Drought Challenges

Current Directions in Water Scarcity Research

VOLUME 2

Series Editors

Robert Mcleman

Henry David Venema

Drought Challenges: Policy Options for Developing Countries

Edited by

Everisto Mapedza

Daniel Tsegai

Michael Brüntrup

Robert McLeman

ELSEVIER

Elsevier
Radarweg 29, PO Box 211, 1000 AE Amsterdam, Netherlands
The Boulevard, Langford Lane, Kidlington, Oxford OX5 1GB, United Kingdom
50 Hampshire Street, 5th Floor, Cambridge, MA 02139, United States

Notices
Knowledge and best practice in this field are constantly changing. As new research and experience broaden our
understanding, changes in research methods, professional practices, or medical treatment may become
necessary.

Practitioners and researchers must always rely on their own experience and knowledge in evaluating and using
any information, methods, compounds, or experiments described herein. In using such information or
methods they should be mindful of their own safety and the safety of others, including parties for whom they
have a professional responsibility.

To the fullest extent of the law, neither the Publisher nor the authors, contributors, or editors, assume any
liability for any injury and/or damage to persons or property as a matter of products liability, negligence or
otherwise, or from any use or operation of any methods, products, instructions, or ideas contained in the
material herein.

Library of Congress Cataloging-in-Publication Data
A catalog record for this book is available from the Library of Congress

British Library Cataloguing-in-Publication Data
A catalogue record for this book is available from the British Library

ISBN: 978-0-12-814820-4

For information on all Elsevier publications visit our website at
https://www.elsevier.com/books-and-journals

Publisher: Candice Janco
Acquisition Editor: Louisa Munro
Editorial Project Manager: Vincent Gabrielle
Production Project Manager: Mohanambal Natarajan
Cover Designer: Matthew Limbert

Typeset by TNQ Technologies

Working together
to grow libraries in
developing countries

www.elsevier.com • www.bookaid.org

Contents

CHAPTER 4 Drought preparedness and livelihood implications in developing countries: what are the options?—Latin America and Northeast Brazil 55

Antonio Rocha Magalhães, Marilia Castelo Magalhães

CHAPTER 5 Drought in the Yucatan: Maya perspectives on tradition, change, and adaptation 67

Miguel Paul Sastaretsi Sioui

Contributors

Stephen Adaawen
Centre for Flight and Migration, Catholic University of Eichstät-Ingolstadt, Eichstätt, Germany

Niranga Alahacoon
International Water Management Institute (IWMI), Colombo, Sri Lanka

Giriraj Amarnath
International Water Management Institute (IWMI), Colombo, Sri Lanka

Michael Brüntrup
Program B, Transformation of Economic and Social Systems, Deutsches Institut für Entwicklungspolitik (German Development Institute) DIE, Bonn, Germany

Brianna Castro
Department of Sociology, Harvard University, Cambridge, MA, United States

Jeganathan Chockalingam
Birla Institute of Technology (BIT), Mesra, India

Olena Dubovyk
Center for Remote Sensing of Land Surfaces (ZFL), University of Bonn, Bonn, Germany; Remote Sensing Research Group (RSRG), University of Bonn, Bonn, Germany

Mesay Kebede Duguma
Consultant, PSNP Donor Coordination Team/World Bank, Addis Ababa, Ethiopia

Moses Duguru
United Nations Office for Outer Space Affairs (UNOOSA), United Nations Platform for Space-based Information for Disaster and Emergency Response (UN-SPIDER), Bonn, Germany

Srinivasa Rao Gattineni
eeMAUSAM, Weather Risk Management Services Private Limited, Hyderabad, India

Gohar Ghazaryan
Center for Remote Sensing of Land Surfaces (ZFL), University of Bonn, Bonn, Germany

Valerie Graw
Center for Remote Sensing of Land Surfaces (ZFL), University of Bonn, Bonn, Germany; Remote Sensing Research Group (RSRG), University of Bonn, Bonn, Germany

Paul Heid
Center for Remote Sensing of Land Surfaces (ZFL), University of Bonn, Bonn, Germany

Lewis Hove
Resilience Hub for Southern Africa, Food and Agriculture Organization of the United Nations, Johannseburg, South Africa

Krishna Reddy Kakumanu
National Institute of Rural Development and Panchayati Raj, Hyderabad, India

Cuthbert Kambanje
Food and Agriculture Organization of the United Nations, Pretoria, South Africa

Ishita Kaushik
University of California, Berkeley, CA, United States

Caroline King-Okumu
Geodata, University of Southampton, Southampton, United Kingdom; The Borders Institute (TBI), Nairobi, Kenya; Centre for Ecology and Hydrology, United Kingdom

Gurava Reddy Kotapati
Acharya N G Ranga Agricultural University, Guntur, India

Rakesh Kumar
National Institute of Hydrology, Roorkee, India

Palanisami Kuppanan
International Water Management Institute, New Delhi, India

Lordman Lekalkuli
National Drought Management Authority, Isiolo, Kenya

Antonio Rocha Magalhães
CGEE-Center for Strategic Studies and Management, Brasilia, Brazil

Marilia Castelo Magalhães
Consultant, Brasília, Brazil

Kane Abdoulah Mamary
Egerton University, Department of Agricultural Economics and Agribusiness Management, Institut d'Economie Rurale-IER, Nakuru, Kenya

Everisto Mapedza
International Water Management Institute (IWMI), Accra, Ghana

Srikanth Maram
National Institute of Rural Development and Panchayati Raj, Hyderabad, India

Karthikeyan Matheswaran
Stockholm Environment Institute (SEI), Bangkok, Thailand

Robert McLeman
Geography & Environmental Studies, Wilfrid Laurier University, Waterloo, Ontario, Canada

Saptarshi Mondal
Birla Institute of Technology (BIT), Mesra, India

Paschal Arsein Mugabe
University of Dar es Salaam (UDSM), Dar es Salaam, Tanzania

Fiona Mwaniki
Kilimo Media International, Nairobi, Kenya

Caroline Mwongera
International Center for Tropical Agriculture (CIAT), Nairobi, Kenya

Chris Miyinzi Mwungu
International Center for Tropical Agriculture (CIAT), Nairobi, Kenya

Udaya Sekhar Nagothu
Norwegian Institute of Bioeconomy Research, Ås, Norway

S. Wagura Ndiritu
Strathmore University Business School, Nairobi, Kenya

H.M. Ngibuini
Independent Natural Resource Management consultant, Nairobi, Kenya

Luxon Nhamo
International Water Management Institute (IWMI), Pretoria, South Africa

Nicholas O. Oguge
Centre for Advanced Studies in Environmental Law & Policy (CASELAP), University of Nairobi, Nairobi, Kenya

Victor A. Orindi
Ada Consortium and National Drought Management Authority (NDMA), Nairobi, Kenya

Rajendra Prasad Pandey
National Institute of Hydrology, Roorkee, India

Peejush Pani
International Water Management Institute (IWMI), Colombo, Sri Lanka; Institute of Remote Sensing and Digital Earth, Beijing, China

Jinelle Piereder
Global Governance at the Balsillie School of International Affairs (BSIA), University of Waterloo, Canada

Joachim Post
German Aerospace Center (DLR), Cologne, Germany

Christina Rademacher-Schulz
Independent Researcher and Consultant, Bonn, Germany

K.V. Rao
Central Research Institute for Dryland Agriculture (CRIDA), Hyderabad, India

Benjamin Schraven
German Development Institute (DIE), Bonn, Germany

Nadine Segadlo
Centre for Flight and Migration, Catholic University of Eichstät-Ingolstadt, Eichstätt, Germany

Kelvin Mashisia Shikuku
International Center for Tropical Agriculture (CIAT), Nairobi, Kenya

Alok Sikka
International Water Management Institute (IWMI), New Delhi, India

Miguel Paul Sastaretsi Sioui
Geography and Environmental Studies, Wilfrid Laurier University, Waterloo, ON, Canada

Vladimir Smakhtin
International Water Management Institute (IWMI), Colombo, Sri Lanka; United Nations University — Institute for Water, Environment and Health (UNU-INWEH), Hamilton, ON, Canada

Jörg Szarzynski
United Nations University, Institute for Environment and Human Security (UNU-EHS), Bonn, Germany; Eurac Research, Bolzano, Italy; Disaster Management Training and Education Centre for Africa (DiMTEC), University of the Free State (UFS), Bloemfontein, Republic of South Africa

Daniel Tsegai
United Nations Convention to Combat Desertification (UNCCD), Bonn, Germany

Juan Carlos Villagrán de León
United Nations Office for Outer Space Affairs (UNOOSA), United Nations Platform for Space-based Information for Disaster and Emergency Response (UN-SPIDER), Bonn, Germany

Yvonne Walz
United Nations University, Institute for Environment and Human Security (UNU-EHS), Bonn, Germany

Lars Wirkus
Data and Geomatics, Bonn International Center for Conversion (BICC), Bonn, Germany

Acknowledgments

The Editors would like to acknowledge several people and organizations without whom this book would not have been possible to realize. We are grateful to Seynabou Gaye for her editorial assistance in the earlier stages of the book's evolution. We would like to acknowledge the United Nations Convention to Combat Desertification (UNCCD) for co-organizing the Africa Drought Conference, which was an opportunity where the editors met and the idea of an edited book later evolved. The German Development Institute (DIE), with the financial support of the German Ministry for Economic Cooperation and Development (BMZ) under its "One World, No Hunger (SEWOH) program," has supported individual chapter authors and the participation of editors in international conferences to gather authors and ideas. The International Water Management Institute (IWMI), through the Mapping the water situation, monitoring the availability of water and the impact on livelihoods in Seven Southern Africa Development Community (SADC) Countries Project during the 2015/16 El Nino-induced drought period, provided material for some of the chapters. This SADC project, led by IWMI, was supported by the Food and Agricultural Organization (FAO) of the United Nations. We would like to thank a number of reviewers who helped to improve the individual chapters. We would also like to acknowledge the individual chapter authors for their persistence and meeting our publication deadlines. We thank our respective institutions for the time that they allowed us to dedicate to this volume. Finally, we thank the Elsevier Team for their support to the publication endeavor.

The designations employed and the presentation of materials in this book do not imply the expression of any opinion whatsoever on the part of the editors and the institutions they are affiliated with concerning the legal or development status of any country, territory, city or area or of its authorities, or concerning the delimitation of its frontiers or boundaries. This book may contain statements, statistics and other information from various sources. The editors and their respective institutions do not represent or endorse the accuracy or reliability of any statement, statistics or other information provided by the authors and contributors contained in this edited volume.

Editor Bios

Everisto Mapedza

Everisto Mapedza is a senior researcher at the International Water Management Institute (IWMI) based in the West Africa office in Accra, Ghana. His interests include institutions, governance, social transformation and gender using the land, water and broadly natural resources lens in the context of developing countries. He is currently a member of the Executive Council of the International Association for the Study of the Commons (IASC). Dr. Mapedza was previously a research fellow at the London School of Economics and Political Science (LSE).

Daniel Tsegai

Daniel Tsegai is a Programme Officer for drought and water scarcity issues at the Secretariat of the United Nations Convention to Combat Desertification (UNCCD). Previously, Dr. Tsegai served as an academic officer with the United Nations University and a senior researcher at the Center for Development Research (ZEF), University of Bonn.

Michael Brüntrup

Michael Brüntrup works at the German Development Institute/Deutsches Institut für Entwicklungspolitik (DIE) as a senior researcher. His interests cover topics related to agriculture and rural development, trade policy, and food security with a geographical focus on sub-Saharan Africa. Since 2017, he is Germany's "Science and Technology Correspondent" for the United Nations Convention to Combat Desertification (UNCCD).

Robert McLeman

Robert McLeman is a professor of Geography and Environmental Studies at Wilfrid Laurier University in Waterloo, Canada. A former diplomat, Dr. McLeman is an internationally recognized expert on environmental migration. He researches the relationship between drought, migration, and population change in rural areas.

Drought risks in developing regions: challenges and opportunities

Everisto Mapedza[a],[*], Robert McLeman[b]

[a]*International Water Management Institute (IWMI), Accra, Ghana,*
[b]*Geography & Environmental Studies, Wilfrid Laurier University, Waterloo, Ontario, Canada*
[*]*Corresponding author*

Introduction

Drought is a hazard that affects most settled areas on a periodic or occasional basis. Even places we associate with being mostly cool and damp—think Ireland, for example (Murphy et al., 2017)—or where the most notorious hazard is flooding as the result of an abundance of precipitation—think Bangladesh (Alam, 2015)—experience sometimes devastating droughts. They can disrupt agricultural production, cause wells and ponds to run dry, force governments to mobilize relief efforts, stimulate migration from affected areas and, in the worst cases, trigger competition and conflict over suddenly scarce resources. Although drought has historically been a rural hazard, disproportionately affecting farmers, pastoralists, and villagers, as cities swell and their water needs grow, drought is becoming an urban hazard as well. Most worryingly, the number of people and places that grapple with drought is expected to grow rapidly this century from the combined effects of continued global population growth, modifications to local eco- and water systems, and changes in the climate caused by continuously growing greenhouse gas emissions (Naumann et al., 2018; Tietjen et al., 2017).

In the last century or so, innovations in agricultural practices, water management systems, detection and early warning technologies, communications, rural financial systems, and insurance regimes have lessened vulnerability to drought—but only for people who can afford them. There is a wide range of proactive strategies and adaptation options that can be implemented across multiple scales, from households to governments, to minimize the direct and indirect impacts of droughts. But the best and most effective strategies are expensive, and necessitate significant adjustments to entrenched practices and systems. They cannot be implemented overnight and require long-term commitments by politicians, officials, and business leaders who, due to the inherent nature of political and market-oriented economic systems, seek short-term/low-cost solutions. The result is that drought adaptation in practice occurs along two tracks: incremental, autonomous adjustments made by farmers, pastoralists, and others whose day-to-day livelihoods are immediately influenced by variations in rainfall, but who often are constrained in their means; and, fitful, reactive measures taken by governments and institutions after drought emergencies occur, with commitment to such measures fading when rains return.

This two-track evolution of drought adaptation reinforces and expands existing societal inequalities, giving a competitive advantage to households that through their own means can afford to purchase irrigation equipment, seeds of drought-resistant crop varieties, and supplemental fodder for livestock. Households that lack cash savings, or have no access to affordable lending or credit, must rely on own scarce resources as well as governments and institutions whose help is available on an ad hoc basis, at best. This is particularly the case in low-income countries in Africa, Asia, Latin America, and the Caribbean, where agricultural populations are large, agricultural systems are transitioning from subsistence to market orientation, formal institutions are often weak and cash-strapped, and public demands for investments in drought adaptation compete with demands for affordable housing, transportation infrastructure, sanitation, health care, and a host of other pressing socioeconomic needs. In many low-income countries, the situation is aggravated by the fact that precipitation falls irregularly and there is simply not enough surface water available to meet local needs on a year-round basis. In less developed countries, economic performance as measured by gross domestic product is strongly correlated to flood and drought events. Even in wealthy nations like the United States, there are systemic barriers that prevent optimal levels of drought adaptation and competing priorities for scarce water supplies (see the chapter on drought risks in Lincoln County, Colorado, this volume). The best, state-of-the-art technologies and the wisest of land management practices cannot entirely eliminate the risk of drought: when the rains do not fall and surface and groundwater reserves become exhausted, crops fail, livestock perish, food shortages ensue, and human systems tip into crisis mode.

The picture we have just painted is bleak, but the situation is not so hopeless as it may appear. The subsequent chapters of this book show that there are many examples of drought risk management that have proven to be effective in low-income countries. The very nature of governing or residing in a low-income country presents unique challenges and barriers to the proactive adoption and implementation of adaptation measures that reduce vulnerability to drought and enhance resilience to variations in water supplies more generally. Yet, through discussion and comparison of best practices, examples of strategies that have worked and those that have not worked, there are valuable lessons to be learned and built upon. Those described in subsequent chapters are not a comprehensive compilation of all the best things we have learned about drought adaptation in developing regions, but are a useful and diverse sampling of the larger body of expertise currently available. The editors will summarize and reflect on lessons learned and future knowledge and research needs in the concluding chapter; in the remainder of the present introduction, we wish to review a number of familiar terms and concepts that may be useful to the reader in framing and interpreting the contributions to this book.

Current and future vulnerability to drought

The potential to experience loss or harm as a result of drought or any other natural hazard has long been described in research and policy in terms of vulnerability. This is, in turn, a function of (1) the nature of the particular hazard to which one is exposed; (2) the sensitivity of livelihoods in a given location to that particular risk; and (3) the capacity of those exposed to adapt (Adger, 2006; Smit and Wandel, 2006). Because it is an outcome of environmental and human systems that are continually changing, vulnerability is never static but is always changing over short- and long-term time scales; and because no two locations or populations are exactly alike, vulnerability differs from one place to another and even within the same community (Cutter et al., 2008). In the case of drought hazards in low-income countries, changes in vulnerability are being accelerated by the impacts of anthropogenic climate change; by high levels of population growth; by people's intensifying use of nature with all its repercussions for the vegetation, the soil, the micro-climate; nature's resilience to drought; and by rapid urbanization. Given such dynamics, a diagnosis of drought vulnerability and adaptation based on current risk factors may not provide a reliable picture of future vulnerability. To identify and anticipate future vulnerability to drought risks in low-income countries requires (1) understanding current exposure, sensitivity, and adaptive capacity; (2) identifying the range of potential adaptation options versus adaptations that are actually taking place; (3) identifying barriers to implementing adaptation options that are not currently being utilized; (4) identifying emergent and potential future risks; and (5) identifying future opportunities for improvement.

Current exposure to drought hazards

A challenge in identifying current or future exposure to drought is that there is no one universally agreed-upon definition or measure of drought. Most tend to fall into one of the three types. The first of these is meteorological drought, which occurs when precipitation levels measured at a given location fall well below average. What constitutes "average" or "well below average" varies from one part of the world to another. Consider the two climate charts shown in Fig. 1.1A and B representing Burkina Faso and Colombia, respectively, two countries featured in chapters that follow. Both countries have seasonal precipitation regimes, in which much less rain falls in the 6 months from November through to April than falls in the remainder of the year. However, in Burkina Faso, the contrast between dry season and wet season is stark: virtually no rain falls during the dry season, whereas Colombia's driest month (January) sees as much rain as the first month of Burkina Faso's rainy season (June). Burkina Faso's wettest month (August) is equal to the seventh wettest month in Colombia.

Given such differences, what constitutes a meteorological drought in Burkina Faso is much different than in Colombia. If no rain is received in Burkina Faso in the month of March, it is of no great concern; hardly any rain falls during that month anyhow. However, in Colombia, were there to be no rainfall in any given month, regardless of the time of year, authorities would start to worry that a meteorological drought might be underway. In Burkina Faso, a 20% shortfall in precipitation in the months of July and August in any given year could be considered a meteorological drought, for relatively little rain will fall in the following 10 months. In Colombia, where rain is received every month of the year, a 20% shortfall in July or August could simply be seen as natural variability and prompt no more than a passing remark in local weather reports.

Meteorological droughts are easy to measure and of greatest interest to those whose job is to monitor climatic conditions on an ongoing basis—weather bureaus, agricultural forecasters, and the like. But in practice, the absolute amount of precipitation that falls from the sky at a given location over a given period is of less practical importance than the actual availability of water on the surface, in the soil, and in accessible groundwater deposits. When precipitation is low over an extended period, water levels drop in rivers, streams, ponds, lakes and wetlands; when levels drop below a critical threshold, the term hydrological drought is invoked. The threshold that triggers such a declaration again varies according to the local context and is usually reflective of the types

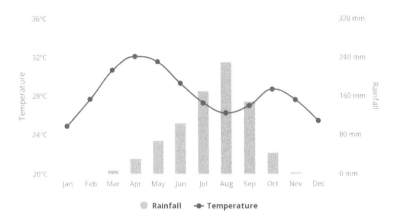

(A) Average Monthly Temperature and Rainfall of Burkina Faso for 1901-2016

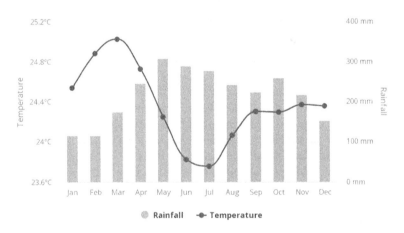

(B) Average Monthly Temperature and Rainfall of Colombia for 1901-2016

FIGURE 1.1

(A) Average temperature and precipitation in Burkina Faso. (B) Average temperature and precipitation in Colombia.

Reproduced from (A and B) World Bank, 2019. World Bank Climate Data Portal. https://climateknowledgeportal.worldbank.org/.

of land uses and the number of people reliant on the water in a particular watershed. Hydrological drought is measured with data collected from a network of gauges placed throughout a watershed and may be combined with estimates of groundwater levels (which may fall when there is not enough precipitation to recharge them). Hydrological drought is an important concern for urban centers in relatively dry areas. The most infamous recent example comes from Cape Town, South Africa, where successive years of below-average rainfall left the city's water supply reservoirs nearly empty by the time rains arrived in the summer of 2018. As a side note, hydrological drought also presents problems where waterways are used for commercial transportation. For example, in 2012, a year when little precipitation fell in the Midwestern US, barges plying several stretches of the Mississippi River started running aground because water levels had dropped too low (Schwartz, 2012). In Germany, in 2018, the river Rhine was so low after several months of low rainfall and high temperatures that river transport had to be drastically reduced with the effect that prices for fuel, usually transported via ship, skyrocketed (Meyer, 2018).

The most familiar category of drought is vegetative drought, which in the context of farming is referred to as agricultural drought. It occurs when there is insufficient moisture in the soil to support the healthy growth of vegetation. Plants self-indicate when a vegetative drought is underway. When a plant's roots cannot find any moisture in the surrounding soil, the inflow of water halts, the internal pressure drops, and the plant begins to wilt. If the lack of water persists, the plant will stop growing, turn from green to brown as chlorophyll becomes inactive, and eventually dies. Some types of agricultural plants, such as perennial grasses on which livestock graze, can survive a lack of soil moisture by deliberately going dormant, ceasing all photosynthesis until rains come. Other plants are much more sensitive, especially when soil moisture shortfalls occur at critical moments of the growth cycle, such as the germination and early seedling period of annual crops, or when the plant is flowering or forming a seed head. When droughts hit during these latter periods, even if the plant survives, the yield of the edible food portion of the plant may be greatly reduced. A variety of standardized measures exist to detect and monitor conditions that can stimulate vegetative drought, well-known examples being the standard precipitation index and the Palmer drought severity index, a distinction being the latter's inclusion of temperature data, with hot temperatures tending to exacerbate evapotranspiration during periods of low precipitation. Such indices are often adjusted by national meteorological organizations to meet their own particular information needs.

However one defines and measures drought, the common denominator is that drought is a period when shortfalls in precipitation cause there to be insufficient water available locally to meet the needs of residents, land users, and nature. What causes precipitation in a particular period to be lower than expected? The potential reasons are many, and a detailed description and analysis is beyond the scope of this chapter. The short version of the story is this: yearly, monthly, and daily variations in precipitation are normal and natural, but unusually severe or extended shortfalls are often associated with fluctuations and anomalies in sea surface temperatures (SSTs), the best known of these being the El Niño Southern Oscillation (ENSO), which is linked to surface temperatures in the mid-Pacific Ocean. An ENSO period occurs when SSTs in the Pacific experience a warming trend and, when they become especially warm, they generate significant changes in the movement of air and moisture through the atmosphere, affecting monsoonal precipitation patterns globally. Regions such as Southeast Asia and Amazonia that are ordinarily quite wet may suddenly tip into drought during a strong ENSO (Räsänen et al., 2016), while other places that are generally more water scarce, such as Pakistan and Nepal, may experience unusually heavy precipitation and consequent flooding (Iqbal and Hassan, 2018).

Regional and local variations in precipitation patterns and departures from expected norms are one component of exposure to drought; another is the number of people living in exposed areas. Researchers have produced a variety of measures of water security that provide a rough barometer

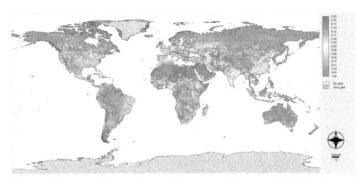

FIGURE 1.2

Spatial index of water scarcity.

Reproduced from Gain, A.K., Giupponi, C., Wada, Y. 2016. Measuring global water security towards sustainable development goals. Environmental Research Letters 11 (12).

of the relative precariousness of countries and their water supplies. One recent example comes from Gain et al. (2016), which combines measures of surface and groundwater availability, population, and drought risks. Not surprisingly, the areas of greatest water insecurity, highlighted in red in Fig. 1.2, are concentrated in and around the regions of dryness in the global climate system—the Saharan and Sudano-Sahelian areas of Africa, the Middle East, South Asia, northeastern China, the Pacific coast of South America, eastern Brazil, the North American Great Plains, and central Australia.

Livelihood sensitivity

Measures of drought occurrence and per capita water resources do not provide a complete picture of drought vulnerability. The impacts of drought on a given population are generally not uniform, but vary considerably according to nonenvironmental factors such as land use and livelihood activities, land tenure arrangements, and labor market and income patterns. At a global level, nearly 70% of all freshwater consumption is used for agriculture (United Nations, 2018), and so people whose livelihoods or wage labor jobs are in the agricultural sector are directly sensitive to changes in the availability and accessibility of water. Cattle and most other livestock (with the exception of camels) can go no more than a day or two without drinking water. Below-average monsoon rainfall or, in a temperate climate, a few dry weeks at planting season can result in a year's worth of lost income for a farm family. For those who work as seasonal agricultural labor, droughts mean less work, less pay, and a greater likelihood they will move to the city in search of better opportunities. These are direct impacts of drought. But there are indirect ways in which nonfarming populations are sensitive to drought, too. Food prices tend to rise when droughts reduce crop yields. Given the interconnected nature of the global food system, a drought in a major food-exporting nation can trigger food price shocks in importing nations thousands of miles away (McLeman, 2013). Urban populations that typically cannot produce their own food are especially sensitive to drought-related food price fluctuations; urban livelihoods may also be affected by local droughts that necessitate conservation measures such as temporary reduction of water use by factories and commercial enterprises.

Adaptive capacity

In principle, there is a wide range of potential actions and strategies that may be implemented proactively and reactively to reduce vulnerability to drought. An often-cited typology by Smit and Skinner (2002) of agricultural adaptations categorizes such options according to actions that are taken by farmers or pastoralists—further categorized as production and financial management adjustments—and adaptations initiated by governments, agri-business, and other higher-level actors. These latter types of adaptation include technological developments (which include such things as irrigation schemes, and development of drought-resistant crop varieties) and institutional programs to support and assist agriculture, such as crop insurance schemes, marketing boards, and maintenance of drought early warning systems. Farm-level adaptations include examples such as switching crop varieties, reallocating existing land, purchasing irrigation equipment, and participating in crop futures schemes.

The wide-ranging adaptation options identified by Smit and Skinner (2002) come primarily from agricultural systems in developed countries in North America and Europe, and many of these are beyond the financial means of farmers and governments alike in low-income countries, where agriculture may have a much greater subsistence orientation, markets are less formal, and insurance regimes are in their infancy. Further, there are many adaptation responses that are quite common in low-income countries but less common today in wealthier countries, such as seasonal labor migration, selling off livestock at distress prices, and simply reducing one's caloric intake when food is scarce or expensive.

In other words, there is a significant adaptation gap that makes low-income countries much more vulnerable to droughts. Many of the chapters that follow in this volume detail existing specific opportunities and barriers for adaptation, providing useful insights into how the adaptation gap might eventually be narrowed.

Future vulnerability to drought in developing regions

Closing the adaptation gap quickly is of considerable importance, for three ongoing processes threaten to increase considerably exposure and sensitivity to drought in low-income countries in coming decades. The first of these is anthropogenic climate change, which will lead to increases in average temperatures over most land regions and, in many regions, will trigger disruptions in prevailing precipitation patterns and stimulate more frequent extreme weather events including droughts. The extent of such changes depends upon future greenhouse gas emissions. Fig. 1.3 shows projections for end-of-century average temperature and precipitation patterns under low and high greenhouse gas emissions scenarios. In both sets of scenarios, average temperatures rise in most low-income countries by at least 1°C, with some regions such as Sudano-Sahelian Africa, the Middle East, Pakistan, Iran, and northern India to expect increases of 4°C or more under high emission scenarios (IPCC, 2014). Should such latter changes come to pass, in the absence of massive improvements in adaptation,

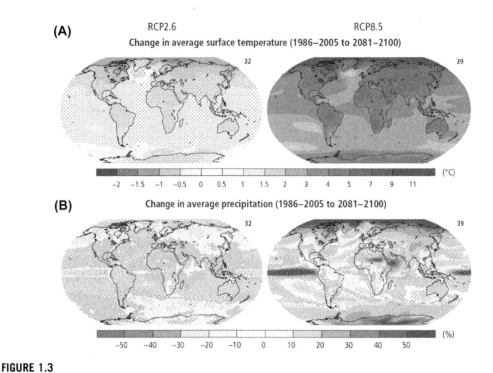

FIGURE 1.3

Future temperature and precipitation patterns under low and high greenhouse gas emission scenarios.
Source: Figure SPM.7 from IPCC, 2014: Climate Change 2014: Synthesis Report. Contribution of Working Groups I, II and III to the Fifth Assessment Report of the Intergovernmental Panel on Climate Change [Core Writing Team, Pachauri, R.K. and Meyer, L.A. (eds.)]. IPCC, Geneva, Switzerland.

low-income countries could expect to experience significant disruptions to food production systems, even those areas such as East Africa and South Asia that might experience a modest increase in annual precipitation (Porter et al., 2014).

A second important change is the rapid increase of population in less developed regions. The United Nations Department of Economic and Social Affairs (UN DESA, 2018) projects that the number of people living in what are currently defined as low-income countries will triple over the course of the present century (Fig. 1.4). Ensuring that there will be sufficient food to support this burgeoning population in an era of rapid climatic change will be one of the most daunting challenges of the present century, with improved drought adaptation being one important part of the solution. This challenge becomes even more daunting when one realizes that, despite impressive improvements in agricultural productivity during the Green Revolution, per capita food production has plateaued as a result of rapid population growth and less dynamic production increases in recent decades (Funk and Brown, 2009).

A third process that creates new challenges and needs for drought adaptation is the rapid urbanization that is currently underway in developing regions. Although low-income countries typically have much larger rural populations than wealthier countries, their urbanization rates have accelerated in recent decades (UN DESA, 2014), creating enormous demands for basic infrastructure

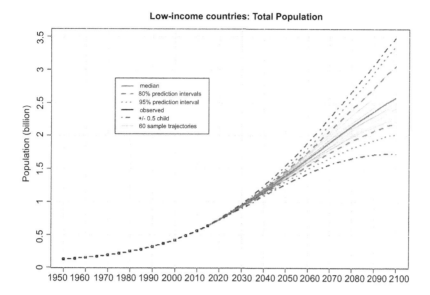

FIGURE 1.4

Projections of future total population of low-income countries (UN DESA, 2018).

such as water, sanitation, and housing. Current urban population growth is already outstripping local water resources in many countries; one recent study estimates that one quarter of all cities globally currently experience significant, ongoing water stress, a disproportionate number of them in low-and middle-income countries (McDonald et al., 2014). Experiences such as that of Cape Town are likely to become more commonplace and lead to greater competition between cities and rural populations over scarce water supplies.

Key topics covered in this volume
Determinants of drought impacts and adaptation opportunities

To some extent, almost every chapter in this book relates evidence that the impacts of drought and the range of possible adaptive responses vary considerably between and within countries. Chapters in the first section of the book have a particularly strong focus on differential impacts and adaptation in developing regions and provide a range of useful insights. The reader will find, for example, chapters that show how drought vulnerability in northeastern Brazil is rising through the combined effects of anthropogenic climate change (which appears to be affecting the severity of droughts and the return rate), environmental degradation through land use/land cover change, and growing population numbers. Improving adaptation in the future requires a multipronged approach to capacity building, with the authors outlining steps being taken in this direction. Readers will in another chapter learn how a single period of El Niño–related droughts led to differential outcomes across Southern Africa due to political and socioeconomic differences across countries. These and other macrolevel determinants of drought impacts highlight the need for closer, ongoing cooperation between countries and between national governments, multilateral organizations, development agencies, and civil societies more generally to reduce vulnerability. In some countries, this high level of cooperation may be necessary to avoid exacerbating competition and conflict in politically unstable areas over fodder and water resources that become scarce during droughts, as is outlined in the next chapter. Migration is often seen as an undesirable outcome of droughts, an indicator of maladaptation, and a potential precursor to civil violence in low-income countries but, as is shown in chapters about drought conflicts and about rural adaptation in Colombia, the linkages are much more complex. Migration by people living in drylands may in many instances be a logical adaptation to drought, one that has been taking place for a very long time, and the greater vulnerability may be among those who lack mobility options. Further, migrants are far more likely to be the victims of violence than the cause of it. In short, as drought hazards become more intense in coming decades and the number of people living in drought-exposed regions grows, policy makers and researchers need to be wary of sweeping generalizations and simplistic narratives about the determinants of drought vulnerability.

Two chapters that stand out in the early section of this book relate to the gender dimensions of drought impacts and adaptation, and to the experience of Indigenous groups. In many societies, women are socially disadvantaged for a variety of cultural, political, and economic reasons, leading them to experience the direct and indirect impacts differently from men. Elderly people, children, and people in poor health also experience drought differently. Consequently, for responses to drought to be effective for all people (and not just some), a gendered approach to identifying determinants and outcomes is necessary. This applies not simply to the narrow goal of drought vulnerability reduction; rather, a holistic approach to addressing gender inequality is required in developing nations, including reconfiguring the socialization process that shapes gender stereotypes from an early age. After all, the factors that make certain societal groups more vulnerable to drought typically also make them more vulnerable to many other risks and hazards as well. As a first step, the authors and editors call here for gender and livelihood considerations to be better integrated into vulnerability assessments and resilience building strategies, starting with better efforts to collect and analyze sex-disaggregated data on drought impacts.

Indigenous groups in many less developed countries are often particularly vulnerable to droughts given their greater likelihood to pursue resource-based livelihoods, their disadvantaged socioeconomic situations and, far too often, the outright discrimination they experience at the hands of governments and fellow citizens. Yet at the same time, having lived on the land for generations, indigenous groups often possess detailed knowledge about drought, its impacts, and the best strategies for mitigating the harm it causes—knowledge governments and institutions pursuing science-based responses often lack and would greatly benefit from. There are many reasons why researchers and policy makers need to engage more deeply with indigenous communities; tapping into traditional ecological knowledge about drought is one more.

Early warning and monitoring systems

The second section of this volume contains chapters that describe ways to improve and expand systems for early detection, warning, and ongoing monitoring of drought in developing regions. Several are quite detailed and technical, and each chapter focuses on experiences and future needs in a particular country or region. As a whole, they provide the reader with a useful sampling of general lessons that may be applied more widely and provide a good indication of where this field is heading. One important future direction is to adapt monitoring and early warning systems for local context; an important reason why such systems have failed to live up to expectations in the past is that the information they generated was too generic in nature, or too broad in spatial scale, to be used as a decision-making input by land users and local authorities. New technologies and innovations in remote-sensing data collection offer great promise for improving drought detection and monitoring in the future, and with refinement of methods for incorporating them into monitoring and assessment systems, will become increasingly valuable to planners and managers in developing regions. Recognizing that drought vulnerability is a product of both physical conditions and socioeconomic factors particular to a given area, chapters in this section also emphasize the utility in incorporating locally specific nonclimatic data into early warning and monitoring systems. A recurrent theme is that drought monitoring is often trapped in particular policy making and decision-making silos, and there is a need for greater consideration of linking drought monitoring assessment with other monitoring and planning activities being undertaken in other sectors and government departments. Conflict monitoring systems in particular need to be better integrated with drought monitoring systems in dryland countries and regions.

Mechanisms for building resilience

Chapters in the third section of this book provide a variety of insights into ways of enhancing drought resilience in developing regions. This section highlights key mechanisms for resilience building against drought. Two chapters here focus on crop insurance and how to make it work better for farmers. Long established in developed countries, insurance regimes are still a relatively new tool in developing countries, and there are many nuances that yet need to be worked out, such as how to encourage farmers to participate, how to ensure payouts are timely and meet farmers needs, and how to ensure the system does not draw too heavily on the public purse. The types of monitoring and assessment data needed to calibrate insurance plans also need to be refined. One important option for insurance regimes is to make sure that they incentivize good land management practices and do not inadvertently encourage risk-taking behaviors that undermine sustainable land management practices.

This section also contains comparisons of the effectiveness of national-level early warning and monitoring systems for drought, their connections with crop and livestock insurance schemes, and the opportunities for improving such systems going forward. One important direction is to make agriculture "climate-smart"—that is, more systemically resilient to a growing range of current and future climate risks—with one of our authors looking at its practical implementation through uptake of drought-tolerant maize and maize—legume intercropping in Uganda. In certain circumstances, these practices enhance farmers' resilience when drought-tolerant varieties of maize are used, but when traditional varieties of maize are intercropped, resilience may actually decline. The lesson in brief: continual innovation in climate-smart agriculture needs to continue, but we must be aware that there are rarely "one-size-fits-all" solutions. An innovation that works well when implemented this way may have the opposite effect when implemented that way. In a later chapter, the importance of improving livestock management practices in dryland areas is highlighted as an important part of making rural areas more drought resilient. Communications plays an important role in doing so, and making sure that farmers and pastoralists receive timely information from authorities about emergent drought risks—and vice versa—may require greater engagement of community leaders.

Other innovations in building drought resilience that appear in this section are rainwater harvesting technologies that enhance not only short-term water availability but improve biomass and contribute to better human health. Although they are not always thought of as innovations in building drought resilience, social protection schemes often underlay successful improvements in livelihood resilience and drought adaptation. Social safety nets for rural people serve a variety of purposes including, as shown in this section, helping farmers and pastoralists in Ethiopia become better able to cope with drought.

Policy options and lessons learned

The last section of chapters in this book looks at the options and opportunities available to policy makers and emphasizes the need for both proactive and reactive planning and strategies for dealing with drought. Far too often, reactive, ad hoc measures have been the main government approach to managing drought risks; this has started to change in some countries, though not all. There are many hurdles to improving national drought planning and implementation in developing countries, such as including planners and decision makers from all levels in national and regional strategies, and coordinating better between governments, nongovernmental organizations, and civil society. Financing drought management strategies remains an ongoing challenge, especially in countries that have a wide range of other, pressing development needs. Yet money spent on drought warning, monitoring, and adaptation is money well spent. Reducing people's exposure to the worst consequences of drought is far less expensive and far better for human well-being than trying to deal with a drought crisis after it occurs, and our authors provide evidence here that confirms how prevention is indeed better than the cure. We also include in this section a chapter that looks at an example of drought risks and resilience building in the United States, as a means of comparing the drought management challenges and opportunities developing countries will face as they transition from developing to developed. This includes evolution of insurance schemes and a look at an institutional option for allocating water rights in water scarce areas, known as the "Colorado Doctrine."

Information gaps and future needs

In each section of this book, authors identify information gaps and future challenges for researchers and policy makers. Here we offer a brief consolidation of these, so that the reader can look for them, and the greater details provided for each, in subsequent chapters.

A first challenge is the lack of coherent and comprehensive drought management policies in many developing countries. In some cases, it is a question of insufficient financial resources, in others it may be a structural issue, with responsibilities being scattered across multiple government departments; in still others, it may simply be a lack of recognition of the importance of drought planning and its benefits. For countries that are developing drought management strategies, or plan to do so, data on drought hazards in developing regions are very limited and usually of poor quality. New remote-sensing technologies will improve the data collection situation, but processing that data and incorporating them into drought management systems will be a major challenge for developing countries.

Data on the gender dimensions of drought vulnerability are also sorely lacking, and once available, they need to be better captured in management systems and responsive actions. Systems need to take into account local conditions and tailor responses accordingly. Local knowledge and observational systems may fill many gaps in formal drought strategies, with more research needing to be done on how to capture and incorporate these. A corollary to this is for information from formal strategies to be made more readily available to people on the ground by taking better advantage of informal, social, and communication networks. Crop, livestock, and drought insurance schemes will be important tools for managers going forward, but there is still a lot of work to be done in terms of data collection, analysis, and identification of best practices in order to make such systems work well for rural populations in developing regions.

Finally, if there is one sentence that summarizes the lessons of this book, it is this: Continual learning for continuous improvement is needed on science, policies, and practices to reduce drought risks in developing countries. Without it, droughts will continue to undermine livelihoods, erode people's trusts in institutions, and reduce our collective chances of meeting the United Nations Sustainable Development Goals.

References

Adger, W.N., 2006. Vulnerability. Global Environmental Change 16, 268–281.

Alam, K., 2015. Farmers' adaptation to water scarcity in drought-prone environments: a case study of Rajshahi district, Bangladesh. Agricultural Water Management 148, 196–206.

Cutter, S.L., Barnes, L., Berry, M., Burton, C.G., Evans, E., Tate, E.C., Webb, J., 2008. A place-based model for understanding community resilience to natural disasters. Global Environmental Change 18, 598–606.

Funk, C.C., Brown, M.E., 2009. Declining global per capita agricultural production and warming oceans threaten food security. Food Security 1 (3), 271–289.

Gain, A.K., Giupponi, C., Wada, Y., 2016. Measuring global water security towards sustainable development goals. Environmental Research Letters 11 (12).

Iqbal, A., Hassan, S.A., 2018. ENSO and IOD analysis on the occurrence of floods in Pakistan. Natural Hazards 91 (3), 879–890.

IPCC, 2014. In: Pachauri, R.K., Meyer, L.A. (Eds.), Climate Change 2014: Synthesis Report. Contribution of Working Groups I, II and III to the Fifth Assessment Report of the Intergovernmental Panel on Climate Change. IPCC, Geneva, Switzerland, 151 pp.

McDonald, R., Weber, K., Padowski, J., Flörke, M., Schneider, C., Green, P., Gleeson, T., Eckman, S., Lehner, B., Balk, D., Boucher, T., Grill, G., Montgomery, M., 2014. Water on an urban planet: urbanization and the reach of urban water infrastructure. Global Environmental Change 27, 96–105.

McLeman, R., 2013. Labor migration and food security in a changing climate. In: Barrett, C. (Ed.), Food Security and Sociopolitical Stability. Oxford University Press, New York, pp. 229–255.

Meyer, A., 2018. Dryness Raise Prices for Heating Oil and Gasoline. www.ostsee-zeitung.de/Nachrichten/MV-aktuell/Niedrige-Flusspegel-treiben-Kosten-fuer-Oel-Transporte-in-die-Hoehe-of-5-November-2018. English translation.

Murphy, C., Noone, S., Duffy, C., Broderick, C., Matthews, T., Wilby, R., 2017. Irish droughts in newspaper archives: rediscovering forgotten hazards. Weather 72 (6), 151–155. https://doi.org/10.1002/wea.2904.

Naumann, G., Alfieri, L., Wyser, K., Mentaschi, L., Betts, R.A., Carrao, H., Feyen, L., 2018. Global changes in drought conditions under different levels of warming. Geophysical Research Letters 45 (7), 3285–3296.

Porter, J.R., Xie, L., Challinor, A.J., Cochrane, K., Howden, S.M., Iqbal, M.M., Lobell, D.B., Travasso, M.I., 2014. Food security and food production systems. In: Barros, C.B.,V.R., Dokken, D.J., Mach, K.J., Mastrandrea, M.D., Bilir, T.E., Chatterjee, M., Ebi, K.L., Estrada, Y.O., Genova, R.C., Girma, B., Kissel, E.S., Levy, A.N., MacCracken, S., Mastrandrea, P.R., White, L.L. (Eds.), Climate Change 2014: Impacts, Adaptation, and Vulnerability. Part A: Global and Sectoral Aspects. Contribution of Working Group II to the Fifth Assessment Report of the Intergovernmental Panel on Climate Change. Cambridge University Press, Cambridge, United Kingdom and New York, NY, USA, pp. 485−533.

Räsänen, T.A., Lindgren, V., Guillaume, J.H.A., Buckley, B.M., Kummu, M., 2016. On the spatial and temporal variability of ENSO precipitation and drought teleconnection in mainland Southeast Asia. Climate of the Past 12, 1889−1905.

Schwartz, J., 2012. After drought, reducing water flow could hurt Mississippi river transport. New York Times A19, 27 November 2012. https://www.nytimes.com/2012/11/27/us/hit-by-drought-mississippi-river-may-face-more-challenges.html.

Smit, B., Skinner, M.W., 2002. Adaptation options in agriculture to climate change: a topology. Mitigation and Adaptation Strategies for Global Change 7 (1), 85−114.

Smit, B., Wandel, J., 2006. Adaptation, adaptive capacity and vulnerability. Global Environmental Change 16, 282−292.

Tietjen, B., Schlaepfer, D.R., Bradford, J.B., Lauenroth, W.K., Hall, S.A., Duniway, M.C., Wilson, S.D., 2017. Climate change-induced vegetation shifts lead to more ecological droughts despite projected rainfall increases in many global temperate drylands. Global Change Biology 23 (7), 2743−2754.

United Nations, 2018. Sustainable Development Goal 6: Synthesis Report on Water and Sanitation. New York.

UN DESA, 2014. World Urbanization Prospects: The 2014 Revision, Highlights (ST/ESA/SER.A/352). United Nations, Department of Economic and Social Affairs, Population Division, New York, 32pp.

UN DESA, 2018. World Population Prospects 2017. United Nations, Department of Economic and Social Affairs, Population Division, New York. https://population.un.org/wpp/.

World Bank, 2019. World Bank Climate Data Portal. https://climateknowledgeportal.worldbank.org/.

Drought, migration, and conflict in sub-Saharan Africa: what are the links and policy options?

Stephen Adaawen[a],*, **Christina Rademacher-Schulz**[b], **Benjamin Schraven**[c], **Nadine Segadlo**[a]

[a]*Centre for Flight and Migration, Catholic University of Eichstät-Ingolstadt, Eichstätt, Germany,*
[b]*Independent Researcher and Consultant, Bonn, Germany,*
[c]*German Development Institute (DIE), Bonn, Germany*
Corresponding author

Introduction

Droughts have become more frequent and severe in many regions due to climate change, with anthropogenic alteration of environmental and hydrological processes aggravating drought impacts (Dai et al., 2004; Giannini et al., 2008; Dai, 2011; Booth et al., 2012; Van Loon et al., 2016; Weber et al., 2018). While the underlying causes, projections, and measurement of droughts are still critical subjects of intense scientific debate among climate scholars (Dai et al., 2004; Lott et al., 2013), there is a general agreement in empirical research on the growing impacts of recurring droughts in sub-Saharan Africa (SSA). The West African Sahel and much of East Africa have particularly been high risk regions for recurrent droughts (Mwangi et al., 2014; Giannini et al., 2017). Aside from well-known Sahelian droughts in the 1970s and 1980s that recorded millions of deaths (Brooks, 2004; Kallis, 2008; Sarr, 2012), recent drought (2012) impacts on food production and water scarcity resulted in food and water crises for more than 1 million people in the Sahelian countries of Niger, Burkina Faso, Mali, Chad, and Mauritania.[1] In the Turkana region of Kenya, for example, drought-induced crop failure and water scarcity led to widespread starvation, loss of livestock and human lives.[2] These impacts have attracted widespread public outcry on the seeming lack of urgency in addressing the food and water crisis in the affected areas.

The impact of climatic extremes in SSA is also reflective of the frequency, spatial extent, and intensity of droughts on the continent. However, African societies have long developed different strategies to cope with climate variability and droughts over the years (Pearson and Niaufre, 2013; Gautier et al., 2016). Climatic extremes have often triggered responses of various kinds including permanent or short-term migration in Africa (Mortimore and Adams, 2001; Liehr et al., 2016). Explanations of climate change impacts and disaster responses in Africa attribute the continent's high vulnerability to droughts as being mainly due to limited adaptive capacities and high poverty levels of rural

---

[1]https://www.wfp.org/stories/drought-returns-sahel.
[2]https://www.bbc.com/news/world-africa-47624616.

Drought Challenges. https://doi.org/10.1016/B978-0-12-814820-4.00002-X

populations. Consequently, interventions to address drought impacts are often reactive rather than pro-active. Mortimore (2010) challenges the subtle endorsement of the poverty and limited adaptive capacity narrative to drought impacts and other climatic extremes on the continent. This is in view of the fact that specification on drought adaptation have often failed to consider the significant changes in key socioeconomic variables that tend to influence the livelihood situations of rural populations. In this light, isolating adaptation from the complex processes of development and change risks distortion (Mortimore, 2010, p.140). He therefore argues for adaptation to droughts in SSA to take into consideration the broader and complex context of development.

Droughts have long been identified as the underlying cause of recurrent famines, loss of livestock, armed conflicts, and large-scale population displacement in many parts of SSA (Homer-Dixon, 1994; Huho and Mugalavaiet al., 2010; Afifi et al., 2012). With ongoing climate change, population growth, and natural resources scarcity, there are suggestions that drought-induced migration will lead to competition and stress in resource-scarce areas, with the potential to degenerate into violent conflicts (Hendrix and Glaser, 2007; Hummel et al., 2012). As already being experienced in the Sahel and East Africa, drought-related farmer–herder conflicts and water tensions have become common and widespread (Benjaminsen et al., 2009; Cabot, 2017). These tensions tend to escalate or ignite latent conflicts, with varying implications for stability and human security in these vulnerable regions (Moritz, 2010). In the case of Darfur, climate change–induced land degradation and droughts have been enumerated as major precursors for the recurrent farmer–herder conflicts; similar observations have also been made in other parts of East and West Africa (Weiss, 2003; UNEP, 2007; Cabot, 2017).

Yet, many other scholars express reservation about the eco-scarcity conflict causation narrative, insisting that there is no strong evidence to support such claims (Barnett and Adger, 2007; Brown, 2010). Despite the growing recognition that climate change events pose significant risks to human security, the attention is also drawn to the many socioeconomic and political factors that act in tandem with environmental factors to trigger conflicts (Raleigh and Urdal, 2007; Gleditsch et al., 2007; Reuveny, 2008a,b; Benjaminsen, 2008).

In general, there has been sustained scientific attention to the questions of drought, human mobility, and conflict in SSA. This notwithstanding, understanding of the linkages is still unclear (Raleigh et al., 2008). Moreover, the policy options at addressing drought impacts on human mobility and conflict remain few. This chapter seeks to contribute to ongoing discussions on the subject by looking beyond linear, cause–effect narratives to highlight and critically examine the multiplicity of complex and interrelated factors that interact with droughts to influence human mobility and conflict. Doing so provides a basis for identifying policy options to address the challenges in drought-vulnerable regions across SSA. This chapter begins with a conceptual overview of climate change, migration, and conflict interlinkages, followed by a review of evidence from the three SSA subregions of West, East, and Southern Africa, and a conclusion that identifies policy options.

Climate change, migration, and conflict nexus: a conceptual overview

Climate change has consistently been identified as a potential global and regional security threat (Brown et al., 2007; Brown and Crawford, 2008; USA Department of Defense, 2019). The droughts and famines that hit the Sahel and much of SSA in the 1970s and 1980s are often cited as examples of the high vulnerability of populations on the continent, and the extent to which local adaptive capacities can be overwhelmed by extreme climatic events (Hulme, 2001; Haarsma et al., 2005). While the humanitarian crisis that resulted from acute food shortages and famines may have prompted strategic interventions and attention to climate-related extremes and disasters, more recent international efforts and research have taken a more proactive view (Batterbury and Warren, 2001; IOM, 2018).

The various IPCC assessment reports (IPCC, 2001, 2007, 2013) have identified that the combined effects of climate change and population growth will accelerate the degradation of natural resources. This will lead to declines in agricultural productivity across many regions, with negative implications for socioeconomic systems. Some other studies have also suggested that climate change—induced human mobility and competition for scarce resources will be the major causes of conflict in vulnerable and fragile regions (Linke et al., 2018). While studies have consistently alluded to the role of environmental/climatic change in inciting violent conflicts, the link remains unclear (Gleditsch et al., 2007). Even for instances of environmentally or natural resource scarcity—induced conflicts that have been reported, it can be observed that most of the conflicts are processual[3] (Moritz, 2010). Moreover, empirical evidence suggests that farmers and herders, or even states that share transboundary water resources, tend to have less conflict and rather cooperate more in the use of resources (Carius et al., 2004; Turton et al., 2006). Further critical assessment of conflict cases has pointed to a multiplicity of complex socioeconomic, political, and environmental factors (direct and indirect) that interact to trigger conflicts (Castles, 2002; Benjaminsen et al., 2009). Reuveny (2007, p. 659) argues, for instance, that environmental migration may trigger conflicts when two or more of the following conditions exist:

a) *Competition:* Environmentally induced population movements have the potential to overstretch resources in receiving areas. The resulting pressure may heighten sentiments and latent tensions between the competing parties.

b) *Ethnic tension:* This channel may be the basis for conflict when climate or environmentally related migrants and host societies are made up of different ethnicities. As may be drawn from the ethnic-induced conflict that erupted between Dayak tribesmen and Madurese migrants within the context of the *transmigrasi* program in Indonesia (Castles, 2002; Tanasaldy, 2009), ethnic differences, fear, or intimidation may cause host residents to be antagonistic to migrants. In instances where there are already longstanding tensions between groups, a little argument or dispute over the use of any resource could inflame passions and a full-blown violent conflict.

c) *Distrust:* Distrust may lead to conflict when people in the place or country of origin suspect that the receiving area is intentionally accepting migrants in order to create an ethnic imbalance between the areas, as in the case of the Bengali—Assamese conflict (Lee, 2001). Alternatively, the government or residents of the receiving area may suspect migrants as being agents sent to foment or incite unrest. Mistreatment of migrants in receiving areas may raise distrust and tensions in the sending area.

d) *Fault lines:* Conflicts may expose frustrations or general discontent with prevailing socioeconomic conditions that predispose people to revolt (Benjaminsen and Ba, 2009). This may happen between pastoralists and herders competing for water and pasture, or when herds destroy farm crops. As was observed in the Tuareg rebellion in Mali, rebels were able to

---

[3]An analytical approach that entails the systematic study of conflict with a focus on the evolution and succession of phases in order unveil patterns of political processes in understanding the complexity of conflicts (see Moritz, 2010).

mobilize disgruntled migrants to challenge the state (Benjaminsen, 2008). Should the state in its quest to curb violence or insurgency respond with force, it may degenerate to armed conflict.

e) *Auxiliary conditions:* Differences in socioeconomic conditions create different potential for conflict. For example, developed economies generally have greater capacity to provide job opportunities for migrants, whereas developing countries generally have more limited employment opportunities and are more highly dependent on environmental resources or agricultural sector. In the face of increasing environmental scarcity, the arrival of environmentally induced migrants may create greater competition for jobs and burden existing resources. With the lack of capacity to absorb the pressure that may evolve, developing countries may therefore be more predisposed to conflicts as compared to developed countries.

Reuveny (2007) suggests that the aforementioned factors apply to both ordinary and climate-related migration. The only difference is the pace and scope of movement, which could in part also determine the likelihood of conflict. In acknowledging the Malthusian undertones and the likely short-comings of his proposal, he draws attention to the fact that climate change—induced migration does not necessarily lead to conflict. Indeed, while climate change—induced migration may cause conflicts, conflicts are often the cause of displacement and mass migration, and there are instances when migration is beneficial to both receiving and source areas in terms of diffusing conflicts.

In short, Reuveny's work emphasizes that the climate change—migration—conflict causation pathway is not simple or linear, it is multicausal, and is highly context dependent—observations that have been made by others as well (e.g., Moritz, 2006; Tanasaldy, 2009).

Droughts, mobility, and conflict in Sub-Saharan Africa
Droughts, mobility, and conflict in West Africa

Precipitation variability and severe droughts are enduring features of climate dynamics in West Africa, with the Sahelian zone having been marked by long periods of dryness since the early 1970s (Hulme, 2001; Giannini et al., 2013). For much of the Sahel, droughts and corresponding decline in rainfall activity have been the cause of crop failure, famines, and loss of livestock, especially in areas predominantly inhabited by pastoralists (Dietz et al., 2001). Droughts in the 1970s and 1980s for food shortages and deaths to both livestock and people, as well as the mass movement of people across the region. Recent drought events have been identified as accounting for the significant loss of farmland, decline in crop yields, and loss of livestock in the dry areas of Mali, Niger, Senegal, Northern Nigeria, and Burkina Faso.

In 2010, for example, 10 million people were affected by hunger as a result of severe droughts across West Africa.[4] Sahelian countries including Niger, Chad, Mali, Burkina Faso, and Mauritania experienced declines in food production of up to 25%, with grain harvests plummeting drastically to 1.4 million metric tons.[5] In the Lake Chad basin, the steady shrinking of the lake due to extended periods of drought, climate change, and extensive water withdrawal threatens nearly 40 million people in Chad, Cameroon, Niger, and Nigeria.[6] Empirical studies suggest the compounding effects of

---

[4]https://www.theguardian.com/environment/2010/jun/03/drought-hunger-west-africa.
[5]http://www.worldwatch.org/drought-west-africa-threatens-food-security.
[6]https://reliefweb.int/report/chad/tale-disappearing-lake.

population growth, farm expansion, loss of vegetation, water scarcity, and land degradation will accelerate desertification and the out-migration of people to more resource-endowed areas (Gautier et al., 2016; De Bruijn et al., 2005). But with most parts of the region facing environmental stress and civil strife, it is argued that recurring droughts have been the major cause of farmer—herder conflicts and interstate tensions on the use of transboundary water resources in the region (Brown and Crawford, 2008; Cabot, 2017).

As is common across West Africa, internal and cross-border migration has been ongoing in Mali and Burkina Faso. In both countries, migration has long been a livelihood strategy to cope with droughts and environmental stress. Most people rely on the production of rain-fed grains like sorghum, maize, millet, and cotton for their livelihoods. But years of debilitating droughts and unreliable rainfall have significantly reduced harvests, and generated severe food shortages in 2012; especially in the Sudano-Sahelian fringes of northern Burkina Faso and Mali (Pearson and Niaufre, 2013).

Given their limited adaptation options, people in these drought-prone areas often resort to preestablished migration routes to other areas as a coping strategy. In analyzing the droughts in the 1980s in Mali, Findley (1994) found a dramatic increase in the number of women and children who migrated during the severe 1983 to 1985 droughts, along with an observed shift to short-duration circular migration involving approximately 64% of migrants. In a more recent research, Pearson and Niaufre (2013) observed that periods of extreme drought often see increased numbers of interregional migration in Mali. In their study, they report that 42% of households intensified seasonal migration during poor harvests, while 17% migrated when there was crop failure and 13% migrated during droughts. Hummel (2016) also found that 40% of people in the Bandiagara District of Mali engaged in migration as a coping strategy to climate shocks like droughts.

Similarly, Burkina Faso experiences a north—south internal migration patterns, as well as long-standing cross-border migration links to Ivory Coast and Ghana. Internal migration in Burkina Faso follows a strongly rural—rural pattern. In the face of environmental deterioration and climate variability, both short- and long-term migration have been identified as prominent off-farm coping strategies in drier regions of northern Burkina Faso (Henry et al., 2004; Nielsen and Reenberg, 2010). As observed among the Rimaiibe ethnic group of northern Burkina Faso, droughts and high rainfall variability lead young adults to migrate to urban areas in Burkina Faso and to Abidjan to seek work and earn money for food (Nielsen and Reenberg, 2010).

In Northern Nigeria, recurring severe droughts and rainfall deficits affect the livelihoods of farmers and herders, causing crop failures, starvation, and disruption of economic activities (Abubakar and Yamusa, 2013). With limited adaptation options, many are forced to migrate to wetter areas in the south of the country. Competition for scarce natural resources and an influx of herders have, in tandem with other socioeconomic and political factors, been major causes of violent conflicts between Fulani herders and farmers (Weiss, 2003; Okeke, 2014).

Like in Northern Nigeria, droughts and competition for scarce natural resources between pastoralists and resident farmers have also been identified as primary precursors for migration and the escalation of conflicts in southwestern Nigeria, Mali and Ghana (Weiss, 2003; Blench, 2004; Tonah, 2006). Based on their findings in the Lake Chad region, Nett and Rüttinger (2016) contend that the shrinking of Lake could further stoke the activities of Boko Haram in the region. Irrespective of the seeming reference to the climate change and eco-scarcity question in explaining the environmentally related conflicts in most parts of the Sahel, other studies also highlight free movement and increasing cooperation between pastoralists and farmers in the use of scarce resources. Indeed, in the cases of the Tuareg rebellion in

Mali and many other violent farmer—herder conflicts in West Africa, studies highlight that migration, exposure to revolutionary discourses, feelings of marginalization, and various other sociopolitical and ecological factors were instrumental in fomenting these violent clashes (Benjminsen, 2008; Raleigh, 2010; Bukari et al., 2018).

Generally, sahelian livelihoods and social systems have adapted to years of climate variability and extremes, as exemplified in livelihood diversification, mobility, and pastoralism. Nevertheless, the impact of drought and water scarcity on food and livestock production will not only continue to contribute to the surge in the number of internally displaced persons, but also worsen the food insecurity and environmental scarcity in the region (Hammer, 2004; Giannini et al., 2017). With violent farmer—herder conflicts or water tensions thus also becoming increasingly common and widespread, the tendency for these tensions to escalate or ignite latent conflicts (ethnic or fault lines) would be great. This will invariably also have implications for the already fragile security situation and stability in the region (see Reuveny, 2007).

Migration is one of the many coping or adaptation options that vulnerable populations may consider in coping with both slow- and rapid-onset extreme events or hazards. However, many complementary factors come into play to precipitate movement during periods of droughts and other environmental change-related stress. At the same time, droughts and extreme events can also lead to immobility, when the affected persons become displaced or overwhelmed such that they may lack the means to undertake any movement (Findley, 1994; Henry et al., 2003). But in considering that ongoing climatic changes may further lead to an increase in frequency and intensity of extreme events in West Africa, there is the need for an integrated approach to drought and natural resource management, food buffers (grain silos), and robust climate forecasts and early warning systems (both drought and conflict). These will prove crucial to managing the effects of droughts and rainfall deficits on food availability, as well as promoting the equitable use of common or shared resources in mitigating potential escalation of eco-scarcity or drought-induced violent clashes in West Africa. Additional strategies in the form of large-scale afforestation programs to curtail water loss, improve soil conditions and water retention, as well as the provision of water points for pastoralists or dams to promote irrigation agriculture could be initiated to minimize the devastating effects of climate extremes, competition, and stress that have the potential to trigger tensions.

Droughts, mobility, and conflict in Eastern Africa

Slow-onset phenomena such as droughts, desertification and soil degradation, linked to changing rainfall patterns, and the resulting scarcity of productive agricultural land and food, have consistently been major issues of concern in Eastern Africa. These challenges have been contributory factors to the periodic large-scale migration and displacement in the region (Ndaruzaniye et al., 2010). As global temperatures rise, drought-induced human relocation is expected to increase. In the East and Horn of Africa Drought Appeal in 2017 (April—December 2017), the International Organization for Migration (IOM) (2017a) reported that Eastern Africa was facing one of the worst droughts in decades and that, combined with political instability due to conflict and armed insurgency in the region, there was a rapid deterioration in food security and an increase in the numbers of displaced persons. In addition to the significant numbers of refugees in the region, approximately 16 million people were under the threat of droughts, with livestock and crop production likely to fall in Ethiopia, Somalia, and Kenya (Afifi et al., 2012; FEWS NET, 2017). At the beginning of 2017, at least 700,000 people were displaced and forced to move due to severe droughts in Somalia, 316,128 people in Ethiopia, and 41,000 in Kenya (IOM, 2017a). The drought-mobility situation in Somalia is, according to the IOM Regional Migrant Response Plan for the Horn of Africa and Yemen 2018–20, multifaceted and complex; being characterized by internal and external displacement due to natural disasters, conflict, man-made crises, forced evictions, and irregular labor migration.

Droughts are the most recurring climate hazard in Ethiopia, especially for pastoralists and agro-pastoralists (GFDRR, 2011). Since the 1970s, the magnitude, frequency, and intensity of droughts have significantly increased with varying impacts the region. These impacts have mainly included: pasture shortages, overgrazing, land degradation, decreased water availability, and livestock diseases. The results have manifested in decreased livestock productivity, crop failure in agro-pastoral areas, food insecurity, increased movements of people and livestock, and increased conflicts over scarce resources. According to IOM (2017a, 2018), Ethiopia has been plagued with the triple challenge of drought, food, and intercommunal conflict since the beginning of 2017.

According to Ezra and Kiros (2001), vulnerability to food crisis had a positive effect on out-migration in rural Ethiopia. While they observed that females were more likely to move in comparison with their male counterparts, mobility for both sexes was facilitated by marriage. Gray and Mueller (2012), in probing the effects of drought on population mobility in the Ethiopian highlands, also noted that migration by men from land-poor households more than doubled during severe droughts. They, however, observed that marriage-related mobility by women was reduced by half, reflecting a decreased ability to finance wedding expenses and setting up a new household. These findings imply that a multiplicity of factors including poverty, high unemployment, land scarcity, environmental degradation, regional income disparities, and a well-established culture of migration in origin regions act in tandem to increase migration propensity.

Djibouti shares long borders with Somalia and Ethiopia and as a result receives large cross-border movements of pastoralists with their livestock during the lean season, and their numbers increasing during droughts (IOM, 2017a). The Djibouti Drought Interagency Contingency Plan for 2017 estimated that the country would receive an additional 5000–10,000 migrants. As Nyaoro et al. (2016) point out, drought-related displacement is frequently associated with pastoralists displacement, which is linked to the loss of livestock and access to land, resources, and markets. Normally, displaced pastoralists often settle along rivers and thus also increasing their vulnerability to floods.

As reported by GFDRR (2017), droughts in Uganda affected close to 2.4 million people between 2004 and 2013. In particular, the drought conditions in 2010 and 2011 caused an estimated loss and damage value of $1.2 billion, equivalent to 7.5% of Uganda's 2010 gross domestic product. The so called "cattle-corridor" stretching from western Uganda through the central region to Teso and Karamoja in the northeast is one of the worst afflicted areas (Egeru, 2015). According to the FAO Resilience Analysis Report (2018), Karamoja is the poorest region in Uganda with a poverty rate that is three times higher than the national average. The high occurrence of droughts, floods, and dry spells undermine the coping capacities of most of the pastoralists and agro-pastoralists, who use internal migration (seasonal) as an integral part of their livelihoods. Based on a drought risk assessment of Karamoja, Jordaan (2015) is of the contention that most drought impacts experienced since 2000 have been exacerbated by lack of preparedness and efficient response, and not necessarily poor climatic conditions. He, therefore, stresses the need for farmers to better adapt by using more drought-resilient agricultural methods.

In Kenya, drought cycles have become shorter, more frequent, and intense. Nyaoro et al. (2016, p.60) reports that the 14 droughts that occurred between 1965 and 2015 affected 48,800,000 people. The consequences of droughts have often included widespread crop failures, acute water shortages, sharply declining terms of trade for pastoralists, and declining animal productivity. These have been the cause of increasing food insecurity, disease outbreaks, severe malnutrition, loss of livelihoods, increased resource-based conflicts, and associated human and livestock displacement in the country (IOM, 2017a).

Regardless of the fact that many countries in East Africa and the Horn of Africa are historically fragile and conflict prone, the impacts of drought on food production and livelihoods have often been the cause of conflict and general insecurity in the region (Meier et al., 2007). Raleigh and Kniveton (2012) in using rainfall as a parameter observed increases in the frequency of conflicts. They found that small-scale conflicts increase with high rainfall variability. It was also detected that high rates of rebel activity intensify with droughts, whereas communal conflicts escalate with anomalously wet conditions. Similar findings by Linke et al. (2018) also showed that Kenyans who relocated or moved in response to drought were consistently more likely to suffer violence relative to those who did not.

With the assessment of the conflict in Darfur, UNEP (2007) also highlighted a strong link between climate change, drought, land degradation, desertification, and the protracted conflict. In contrast to these findings, Kevane and Gray (2008) reject the characterization of the Darfur conflict as a climate change—inspired conflict. Based on their analyses of long-term rainfall datasets across SSA between 1881 and 2006, they observed that there was no correlation between structural breaks in rainfall and incidence of conflict in the area. They point instead to weak governance and political marginalization in Sudan as being largely responsible for the violence.

Although droughts may help fuel farmer—herder conflicts and rebel insurgencies in East Africa, other socioeconomic and political factors cannot be discounted. In the Kilosa District of Tanzania, for example, Benjaminsen et al. (2009) explain that violent clashes between local farmers and pastoralists were not only due to competition for scarce resources but also triggered by government villagization programs, uncertain land tenure, antipastoral policies, and corruption. These findings highlight the potential danger in narrowly focusing on drought, environmental stress, and climatic extremes in addressing environmentally related conflicts. The findings also lend credence to the need for integrated early warning systems that concurrently monitor the socioeconomic, political, and environmental dynamics as complementary triggers of violent conflicts in the region (Meier et al., 2007).

Droughts, mobility, and conflict in Southern Africa

The IPCC (2014) predicts that Southwestern Africa will likely experience severe droughts during the 21st century, leading to water shortages and food insecurity. An IOM examination of the SADC (Southern African Development Community) region concludes that rising uncertainty in agricultural cultivation, livestock production, and subsistence farming will trigger environmental migration in search for alternative livelihood strategies (Pourazar, 2017). The SADC region experienced its worst drought in 35 years during the 2015—16 El Niño. As a result of the drier than normal conditions, there was a decline in precipitation levels, which led to poor crop harvests in the region. This resulted in the corresponding declaration of a regional drought disaster in March 2016 by SADC Council of Ministers (RIASCO, 2016; UN-OCHA, 2016; Egeland et al., 2016). In total, about 32 million people were considered food insecure, with harvests of main staples such as maize, wheat, and sorghum being especially affected (UN-OCHA, 2016; World Bank, 2013; Kassie et al., 2013).

Drought-induced water scarcity is a growing challenge in South Africa. The country's dependence on surface water resources has seen a significant increase in demand for both domestic and agricultural use (Lötter, 2017; Touter and Mauck, 2017; Conrad and Carstens, 2017). An ongoing lack of rain in the Western Cape Province since 2015 led to severe water shortages in Cape Town and the surrounding areas in 2018. This necessitated water rationing by city authorities where Cape Town residents were restricted to use 50 L of water per day (Perine and Keuck, 2018; Donnenfeld, 2018). Although the city managed to half its annual water consumption from 2016 to 2017, Cape Town water authorities announced April 12, 2018, as "day zero." The day meant that water supply would be cut off if water consumption was not reduced drastically (Vogt and Barbosa, 2018). Better rainfall in mid-2018 allowed authorities to relax restrictions, but the need for careful usage of scarce water resources persists (Pérez-Pena, 2018; McKenzie and Swails, 2018).

Drought is also a significant challenge in Madagascar's arid "Grand Sud" region. The region is home to 1.8 million people, of which roughly 80% drive their livelihood from rain-fed agriculture (UN Office of the Resident Coordinator Madagascar, 2016). Since 2013, the region suffered an extreme drought period that was reinforced by the impacts of the 2015 El Niño. The situation degenerated into a humanitarian crisis in February 2016 that left more than one million people food insecure (UN-OCHA, 2016). Affected communities use migration as a strategy to cope with drought (IOM, 2017b). In the Androy District, drought and subsequent lack of revenue from farming precipitated high levels of out-migration from rural areas to the district capital in search of income. Unlike previous times, when people would return to their home areas when rain was expected, people who migrated during the drought were observed to be staying much longer or permanently.

Drought is also a significant stimulus for migration in Botswana, where farmers and pastoralists in arid and semiarid areas experience erratic rainfall, water shortages, dry spells, and desertification (Heita, 2018; Pourazar, 2017; Kgosikoma, 2006; UNDP, 2018). Pastoralists especially use mobility to cope with drought and to secure sufficient water and forage for their livestock (Kgosikoma, 2006). Increasing rural–urban migration to search for alternative economic opportunities puts pressure on rapidly growing urban areas such as Gabarone, Francistown, Kweneng East, and Selebi-Phikwe (Pourazar, 2017).

At the regional level, Ashton (2000) suggests that conflicts and disputes over water will continue well into the future, as increasing demand and pressure continue on finite water resources in Southern Africa. But, given that water scarcity–induced conflicts have generally been secondary to concerns of territorial integrity or sovereignty, he is of the contention that it is highly unlikely that conflicts over water would degenerate into any full-scale military conflict or "water wars" in Southern Africa. Nevertheless, recurring and severe droughts have rekindled latent conflicts relating to the use of shared water resources in the region. The Chobe River, which drains Botswana, Namibia and Zambia, has long been a subject of dispute between the countries. Although Botswana won an international court case as the owner of the Sedudu Island, situated in the middle of the Chobe River, drought-induced declines in water levels have pushed Namibian fishermen to venture in to the Botswana side of the Island. The increasing tensions have led to shootings and subsequent arrests of Namibian fishermen by the Botswana Army (Kings, 2016). Similar disputes arising from the use and ownership of the Orange and Limpopo Rivers have also led to both overt and covert tensions across countries that share the rivers.

For the SADC region, there are long-established treaties between countries on the use of shared water resources. But due to a seeming weakness in enforcing or adhering to these treaties, increasing frequency and severity of droughts in the region could ignite simmering tensions with the potential to escalate into large-scale violent conflicts in the region. In view of the likely negative implications this would have for socioeconomic and political stability in the region, a recommendation is made for a more integrated and fairer water agreements (Kings, 2016).

Summary and conclusions: what are the policy options?

The evidence concerning the relationship between droughts, human mobility, and conflict as described above suggests a highly complex and nonlinear set of interactions. As widely noted in the empirical research, and in Reuveny's (2007) highlighting of the linkages between climate/environmental change, migration, and conflicts, it can be discerned that the context (structural conditions) and other intervening factors interact to ignite conflicts during droughts. As compared with other environmental hazards, droughts seem to trigger different forms of human mobility depending on certain political, economic, and sociocultural conditions, with great variations among the three regions of SSA. Some important summary points to be taken away from each region are as follows:

1) In West Africa, cooperation in the use of resources has long been integral to farmer–herder relations. But recurring droughts have contributed to the loss of farmland, declining crop yields, and loss of livestock in semiarid areas (e.g., in Mali, Niger, Senegal). Aside from the impact on food security, drought-induced water scarcity, mobility and competition for the use of scarce natural resources, as observed in the Lake Chad region, droughts have become major causes of farmer–herder conflicts and interstate tensions for transboundary water resources. Alongside the tendency of droughts to further accentuate already existing mobility patterns, there is also a potential for families overwhelmed by loss of livestock, livelihoods, and/or lack of institutional or social support systems, to become trapped and unable to move as a coping strategy. Increasing numbers of violent farmer–herder conflicts and tensions over water have the potential to escalate latent conflicts, with implications for the already fragile security situation and stability in the region—with an increasing likelihood of (further) forced displacement.

2) In Eastern Africa, countries continue to grapple with the devastating effects of more frequent and intensive droughts in the region. The subsequent impacts on food and pastureland tends to draw competing parties into conflicts. Existing ethnic and political tensions serve to fuel pervasive farmer–herder conflicts, with droughts having further heightened food insecurity and economic losses. The increased resource-based conflicts and massive drought-induced displacement are particularly suffered by pastoralists and rural small-scale farmers who inhabit highly drought-prone areas. Rural-to-urban migration is more related to general livelihood diversification but is used as an emergency response to drought as well.

3) Arid and semiarid areas of Southern Africa have experienced major droughts in the recent past. These have largely contributed to severe impacts on water supplies and production of major crops like maize and sorghum. Urban centers in South Africa are especially at risk. Prolonged dry spells and desertification affect cattle and livestock farming. Mostly, the affected people migrate to search for alternative livelihood strategies. Although migration was previously seasonal or temporary, observations from recent studies in Botswana and Madagascar reveal that migration is increasingly becoming more permanent in nature. Increasing demand and pressure on shared water resources during droughts have raised tensions between countries in the SADC region. Consequently, well-managed water resources and use is critical to stemming the potential of any transboundary armed conflicts in the region.

Generally, in SSA, drought-related migration is often seasonal or circular labor within countries or between neighboring countries. Mostly, individual household members—rather than the entirety of households or families—are migrating for a rather limited period to urban areas, centers of commercial agriculture, or mining sites. A main goal of these migrants is to earn money to remit back home in order to mitigate hardships that their families back home are facing. In this way, translocal spaces of financial, communicational, or other exchange between migrants and their families are being generated, laying the ground for what is often called "migration as adaptation" (McLeman and Smit, 2006; Gemenne and Blocher, 2017).

A key political emphasis should thus be on governing human mobility in a way that makes it possible to avoid forced displacement and any foreseeable conflicts. This will contribute to maximizing positive mechanisms of migration (e.g., financial or other remittances) and to minimizing negative aspects, such as labor exploitation or human trafficking. The migration policy community often speaks favorably of "migration as adaptation," and emphasizes governance challenges when it comes to human mobility in the context of climate. By contrast, the environmental or climate policy community is still struggling with the issue of "climate migration" and rather perceives it as highly sensitive, a phenomenon that should be prevented. It is necessary to foster a dialogue between different policy fields and communities: climate change, migration, development cooperation, urban planning, humanitarian assistance, rural development, and agriculture. Scientists as well as practitioners need to facilitate such a knowledge-based dialogue. The goal has to be overcoming the predominant "sedentary bias" of preventing migration (instead of actively setting a framework for it), generating a common understanding of the challenges related to human mobility in the context of climate change, and creating an awareness of it being a cross-cutting problem, which can only be addressed in a joint effort.

In terms of drought and conflict management, there is the need for a more proactive approach to addressing drought impacts. Alongside the integration of drought and conflict into early warning systems, national policies should focus on mainstreaming drought and conflict management into long-term national development plans. Emphasis should be placed on incorporating local knowledge and participation of local authorities in conflict mediation and efficient early warning systems. This would help drive more thorough consideration of early warning systems and comprehensive appreciation of climate change—migration—conflict linkages in stemming resource use and farmer—herder conflicts in SSA.

Efforts at disaster risk reduction and mitigation should also consider the efficiency of weather-based agriculture or crop insurance schemes to stem the impacts of droughts on the continent. Besides encouraging the cultivation of drought-resistant crops and the promotion of climate-smart agriculture, there is a need for governmental policy to focus on equipping the appropriate institutions to raise awareness. Additional support in the form of subsidies or premiums for poor farmers and pastoralists would encourage the consideration of weather-based insurance schemes. These efforts could be complemented by drought mitigation and land degradation strategies like rainwater harvesting, underground water irrigation, construction of dug-outs at strategic locations, shared management of designated pasturelands, and the cultivation of cover crops to promote soil—water retention and prevention of erosion. Many of the following chapters of this volume provide further details on the implementation of such strategies. Such efforts would greatly contribute to addressing the role of drought in triggering large-scale displacements and the potential for conflicts.

References

Abubakar, I.U., Yamusa, M.A., 2013. Recurrence of drought in Nigeria: causes, effects and mitigation. International Journal of Agriculture and Food Science Technology 4 (3), 169−180.

Ashton, P., 2000. Southern African Water Conflicts: Are they Inevitable or Preventable? The African Dialogue Lecture Series. Pretoria University, South Africa.

Afifi, T., Govil, R., Sakdapolrak, P., Warner, K., 2012. Climate change, vulnerability and human mobility: perspectives of refugees from East and Horn of Africa. In: UNU-EHS Report No. 1. Partnership Between UNU and UNHCR. United Nations University Institute for Environment and Human Security (UNU-EHS), Bonn.

Barnett, J., Adger, W.N., 2007. Climate change, human security and violent conflict. Political Geography 26, 639−655.

Batterbury, S., Warren, A., 2001. The African Sahel 25 years after the great drought: assessing progress and moving towards new agendas and approaches. Global Environmental Change 11, 1−8.

Benjaminsen, T.A., Maganga, F.P., Abdallah, J.M., 2009. The Kilosa killings: political ecology of a farmer−herder conflict in Tanzania. Development and Change 40 (3), 423−445.

Benjaminsen, T.A., Ba, B., 2009. Farmer−herder conflicts, pastoral marginalisation and corruption: a case study from the Inland Niger Delta of Mali. The Geographical Journal 175 (1), 71−81.

Benjaminsen, T., 2008. Does supply-induced scarcity drive violent conflicts in the African Sahel? The case of the Tuareg rebellion in northern Mali. Journal of Peace Research 45 (6), 819−836.

Blench, R., 2004. Natural Resource Conflicts in North-Central Nigeria: A Handbook and Case Studies. Mandaras Publishing, London.

Booth, B.B.B., Dunstone, N.J., Halloran, P.R., Andrews, T., Bellouin, N., 2012. Aerosols implicated as a prime driver of twentieth century north Atlantic climate variability. Nature 484 (7393), 228−232.

Brooks, N., 2004. Drought in the African Sahel: long term perspectives and future prospects. Tyndall Centre for Climate Change Research 61, 1−31.

Brown, I.A., 2010. Assessing eco-scarcity as a cause of the outbreak of conflict in Darfur: a remote sensing approach. International Journal of Remote Sensing 31 (10), 2513−2520.

Brown, O., Crawford, A., 2008. Assessing the Security Implications of Climate Change for West Africa: Country Case Studies of Ghana and Burkina Faso. International Institute for Sustainable Development, Manitoba, Canada.

Brown, O., Hammill, A., McLeman, R., 2007. Climate change as the 'new' security threat: implications for Africa. International Affairs 83 (6), 1141−1154.

Bukari, K.N., Sow, P., Scheffran, J., 2018. Cooperation and Co-existence between farmers and herders in the midst of violent farmer-herder conflicts in Ghana. African Studies Review 61 (2), 78−102.

Cabot, C., 2017. Climate change and farmer−herder conflicts in West Africa. In: Cabot, C. (Ed.), Climate Change, Security Risks and Conflict Reduction. Springer.

Carius, A., Dabelko, G.D., Wolf, A.T., 2004. Water, conflict and cooperation. ECSP Report, Issue 10 (Policy Brief), pp. 60−66.

Castles, S., 2002. Environmental Change and Forced Migration: Making Sense of the Debate. New Issues in Refugee Research. Working Paper No. 70. UNHCR, Geneva, pp. 1−14.

Conrad, J., Carstens, M., 2017. Groundwater. In: Mambo, J., Faccer, K. (Eds.), Understanding the Social and Environmental Implications of Global Change. Second Edition of the Risk and Vulnerability Atlas. Pretoria.

Dai, A., 2011. Drought under global warming: a review. Advanced Review 2, 45−65.

Dai, A., et al., 2004. Comment: the recent Sahel drought is real. International Journal of Climatology 24, 1323−1331.

De Bruijn, M., Van Dijk, H., Kaag, M., Van Til, K. (Eds.), 2005. Sahelian Pathways: Climate and Society in Central and South Mali. African Studies Centre Research Report, 78/95.

Dietz, A.J., Ruben, R., Verhagen, A. (Eds.), 2001. The Impact of Climate Change on Drylands with a Focus on West Africa. NOP-ICCD Research Project 952240, Wageningen.

Donnenfeld, Z., 2018. South Africa's Water Crisis is Bigger than the Cape. https://issafrica.org/iss-today/south-africas-water-crisis-is-bigger-than-the-cape.

Egeru, A., 2015. 'Mental Drought' Afflicts Uganda's Cattle Corridor. https://www.researchgate.net/publication/305392400_'Mental_drought'_afflicts_Uganda's_cattle_corridor.

Egeland, J., Bilak, A., Rushing, E., 2016. Global Report on Internal Displacement 2016. http://www.internal-displacement.org/globalreport2016/.

Ezra, M., Kiros, G.-E., 2001. Rural out-migration in the drought prone areas of Ethiopia: a multilevel Analysis. International Migration Review 35 (3), 749−771.

Famine Early Warning Systems Network (FEWS-NET), 2017. Prolonged drought drives a food security Emergency in Somalia and southeastern Ethiopia, *EAST AFRICA Food Security Alert*. http://fews.net/sites/default/files/documents/reports/EA_Alert_06_2017_final.pdf.

Findley, S.E., 1994. Does drought increase migration? A study of migration from rural Mali during the 1983-1985 drought. International Migration Review 28 (3), 539−553.

Food and Agricultural Organization of the United Nations (FAO), 2018. Resilience analysis in Karamoja, Uganda report. http://www.fao.org/3/i8365en/I8365EN.pdf.

Gautier, D., Denis, D., Loactelli, B., 2016. Impacts of drought and responses of rural populations in West Africa: a systematic review. Climate Change 7, 666−681.

Gemenne, F., Blocher, J., 2017. How can migration serve adaptation to climate change? Challenges to fleshing out a policy ideal. The Geographical Journal 1−12.

Giannini, A., Krishnamurthy, P.K., Cousin, R., Labidi, N., Choularton, R.J., 2017. Climate risk and food security in Mali: a historical perspective on adaptation. Earth's Future 5, 144−157.

Giannini, A., Salack, S., Lodoun, T., Ali, A., Gaye, A.T., Ndiaye, O., 2013. A unifying view of climate change in the Sahel linking intra-seasonal, interannual and longer time scales. Environmental Research Letters 8, 1−8.

Giannini, A., Biasutti, M., Held, I.M., Sobel, A.H., 2008. A global perspective on African climate. Climate Change 90, 359−383.

Gleditsch, N.P., Nordås, R., Salehyan, I., 2007. Climate Change and Conflict: The Migration Link. In: Coping with Crisis Working Paper Series. International Peace Academy.

Global Facility for Disaster Reduction and Recovery (GFDRR), 2017. Natural Hazard Risk: Uganda. https://www.gfdrr.org/en/uganda.

Global Facility for Disaster Reduction and Recovery (GFDRR), 2011. Vulnerability, Risk Reduction, and Adaptation to Climate Change: Climate Risk and Adaptation Country Profile (Ethiopia). https://www.gfdrr.org/sites/default/files/publication/climate-change-country-profile-2011-ethiopia.pdf.

Gray, C., Mueller, V., 2012. Drought and population mobility in rural Ethiopia. World Development 40 (1), 134−145.

Haarsma, R.J., Selten, F.M., Weber, S.L., Kliphuis, M., 2005. Sahel rainfall variability and response to greenhouse warming. Geophysical Research Letters 32 (L17702), 1−4.

Hammer, T., 2004. Desertification and migration: a political ecology of environmental migration In West Africa. In: Unruh, J.D., Krol, M.S., Kliot, N. (Eds.), Environmental Change and its Implications for Population Migration. Kluwer, Dordrecht, pp. 231−246.

Hendrix, C.S., Glaser, S.M., 2007. Trends and triggers: climate, climate change and civil conflict in sub-Saharan Africa. Political Geography 26 (6), 695−715.

Henry, S., Schoumaker, B., Beauchemin, C., 2004. The impact of rainfall on the first out-migration: a multi-level event-history analysis in Burkina Faso. Population and Environment 25 (5), 423−460.

Henry, S., Boyle, P., Lambin, E.F., 2003. Modelling inter-provincial migration in Burkina Faso, West Africa: the role of socio- demographic and environmental factors. Applied Geography 23, 115−136.

Heita, J., 2018. Assessing the Evidence: Migration, Environment and Climate Change in Namibia. Geneva. https://environmentalmigration.iom.int/assessing-evidence-migration-environment-and-climate-change-namibia.

Homer-Dixon, T.F., 1994. Environmental scarcities and violent conflict: evidence from cases. International Security 19 (1), 5−40.

Huho, J., Mugalavai, E., 2010. The effects of droughts on food security in Kenya. The International Journal of Climate Change: Impacts and Responses 2 (2), 61−72.

Hulme, M., 2001. Climatic perspectives on Sahelian desiccation: 1973-1998. Global Environmental Change 11, 19−29.

Climate change, environment and migration in the Sahel: Selected issues with a focus on Senegal and Mali. In: Hummel, D., Doevenspeck, M., Samimi, C. (Eds.), 2012. Micle Working Paper, 9.

Hummel, D., 2016. Climate change, land degradation and migration in Mali and Senegal: some policy implications. Migration and Development 5 (2), 211−233.

Intergovernmental Panel on Climate Change (IPCC), 2007. Climate change 2007: impacts, adaptation and vulnerability. In: Parry, M.L., Canziani, O.F., Palutikof, J.P., van der Linden, P.J., Hanson, C.E. (Eds.), Contribution of Working Group II to the Fourth Assessment Report of the Intergovernmental Panel on Climate Change. Cambridge University Press, Cambridge.

Intergovernmental Panel on Climate Change (IPCC), 2014. Climate change 2014: synthesis report. In: Contribution of Working Groups I, II, and III to the Fifth Assessment Report of the Intergovernmental Panel on Climate Change, Geneva.

IPCC, 2013. Climate change 2013: the physical science basis. In: Stocker, T.F., Qin, D., Plattner, G.-K., Tignor, M., Allen, S.K., Boschung, J., Nauels, A., Xia, Y., Bex, V., Midgley, P.M. (Eds.), Contribution of Working Group I to the Fifth Assessment Report of the Intergovernmental Panel on Climate Change. Cambridge University Press, Cambridge, United Kingdom and New York, NY, USA, p. 1535.

Intergovernmental Panel on Climate Change (IPCC), 2001. Climate change 2001: impacts, adaptation and vulnerability. In: McCarthy, J.J., Canziani, O.F., Leary, N.A., Dokken, D.J., White, K.S. (Eds.), Contribution of Working Group II to the Third Assessment Report of the Intergovernmental Panel on Climate Change. Cambridge University Press, Cambridge, UK, and New York, USA, p. 1032.

International Organization for Migration (IOM), 2018. Mapping human mobility (migration, displacement and planned relocation) and climate change in international processes, policies and legal frameworks. In: Pillar II: Policy − International/Regional Activity II.2. IOM Migration, Environment and Climate Change Division. Available at: https://unfccc.int/sites/default/files/resource/WIM%20TFD%20II.2%20Output.pdf.

International Organization for Migration (IOM), 2017a. IOM East and Horn of Africa Drought Appeal. https://www.iom.int/sites/default/files/country_appeal/file/East_Africa_Drought_Appeal-apr-dec2017.pdf.

International Organization for Migration (IOM), 2017b. Evidencing the Impacts of the Humanitarian Crisis in Southern Madagascar on Migration, and the Multisectoral Linkages that Drought-Induced Migration has on Other Sectors of Concern. https://environmentalmigration.iom.int/evidencing-impacts-humanitarian-crisis-southern-madagascar-migration-and-multisectorial-linkages.

Jordaan, A.J., 2015. Karamoja, Uganda Drought Risk Assessment: Is Drought to Blame for Chronic Food Insecurity? http://www.academia.edu/11756290/Karamoja_Drought_Risk_Assessment_Is_drought_to_blame_for_chronic_famine_and_food_insecurity.

Kallis, G., 2008. Droughts. Annual Review of Environment and Resources 33, 85−118.

Kassie, B.T., Hengsdijk, H., Rötter, R., Kahiluoto, H., Asseng, S., Van Ittersum, M., 2013. Adapting to climate variability and change: experiences from cereal-based farming in the central rift and Kobo Valleys, Ethiopia. Environmental Management 52.

Kgosikoma, O.E., 2006. Effects of Climate Variability on Livestock Population Dynamics and Community Drought Management in Kgalagadi, Botswana. http://www.ccardesa.org/knowledge-products/effects-climate-variability-lifestock-population-dynamics-and-and-community.

Kings, S., 2016. Climate Change is Testing Southern Africa Water Agreements. https://www.climatechangenews.com/2016/12/02/climate-change-is-testing-southern-africa-water-agreements/.

Kevane, M., Gray, L., 2008. Darfur: rainfall and conflict. Environmental Research Letters 3, 1−10.

Lee, S.W., 2001. Emerging threats to international security: environment, refugees, and conflict. Journal of International and Area Studies 8 (1), 73−90.

Liehr, S., Drees, L., Hummel, D., 2016. Migration as societal response to climate change and land degradation in Mali and Senegal. In: Yaro, J.A., Hesselberg, J. (Eds.), Adaptation to Climate Change and Variability in Rural West Africa. Springer International Publishing, Switzerland.

Linke, A.M., et al., 2018. The consequences of relocating in response to drought: human mobility and conflict in contemporary Kenya. Environmental Research Letters 13, 1−9.

Lott, F.C., Christidis, N., Stott, P.A., 2013. Can the 2011 east African drought be attributed to human-induced climate change? Geophysical Research Letters 40, 1177−1181.

Lötter, D., 2017. Risk and vulnerability in the South African farming sector. Implications for sustainable agriculture and food security. In: Mambo, J., Faccer, K. (Eds.), Understanding the Social and Environmental Implications of Global Change, Second Edition of the Risk and Vulnerability Atlas. Pretoria.

McKenzie, D., Swails, B., 2018. Day Zero Deferred, But Cape Town's Water Crisis is Far from Over. https://edition.cnn.com/2018/03/09/africa/cape-town-day-zero-crisis-intl/index.html.

McLeman, R., Smit, B., 2006. Migration as an adaptation to climate change. Climate Change 76 (1−2), 31−53.

Meier, P., Bond, D., Bond, J., 2007. Environmental influences on pastoral conflict in the Horn of Africa. Political Geography 26, 716−735.

Mortimore, M., 2010. Adapting to drought in the Sahel: lessons for climate change. WIREs Climate Change 1 (1), 134−1143.

Mortimore, M.J., Adams, W.M., 2001. Farmer adaptation, change and 'crisis' in the Sahel. Global Environmental Change 11 (1), 49−57.

Moritz, M., 2010. Understanding herder-farmer conflicts in West Africa: outline of a processual approach. Human Organization 69 (2), 138−148.

Moritz, M., 2006. Changing contexts and dynamics of farmer-herder conflicts across West Africa. Canadian Journal of African Studies 40 (1), 1−40.

Mwangi, E., Wetterhall, F., Dutra, E., Di Giuseppe, F., Pappenberger, F., 2014. Forecasting droughts in East Africa. Hydrology and Earth System Sciences 18, 611−620.

Ndaruzaniye, V., et al., 2010. Climate Change and Security in Africa. https://www.africa-eu-partnership.org/sites/default/files/documents/doc_climate_vulnerability_discussion_paper.pdf.

Nett, K., Rüttinger, L., 2016. Insurgency, Terrorism and Organised Crime in a Warming Climate: Analysing the Links Between Climate Change and Non-State Armed Groups. Climate Diplomacy Report. Adelphi, Berlin.

Nielsen, J.Ø., Reenberg, A., 2010. Temporality and the problem with singling out climate as a current driver of change in a small West African village. Journal of Arid Environments 74 (4), 464−474.

Nyaoro, D., Schade, J., Schmidt, K., 2016. Assessing the Evidence: Migration, Environment and Climate Change in Kenya. https://publications.iom.int/books/assessing-evidence-migration-environment-and-climate-change-kenya.

Okeke, O.E., 2014. Conflicts between Fulani herders and farmers in Central and Southern Nigeria: discourse on proposed establishment of grazing routes and reserves. AFRREV IJAH 3 (1), 66−84. S/9.

Pearson, N., Niaufre, C., 2013. Desertification and Drought Related Migrations in the Sahel: The Cases of Mali and Burkina. The State of the Environment, IDDR, pp. 79−98.

Perine, C., Keuck, H., 2018. Building Urban Resilience to Climate Change. A Review of South Africa. https://www.climatelinks.org/sites/default/files/asset/document/180327_USAID-ATLAS_Building%20Urban%20Resilience%20to%20CC_South%20Africa_to%20CL_rev.pdf.

Pérez-Pena, R., 2018. Cape Town Pushes Back 'day Zero' as Residents Conserve Water. https://www.nytimes.com/2018/02/20/world/africa/cape-town-water-day-zero.html.

Pourazar, E., 2017. Spaces of Vulnerability and Areas Prone to Natural Disaster and Crisis in Six SADC Countries. https://publications.iom.int/books/spaces-vulnerability-and-areas-prone-natural-disaster-and-crisis-six-sadc-countries.

Raleigh, C., 2010. Political marginalization, climate change and conflict in African Sahel states. International Studies Review 12 (1), 69−86.

Raleigh, C., Kniveton, D., 2012. Come rain or shine: an analysis of conflict and climate variability in East Africa. Journal of Peace Research 49 (1), 51−64.

Raleigh, C., Jordan, L., Salehyan, I., 2008. Assessing the Impact of Climate Change on Migration and Conflict. Social Development Department, The World Bank, Washington DC.

Raleigh, C., Urdal, H., 2007. Climate change, environmental degradation and armed conflict. Political Geography 26, 674−694.

Regional Inter-Agency Standing Committee (RIASCO), 2016. RIASCO Action Plan for Southern Africa. Revised Regional Response Plan for the El Nino-Induced Drought in Southern Africa. https://reliefweb.int/report/world/riasco-action-plan-southern-africa-response-plan-el-ni-o-induced-drought-southern.

Reuveny, R., 2007. Climate change-induced migration and violent conflict. Political Geography 26, 656−673.

Reuveny, R., 2008a. Ecomigration and violent conflict: case studies and public policy implications. Human Ecology 36, 1−13.

Reuveny, R., 2008b. Climate change-induced migration and violent conflict. Political Geography 26, 656−673.

Sarr, B., 2012. Present and future climate change in West Africa: a crucial input for agricultural research prioritization for the region. Atmospheric Science Letters 13, 108−112.

Tanasaldy, T., 2009. Ethnic geography in conflicts: the case of West Kalimantan, Indonesia. Review of Indonesian and Malaysian Affairs 43 (2), 105−130.

Tonah, S., 2006. Migration and farmer-herder conflicts in Ghana's Volta basin. Canadian Journal of African Studies /Revue canadienne des études africaines 40 (1), 152−178.

Touter, M., Mauck, B., 2017. Surface water. In: Mambo, J., Faccer, K. (Eds.), Understanding the Social and Environmental Implications of Global Change, Second Edition of the Risk and Vulnerability Atlas. Pretoria.

Turton, A.R., Patrick, M.J., Julien, F., 2006. Transboundary water resources in southern Africa: conflict or cooperation? Development 49, 1−10.

United Nations Development Programme (UNDP), 2018. Climate Change Adaptation in Africa. UNDP Synthesis of Experiences and Recommendations 2000−2015. http://www.undp.org/content/dam/undp/library/Climate%20and%20Disaster%20Resilience/Climate%20Change/CCA-Africa-Final.pdf.

United Nations Environment Programme (UNEP), 2007. Sudan: Post-Conflict Environmental Assessment. United Nations Environment Programme (UNEP), Nairobi-Kenya.

United Nations Office for the Coordination of Humanitarian Affairs (UNOCHA), 2016. El Niño: Overview of Impact, Projected Humanitarian Needs and Responses. https://reliefweb.int/report/world/el-ni-o-overview-impact-projected-humanitarian-needs-and-response-02-june-2016.

United States of America Department of Defense, 2019. Report on Effects of a Changing Climate to the Department of Defense. https://climateandsecurity.files.wordpress.com/2019/01/sec_335_ndaa-report_effects_of_a_changing_climate_to_dod.pdf.

Van Loon, A.F., et al., 2016. Drought in the anthropocene. Nature Geoscience 9, 89−91.

Vogt, J., Barbosa, P., 2018. Drought and Water Crisis in Southern Africa. http://publications.jrc.ec.europa.eu/repository/bitstream/JRC111596/drought_water_crisis_in_southern_africa2018_doi_isbn.pdf.

Weber, T., Haensler, A., Rechid, D., Pfeifer, S., Eggert, B., Jacob, D., 2018. Analyzing regional climate change in Africa in a 1.5, 2, and 3°C global warming world. Earth's Future 6, 643−655.

Weiss, H., 2003. Migrations during times of drought and famine in early colonial northern Nigeria. Studia Orientalia 95, 1−29.

World Bank, 2013. Turn Down the Heat. Climate Extremes, Regional Impacts, and the Case for Resilience. Washington D.C. http://www.worldbank.org/en/topic/climatechange/publication/turn-down-the-heat-climate-extremes-regional-impacts-resilience.

Further reading

Baudoin, M.-A., Vogel, C., Naik, M., 2017. Living with drought in South Africa: lessons learnt from the recent El Nino drought period. International Journal of Disaster Risk Reduction 23, 128–137.

Collier, P., Conway, G., Venables, T., 2008. Climate change and Africa. Oxford Review of Economic Policy 24 (2), 337–353.

International Organization for Migration (IOM) Madagascar, 2017. IOM Madagascar Annual Report 2017. https://www.iom.int/sites/default/files/country/docs/Madagascar/iom-madagascar-annual-report-2017.pdf.

Lesolle, D., 2012. SADC Policy Paper on Climate Change: Assessing the Policy Options for SADC Member States. https://www.sadc.int/files/9113/6724/7724/SADC_Policy_Paper_Climate_Change_EN_1.pdf.

Nicholson, S., 2005. On the question of the "Recovery" of the rains in the West African Sahel. Journal of Arid Environments 63, 615–641.

United Nations Children Emergency Fund (UNICEF), 2018. Djibouti Humanitarian Situation Report: Situation in Numbers. https://www.unicef.org/appeals/files/UNICEF_Djibouti_Humanitarian_Situation_Report_MidYear_2018.pdf.

Lessons from the El Nino–induced 2015/16 drought in the Southern Africa region

3

Lewis Hove[a,*]**, Cuthbert Kambanje**[b]

[a]*Resilience Hub for Southern Africa, Food and Agriculture Organization of the United Nations, Johannseburg, South Africa,*
[b]*Food and Agriculture Organization of the United Nations, Pretoria, South Africa*
[*]*Corresponding author*

Introduction

The Southern Africa region experienced a severe El Nino–induced drought during the 2015/16 season. This drought adversely affected many sectors of the regional economy including agriculture, food and nutrition security, health, water, sanitation, and energy. The drought resulted in reduced crop yields, with the region recording a 15% decrease in yield compared to 5-year average for maize which is the main staple cereal of the region (SADC, 2016). Overall, the region had a cereal deficit of 9.1 million tons that had to be procured from outside the region (SADC, 2016). The drought also depleted above and below groundwater resources for human consumption and for crop and livestock production. As a result of the drought, 634,000 cattle were reported to have died because of disease outbreaks and inadequate pasture and water (UNICEF, 2017). The poor sanitation and disrupted livelihoods resulted in increased reports of communicable diseases. Energy generation at Kariba Dam was reduced to 50% of potential capacity, with adverse effects on the Zambia and Zimbabwe industries (Goussard, 2016). In a region where more than 70% of the population depends on rain-fed agriculture for livelihoods and income (SADC, 2018), the drought affected more than 40 million people with 23 million requiring emergency humanitarian support by June 2016 (SADC, 2016; OCHA, 2016). In this paper, drought is defined as the naturally occurring phenomenon that exists when precipitation has been significantly below normal recorded levels, causing serious hydrological imbalances that adversely affect land resource production systems (WMO, 1992)

Development practitioners and scholars alike generally concede that droughts are becoming more and more frequent in the region. There is also a general belief that the El Niño Southern Oscillation (ENSO) phenomenon which affects the Southern African region's weather is becoming stronger and bringing more extreme weather events to the region. That the 2015/16 drought had been forecast in August 2015 and most countries developed contingency plans which were not timeously implemented, is an indication that there are challenges with respect to political will and therefore affecting resources

Drought Challenges. https://doi.org/10.1016/B978-0-12-814820-4.00003-1

33

allocation to prevent disasters. It would also appear that the focus is more on responding to drought rather than acting to prevent the usually drastic consequences.

Given the foregoing, the purpose of this chapter is to derive some lessons from the occurrence of the 2015/16 drought, i.e., in terms of the emerging pattern of natural phenomenon of the drought to the associated response frameworks at various levels (regional, national, and community), to inform future drought policy, program, and project development. The chapter pursues three objectives beginning by a characterization of the phenomenon of droughts in Southern Africa focusing on the frequency and impact. This is followed by a review of the responses to drought at various levels including at national, regional, and local levels examining the policy, program, project, institutional characteristics of the responses. The chapter concludes by a synthesis of the lessons that can be learned from both the emerging character and response frameworks to drought in Southern Africa.

Materials and methods

The chapter relies on the review of secondary data and information and presents it in a descriptive format to achieve its objectives. Data and information were collected through a review of literature and interviews with selected key informants in the agriculture, disaster management and humanitarian sectors in the Southern African region. For information related to Southern Africa's response to the El Nino drought, the chapter is mainly informed by an in-depth review of response plans as well as reports on the review of the responses for the Southern African Development Community (SADC), Food and Agriculture Organization of the United Nations (FAO), United Nations Children's Education Fund (UNICEF), World Food Program (WFP), Office for Coordination of Humanitarian Affairs (OCHA), a number of bilateral and multilateral development partners and international nongovernmental organizations. A limited number of interviews were held with the staff members that were coordinating the regional response including those from the SADC Disaster Risk Reduction Unit and the SADC Climate services Center to the regional level to follow up on some issues. To ground the analysis in theory, the authors also use information from and reviews of selected scholarly articles on droughts, their impacts and response. Although the authors cite from scientific sources, the major limitation is limited availability of scientific articles on this emerging and contemporary issue, thus explaining the overreliance on reports from UN agencies, regional institutions, and other organizations that are actually involved in implementation of response programs to deal with emergencies in the region.

A characterization of the drought phenomenon in Southern Africa
Theoretical foundations—what constitutes a drought?

To begin characterizing the drought phenomenon in Southern Africa and due to the many definitions that many scholars have put forward, it is important to highlight that in this chapter the authors go by the definitions put forward by the WMO (1992). Drought is defined as "prolonged absence or marked deficiency of precipitation" or "period of abnormally dry weather sufficiently prolonged for the lack of precipitation to cause a serious hydrological imbalance." In addition to this definition, according to Zoltan (2008), it is different from dry seasons which are the period of the year characterized by almost complete absence of rainfall and dry spells which are periods of abnormally dry weather. Zoltan (2008) goes further to highlight various types of droughts, namely atmospheric drought which occurs when there is "too high saturation deficit"; meteorological drought which is when there is "a longer period of time with considerably less than average precipitation amounts"; agricultural drought which occurs when "available soil moisture is inadequate and yield is considerably less than the average because of water shortage"; hydrological drought which refers to "a period of below-normal streamflow"; physiological drought which is when "plants are unable to take up water in spite of the present sufficient soil moisture"; and, socioeconomic drought which affects the supply and demand of some economic goods with elements of met, hydrological, and agricultural droughts. Gleaning from these definitions, it is clear that the 2015/16 drought that occurred in Southern Africa had strong elements of atmospheric, meteorological, agricultural, hydrological, physiological,[1] and socioeconomic droughts, with the various types being manifestations of the prolonged absence or deficiency of precipitation.

Retrospective analysis of drought occurrence in Southern Africa —nature and impact of past droughts

In this section, the authors briefly analyze the nature and impact of droughts that have been experienced in the Southern Africa region from the 1900 up to the El Nino—induced drought during the 2015/16 agricultural season. A number of approaches have been used by different scholars to characterize the severity and impact, while others attempt to retrofit some patterns in the occurrences of drought in the region.

To identify and classify drought events in Southern Africa from 1980, Chikoore (2016) applies a drought index based on the precipitation—evapotranspiration (P—E) anomaly. The P—E anomaly uses the Global Precipitation Climatology Project (GPCP) as the precipitation reference and seasonal cycle and the Modern Era Retrospective Analysis for Research and Analysis (MERRA) sensible heat flux to denote potential evapotranspiration. This drought index identifies extreme drought events during 1982 and 1992, severe drought during 1983 and 1995, and moderate drought conditions during 1984, 1987, and 2003 and concludes that drought events were most frequent and more severe during the 1980s, but fewer compared to the 2000s, while successive droughts occurred during 1982—83.

Although noting the existence of weak long-term trends and higher/increased interannual rainfall variability in Southern Africa since 1970s, Jury (2013) appears to be highlighting the strong influence of the Pacific ENSO as well as interactions with Indian Ocean and Atlantic climates on the intradecadal oscillations, giving some pattern to the occurrence of drought in Southern Africa. According to Chikoore (2016), it is argued that a positive phase of the DMI coupled with warm ENSO act to produce severe drought over Southern Africa highlighting, for instance, that the two co-occurred during 1982/83, 1986/87, 1987/88, 1991/92, 1994/95, and 1997/98 to produce droughts in these years. Table 3.1 shows El Nino and La Nina seasons from 1979 in Southern Africa.

Many scholars, for instance, Jury (2013), Chikoore (2016), Obasi (2005) are in agreement with the patterns that shape the resultant climatic outcomes in Southern Africa with the ENZO modulating year-to-year climate variability among other factors. These include the north—south displacement of the Hadley cell and changes in solar insolation that results in large swings in rainfall over the annual cycle in Southern Africa; the role of zonal circulations in summer causing easterly winds drawing

---

[1]Some authors for example Wilhite, 2016 prefer to have "physiological" drought covered under "agricultural" drought.

Table 3.1 Weak, moderate and strong El Nino and La Nina seasons, 1979–2017.

El Niño seasons			La Niña seasons		
Weak	**Moderate**	**Strong**	**Weak**	**Moderate**	**Strong**
2004/05	1986/87	1982/83	1983/84	1998/99	1988/89
2006/07	1987/88	1997/98	1984/85	2007/08	1999/2000
1991/92		1995/96		2010/11	
1994/95			2000/01		
2002/03			2005/06		
2009/10			2008/09		
2011/12			2012/13		
2013/14		2015/16*	2014/15		
			2016/17		

Source: Modified from Chikoore (2016), except *, inserted by authors.

moisture from the Southwest Indian Ocean and warm Agulhas Current; and in winter, westerly winds bring dry air from the South Atlantic Ocean and cool Benguela Current.

According to Maish et al. (2014), based on the analysis of droughts during 1900–2013, droughts have intensified in terms of frequency, severity, and geospatial coverage over the last few decades—droughts that occurred in 1972–73, 1983–84, and 1991–92 were most intense and widespread. The impact of drought is normally felt across sensitive sectors such as water, food, and energy with a widespread impact on agricultural produce. Due to the high dependence of rain-fed agriculture, agricultural systems in Southern Africa are particularly vulnerable to drought. In the Southern African region, about 80% of rural people, who constitute 70% of the population derive their livelihoods directly from agricultural and natural resources are directly affected by climate-related shocks.

Economies of many countries in sub-Saharan Africa are particularly vulnerable to the effects of drought because they depend on rain-fed agriculture and have low levels of income per capita. Table 3.2 shows frequencies and estimated numbers of affected people and economic damage for disasters in the Southern Africa region between 2000 and 2016. It shows that droughts have affected the highest number of people, approximately 74 million people, and with a high estimated economic damage of US$2.1 billion.

The analysis of the impact of drought on African economies done by Benson and Clay (1994) bring to the fore some classical conclusions which still apply in the contemporary situation. Of particular relevance, based on the analysis of the more severe droughts which affected many African countries in the 1980s, the study concludes that (1) drought shocks have large, but highly differentiated economy-wide impacts; (2) the likely frequency, scale, and character of these impacts depends on the interaction between economic structure and resource endowments, as well as more immediate short-term effects; (3) relatively more developed economies in Africa may be more vulnerable to drought shocks than least developed or arid countries in terms of macroeconomic aggregates and rates of economic recovery; and (4) different regions of Africa are experiencing different longer-term climatic trends which imply that regional strategies are required for mitigation and relief of droughts.

Table 3.2 Frequency, estimated numbers of affected people, and economic damage for five most important disasters in Southern Africa (2000–16).

Disaster type	Occurrence/ frequency	Total affected	Estimated total damage ($)
Flood	198	16,142,359	2,424,204,000
Drought	46	73,842,258	2,108,000,000
Storm	87	5,397,912	858,722,000
Earthquake	15	196,444	515,000,000
Wildfire	11	68,796	440,000,000

Source: From D. Guha-Sapir, R. Below, Ph. Hoyois, Emergency Events Database (EM-DAT). Available from www.emdat.be. EM-DAT, CRED / UCLouvain, Brussels, Belgium - www.emdat.be (D. Guha-Sapir).

The 2015/16 El Nino—induced drought in Southern Africa

As shown in the preceding section, the Southern Africa region weather system is susceptible to two phenomena namely the El Niño and the La Niña which are opposite phases in the ENSO. According to the WMO (2014), the ENSO is a naturally occurring phenomenon involving fluctuating ocean temperatures in the central and eastern equatorial Pacific, coupled with changes in the atmosphere. The El Niño originates from warmer-than-average sea surface temperatures in the eastern Equatorial Pacific and is generally associated with reduced total rainfall that is received in a shorter period than normal across the region; while the La Niña represents below-average sea surface temperatures across the east-central Equatorial Pacific and is associated with above normal rains that usually result in flooding of most of the major rivers of the region (WMO, 2014).

The severity and impact of the 2015/16 El Niño—induced drought creates the need to revisit the whole drought response architecture in the region and enriches the impetus for this chapter. While the natural traditional El Nino phenomenon results in lower rainfall in the region, there are clear indications that 2015/16 season in the Southern Africa region was hit by one of the strongest El Niño's characterized by late and below normal rains which resulted in a drought (Relief Web, 2015, FAO, 2016[2]). A number of scientific analyses have conceded that the 2015/2016 El Niño was one of the strongest extreme events ever recorded and has been generally considered to be of the magnitude of the 1997/1998 El Niño (Engelbrech and Engelbrecht, 2017; Paek et al., 2017; Relief Web, 2015). FAO (2016) highlighted that the 2015/16 agricultural season drought was driven by one of the strongest El Niño events of the last 50 years. It is also important to note that 2015/16 drought conditions followed the 2014–15[3] agricultural season that was similarly characterized by hot and drier-than-normal season for many countries in the region (FAO, 2016) (Fig. 3.1).

Most of the Southern Africa region experienced below normal rainfall which has led to the drought during 2015/16 El Niño (see map). The first half of the 2015/16 season (September to December) was characterized by slow onset of rain, lower than normal rain and higher than normal temperatures

---

[2]Regional Response Plan.

[3]Some countries such as Madagascar had experienced up to three consecutive drought seasons.

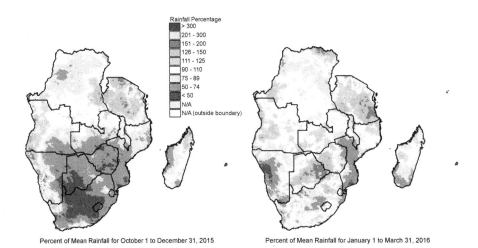

Rainfall Percentage
> 300
201 - 300
151 - 200
126 - 150
111 - 125
90 - 110
75 - 89
50 - 74
< 50
N/A
N/A (outside boundary)

Percent of Mean Rainfall for October 1 to December 31, 2015   Percent of Mean Rainfall for January 1 to March 31, 2016

FIGURE 3.1

Mean rainfall for Southern African countries during the period of January to March, 2016.

Reproduced from FAO (2016).

(SADC, 2015). This has resulted in developing dryness across large portions of the subregion, delaying planting activities by up to 50 days and negatively affecting the establishment of the early-planted crops.

During the second half of the season, SADC (2016a,b,c) highlights that the regional drought slightly eased in January with some areas receiving well-above normal and in some cases over twice the normal amount resulting in some areas experiencing flooding (e.g., in Tanzania and northern Mozambique). The drought conditions, however, restrengthened through to mid-February. In large parts of the region, the maximum rainfall anomaly (difference from normal) was experienced and for the second year in succession, rainfall was widely 20%—60% below average for the summer rainy season (October to April) in 2015/2016 (WMO, 2017).

During the first half of the 2015/16 season (September to December), very high temperatures persisted especially in the southern half of the region, driving up rates of water loss from surface water bodies (SADC, 2015). The year 2016 started with an extreme heat wave in Southern Africa in the first week of January. On January 7, 2016, it reached 42.7°C in Pretoria and 38.9°C in Johannesburg, both of which were 3°C or more above the all-time records at those sites (WMO, 2017) (Fig. 3.2). However, scholarly articles demonstrate the complexity of attaching conditional probabilities of heat wave occurrence given drought (in this case the El Nino—induced drought) with models exhibiting less ability in reproducing the observed conditional probability of a heat wave given El Niño conditions (Lyon, 2009). What it implies is that more scientific modeling and analysis would be required to fully attribute the recorded temperatures to the stronger El Ninos. Large areas of Southern Africa exhibited below-normal vegetation conditions in cropped areas, indicating low soil moisture and retarded development of the early-planted crops (FAO, 2016). Fig. 3.3 shows Agricultural Stress Index percentage of cropped areas suffering from water stress in first dekad and second dekad.

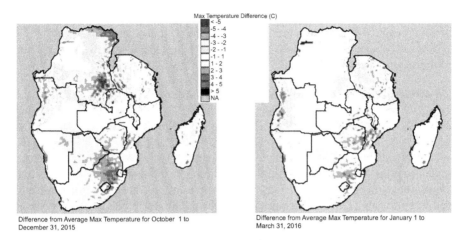

Difference from Average Max Temperature for October 1 to December 31, 2015

Difference from Average Max Temperature for January 1 to March 31, 2016

FIGURE 3.2

Mean air temperature for Southern African countries during the period of January to March, 2016.

Source: CHG EWX.

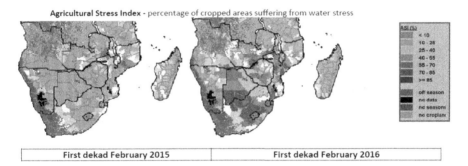

FIGURE 3.3

Agricultural Stress Index percentage of cropped areas suffering from water stress in first dekad and second dekad.

Source: Food and Agriculture Organization of the United Nations, [2016], [Southern Africa El Niño Response Plan (2016/17)].

Reproduced with permission.

Impacts of the drought on the various sectors, economic segments, and population groups

The drought impacted on agricultural productivity and production as drought affected most of the countries. A number of international organizations, United Nations technical agencies, and regional bodies and scientific organizations, for example, FAO, WMO, UNICEF, OCHA, USAID, SADC, World Vision working in the Southern Africa region have reported on the impact of the 2015/16 drought. Across the region 40 million rural people, or 22 per cent of total rural population, were affected by the drought, with 23 million requiring emergency humanitarian assistance.

The impacts can be summarized following Benson and Clay (1994)'s classical characterization of the impact of drought as follows:

Impacts on crops, livestock, and food and nutrition security

According to FAO (2017), the El Nino–induced drought resulted in 40 million people being acutely food insecure as a result of widespread crop failure, low production, and loss of livelihoods through income losses and assets in the worst-affected areas. UNICEF (2017) reported that the El Niño event contributed to severe acute malnutrition among 237,000 children under the age of 5 years, while in some countries the percentage of children suffering from severe acute malnutrition reached 7%–10%, well above the international emergency threshold of 3%. This according to UNICEF was compounded by preexisting chronic malnutrition, expressed through levels of stunting ranging from 8% to 47%.

FAO (2017) reported that livestock production was significantly affected by lack of pasture and water resulting in significant deaths in Lesotho, South Africa, Swaziland, and Zimbabwe and increased incidences of diseases such as foot-and-mouth disease as reported in northern Namibia and southern Angola. According to the RIASCO Action Plan for Southern Africa: Response Plan for the El Niño–induced drought in Southern Africa May 2016–April 2017, there were more than 634,000 livestock deaths. In addition, some pastoral people migrated in search of grazing land, complicating the social impact and rendering assistance more complex (UNICEF, 2017).

The price of commodities such as the staple maize increased to levels that were beyond the reach of most affected communities in these countries. According to UNICEF (2017), food prices increased (in some cases almost doubling). Prices of maize, which is the main staple food in the region, in surplus producing areas were significantly above international prices, and compared with the previous season, maize prices in Malawi and Mozambique had doubled (SADC, 2017). Access to food by household was further constrained by low purchasing power as a result of the devaluation of some currencies, notably the South African rand and the Zambian kwacha against major currencies. According to FAO (2016), the situation was further exacerbated by the loss of national revenues in the region, due to the fall in global commodity prices of their main exports, including minerals and oil.

Impacts on household access to water and sanitation

According to the SADC (2017), the drought resulted in widespread water shortages which reduced access to clean water for many rural communities with the attendant increase in the incidents of communicable diseases such as cholera in Angola, Malawi, Mozambique, Zambia and Zimbabwe; typhoid in Malawi and Zimbabwe; diarrhea and dysentery in Zimbabwe. Botswana, Namibia, and South Africa experienced increased incidents of Malaria. Scarcity of water also affected the operations at most rural health facilities and resulted in reduced school attendance as parents kept their children away from school. SADC (2017) also reported that this was a common problem in Swaziland where more than

80% of schools were affected by water shortages, and women and children suffered the most due to the drought as they spent most of their time sourcing family requirements such as water. According to UNICEF (2017), more than 4 million people in the region lacked access to clean water and basic sanitation.

During 2016, OCHA (2016) reported significant gaps in seed supply and demand in drought-affected countries. The shortage of seed in the formal market is likely to negatively affect the capacity of governments and development partners to quickly and effectively respond to the crisis at scale through provision of certified quality seed and other agricultural inputs.

Other social and economic impacts of the drought

Some Scholars such as Laosuthi and Selover (2007) argue that the positive and negative impacts of the El Nino phenomenon often cancel each other based on analysis presented in the *Eastern Economic Journal* that analyzed the economic effects of the El Niño weather pattern on 22 economies. However, the conclusion from the listed scholars was based on the argument that "The El Niño acts to redistribute heat and precipitation and it 'bestows beneficial effects on economies as well as negative effects'." It is further noted that the cancellation effect is likely to be more complete the larger the geographical area of a country and/or the more diversified the economy. However, the situation for the Southern Africa region appears to be differing with the conclusion by Laosuthi and Selover as the livelihood of more than 60% of the region's populations is dependent on rain-fed agriculture whose performance is directly linked to the distribution and total amount of rainfall. The 2015/16 drought resulted in reduced crop yields, loss of assets in particular livestock and incomes, and increased surface and groundwater abstraction and also reduced power generation. According to FAO (2017), the negative economic impact was further multiplied by weak economies of some of the countries in the region associated by unfavorable exchange rates and slow economic growth due to loss of national revenues resulting from low commodity prices of main exports such as minerals and oil. For example, the South Africa rand and Zambian kwacha hit record lows in December 2015.

The drought caused low water levels at Kariba Dam that resulted in 1,000 MW less power generation than usual, resulting in citizens and businesses in Zambia and Zimbabwe enduring 8- to 10-hour daily load shedding (Goussar, 2016). In addition, Goussar (2016) also noted that the volatility and shortage of power supply could have led to shying away of foreign direct investors especially to invest in heavy industry which manifested in many delayed or canceled mining projects.

During 2016, OCHA (2016) asserted that increased migration was due to lack of food and water during the El Niño—induced drought. The search for employment also resulted in increased numbers of separated and unaccompanied children. Child labor was believed to have increased since the start of El Niño, which affected more boys than girls. The study reports a rising number of school dropouts, including 6000 children dropping out of schools in Zimbabwe due to hunger or the need to help their families.

Responding to the droughts with a focus on 2015/16 drought

In this section, a review of the response to the 2015/16 drought is presented. The authors first examine how the early warning systems and process at regional and national levels were organized. This is followed by a synthesis of the response mechanisms and processes at regional, national, and organizational level.

The effectiveness and efficiency of stakeholders' response to the drought were assessed with respect to the following three pillars for effective drought risk reduction that were proposed by Bruntrup and Tsegai (2017):

a. Monitoring and early warning systems
b. Vulnerability and impact assessment
c. Risk mitigation measures

Drought monitoring and early warning systems

The Southern Africa Regional Climate Outlook Forum (SARCOF) spearheads the most widely adopted seasonal early warning mechanism in the region. The SARCOF is a regional climate outlook prediction and application process adopted by the 14 countries comprising the Southern African Development Community (SADC) Member States in conjunction with other partners. It is one of the World Meteorological Organization (WMO) Regional Climate Outlook Forums, developed under World Climate Services Program (WCSP) of the World Climate Program, active in several parts of the world, which routinely provide real-time regional climate outlook products. The SADC Climate Services Center (SADC—CSC) in Botswana coordinates the SARCOF.

At the national level, there is considerable variation in the institutional architecture for disaster management and response including early warning systems across Southern Africa. In some countries, National Disaster Management Agencies (NDMAs) have been established as statutory bodies under an Act of Parliament, while in others disaster management is coordinated from the offices of the Heads of State and Government. Madagascar, Lesotho, and Swaziland are examples of countries where NDMAs are established under legislative provisions. However, the national level early warning systems in SADC appear to draw heavily upon the outcome of the SARCOF. As SARCOF is a regional grouping of weather and climate experts, national weather and climate institutions in Southern Africa appear to be relying more on its outlook projections. At the national level, similar forums provided outlook projections, for example, the National Climate Outlook Forum (NACOF) in Zimbabwe, Meteorological Observation Network in Zambia, and National Meteorological Services (NMS) in Malawi.

A debatable issue whether the national level early warning systems are adequately providing the services of early warning to the government and the affected communities by ensuring that (user friendly) climate information products are understood by and communicated to user and allows interagency coordination of policies, sectoral plans, and an ongoing process for understanding and responding to risks posed/opportunities brought about by past, current, and future climate. Ideally, early warning systems should be complete with a clear risk and hazard classification system and profiling; escalation protocols (alert levels), and response protocols/processes.

Additional early warning for the 2015/16 drought was also provided by the Famine Early Warning Systems Network (FEWS NET) which provides early warning and analysis on food insecurity to eight countries in Southern Africa. From October 2015 to September 2016, FEWS NET projected that emergency food assistance in the countries FEWSNET covers would be roughly 30% higher than previous year's estimates. In about half of the countries covered, El Niño's impacts on climates are a primary driver of acute food insecurity (FEWSNET, 2018).

The reviews done by FAO (2016) and SADC (2016a,b,c) provide the most elaborate account of how early warning and response to 2015/16 drought was organized. In August 2015, SARCOF presented a consensus outlook for 2015/2016 indicating normal to below normal rainfall for the periods of October to December 2015 and January to March 2016 over most of the Southern African region due to the predicted El Niño weather phenomenon (SADC, 2016a,b,c). Following this forecast, SADC embarked upon a series of early activities to prepare for a comprehensive response. The SADC Secretariat mobilized financial and technical support from RIASCO (see proceeding section) to organize meetings at regional and national levels to assess the situation and respond. With support from RIASCO, SADC Secretariat convened a regional meeting in February 2016 where the regional drought emergency was officially confirmed, and SADC and its Member States agreed to recommend the declaration of a regional drought emergency to the Council of Ministers. The SADC Council of Ministers recommended the declaration of a regional drought emergency at its meeting of March 2016 and mandated the SADC Secretariat to establish the SADC El Niño Response and Coordination team at the Secretariat in Gaborone, Botswana, to coordinate a regional response to the drought in close collaboration with Member States. The team, made up of representatives from the SADC Secretariat and UN Agencies (Food and Agriculture Organization (FAO), the UN Office for the Coordination of Humanitarian Affairs (OCHA), World Food Program (WFP), United Nations Children's Emergency Fund (UNICEF), and the United Nations Development Program (UNDP)) was set up in May 2016. The main task of the coordination unit was to help member states and the SADC Secretariat undertake vulnerability and impact assessments as well as develop response plans for the El Nino—induced drought in a coordinated manner.

According to SADC (2017), the team's operational budget was estimated at US$ 1,439,600, but only US$ 918,000 was mobilized for its operations leaving a funding gap of US$ 521,600. The review also highlights that the Coordination team's first action was to prepare the Regional Humanitarian Appeal which was launched by the SADC Chairperson in June 2016. In addition, the Transport and Logistics Unit of the Coordination Team assessed the regional transport facilities including ports and transport corridors to establish their capacity and adequacy for moving the large volumes of food aid which were required to provide emergency relief to the estimated 23 million SADC citizens who needed food relief. These facilities were found to have adequate capacity. The results of this assessment were used to develop a regional transport plan aimed at smoothening the transportation of cargo across the region. The plan provided for the granting of waivers of sabotage restrictions for drought relief imports, the issuance of cross-border permits and visas for railway and road transport crews, expedited customs clearance and provision of security escorts for relief cargo in transit from the ports to landlocked states. The plan also provided a framework for improved coordination between national disaster management agencies and the logistics cell that was embedded within the SADC El Niño Logistics and Coordination team at the SADC Secretariat.

Drought vulnerability and impact assessments

The regional and national vulnerability committees that were established by SADC in 1999 undertake the assessment of vulnerability on impacts of droughts and other hazards. The National Vulnerability Assessment Committees (NVACs) coordinate the annual vulnerability assessment and analysis. According to SADC (2018a,b,c), NVACs are multisectional committees led by relevant government ministries with wide ranging memberships, which includes different government departments, nongovernmental organizations (NGOs), and international organizations involved in poverty reduction and socioeconomic development. NVACs carry out annual and periodic vulnerability assessments, in addition to studies on selected topics such as nutrition, climate change, and related themes that are critical in VAA. The RVAC did not deviate from its usual annual timelines and produced its report on the 2015/16 season in July 2016.

The NVACs are acknowledged as the main system to monitor, report, and respond to food insecurity in the region, despite the fact that they are not legal entities. The NVACs use different tools and methodologies to monitor and assess impacts of hazards, including the drought, which makes intercountry comparisons impossible. An attempt to standardize the monitoring, classification, and reporting on food insecurity was the adoption by SADC at the February 2016 meeting, of the Integrated Phase Classification systems in Southern Africa. The Integrated Food Security Phase Classification (IPC) is a set of standardized tools that aims at providing a "common currency" for classifying the severity and magnitude of food insecurity. This evidence-based approach uses international standards, which allow comparability of situations across countries and over time. It is based on consensus-building processes to provide decision makers with a rigorous analysis of food insecurity along with objectives for response in both emergency and development contexts. However, IPC as of 2016 was only used in a few countries due to limited technical capacity.

Mitigation measures implemented

The SADC launched a regional appeal that included interventions at both the national and regional levels in July 2016. The emergency appeal raised the profile of the challenge and the impending impacts of the drought. According to the SADC Humanitarian Appeal, the drought emergency had impacted the lives of up to 40 million people with 23 million in need of immediate emergency assistance across the region. A total of US$ 2.7 billion was estimated as the level of financial support that would be needed to provide assistance to those in need. The regional appeal recognized the need for multisectoral approaches to addressing the impacts of the drought and included interventions that protected livelihoods and productive assets, as well as those that build resilience to future droughts.

According to SADC (2017), the SADC response to the El Niño drought emergency was developed through an analysis of information and plans that were provided by Member States with the objective of creating a coherent regional approach to the emergency. The Appeal took into consideration the need to bridge the gap between humanitarian response and resilience building. The response developed by SADC would be grafted onto already ongoing SADC programs that include long-term resilience building. Wherever possible, SADC also encouraged international development partners to build upon the responses developed by Member States as foundations for building resilience over the long term. Resilience building was defined in broad terms to include provision of support services such as water and sanitation and nutrition services. Collaboration across sectors in addressing community needs was also factored into the crafting of approaches to resilience building. SADC (2017), however, notes that country differences in tools and approaches made it difficult for developing partners and donors to compare and prioritize across the countries—this naturally creates the need for stronger harmonization.

The RIASCO developed a complementary response plan that targeted the seven most affected countries of Angola, Lesotho, Madagascar, Malawi, Mozambique, Swaziland, and Zimbabwe. The plan targeted 1,035,186 households (5,175,930 people) with livelihood assistance to enable affected households to produce their own food and protect their assets (Table 3.3). Like the SADC plan, the RIASCO plan recognized that focusing on immediate humanitarian needs without considering recovery and resilience was unsustainable nor desirable. The plan therefore also focused on bridging the gap between response and resilience over the long term as a way of building communities that are better equipped to handle shocks in the future.

The plan was developed around three pillars:

1. **Pillar 1: The Humanitarian Response:** Focusing on critical humanitarian needs, response plans, and funding needs broken down into the focus sectors of food security, agriculture and livelihoods, health and nutrition, water, sanitation and hygiene, education, and protection in the seven countries that were assessed as requiring the greatest need for international assistance.
2. **Pillar 2: Building Resilience:** Advocating for the region to break the cycle of short-term response to emergencies which have now become part of the development landscape in Southern Africa.
3. **Pillar 3: Potential Economic Impacts and Solutions to Impacts of Droughts:** Placing disaster response and mitigation within the context of economic development realities in the region

Table 3.3 Humanitarian drought response funding in RIASCO priority countries.

Country	Number of people affected (million)	Funding requirements ($)	Funding received ($)	Gap ($)	Gap (%)
Malawi	6.50	395,361,811	340,372,318	40,197,927	10
Zimbabwe	4.10	352,304,020	218,440,361	133,863,659	38
Mozambique	2.00	179,070,000	153,976,374	25,093,626	14
Madagascar	1.10	154,934,800	95,538,735	59,396,065	38
Swaziland	0.64	95,360,000	39,748,100	55,611,900	58
Angola	0.78	70,409,614	11,280,049	59,129,565	84
Lesotho	0.71	52,641,594	40,729,865	11,911,729	23
Total	15.83	1,300,081,839	900,085,802	385,204,472	30

Source: From Southern Africa Regional Inter-Agency Standing Committee (RIASCO)-United nations ©2016 United Nations. Reprinted with the permission of the United Nations.

also citing resilience building interventions supported by development agencies including the World Bank (WB), African Risk Capacity (ARC), and the African Development Bank (AfDB).

As of July 2017, the RIASCO Plan had realized 70% of its total funding requirements across all sectors with the Food Security and Agriculture the best supported. The water, sanitation, and hygiene, and nutrition and health sectors were the least funded with funding gaps of 68% and 48%, respectively. The high level of support for food security and agriculture was indicative of the focus of the initial response on immediate lifesaving activities with less attention to early recovery and resilience building. This is clearly indicated by the fact that of the US$ 1.3 billion funding requirements for disaster response, only $ 2,500,000 was earmarked for recovery and resilience building.

The fact that the SADC regional appeal only received about 12% of the required funding and there were no reported deaths due to the drought could mean that the funding requirements were overestimated, which raises questions on the quality of data collected through the national and regional mechanisms. There is need to improve the coordination of stakeholders as well as the quality of data routinely collected at national level. It is apparent that funding from the external sources is not always timely available, hence the need to governments to fund their own response plans. The development of the SADC disaster fund and the use of insurance products in responding to droughts are welcome recent developments.

SADC Council of ministers approved a Regional Disaster Preparedness and Response Strategy and Fund in August 2017 for use in coordinating the region's response to disasters at Member State level (SADC, 2018a,b,c). The fund is seen as a potentially effective mechanism to support member states to facilitate rapid response to disasters which overwhelm member states and provide immediate lifesaving support while resources for more comprehensive responses are being mobilized. In addition, according to SADC (2016b), the SADC has a Financial Food Reserve Facility that is meant to provide a reserve to feed 25% of the region's population over a period of 6 months in the event of a disaster while the affected countries are preparing longer-term comprehensive responses. Required funding levels for the Disaster Response Fund and the Financial Food Reserve Facility, which will be sourced from Member States and development partners, are estimated at US$ 1,900,000 and US$ 360,000,000, respectively. These financial modalities, however, are still in their infancy, and their effectiveness remains to be seen.

The African Union has also established the African Risk Capacity aimed at improving the capacities of African countries to plan and respond to extreme weather events and natural disasters as a way of mitigating the risk of food insecurity. An ARC Insurance Company Limited (ARC Ltd.) was established under this initiative to serve as a financial affiliate and licensed insurer under ARC. More than 32 AU Member States have signed up to this continent-wide insurance scheme. Malawi which joined the scheme in 2015 was one of the countries which benefitted from payouts from this insurance scheme following the 2015—16 drought emergency. A total of US$ 8.1 million was paid to the Government of Malawi in addition to support with the formulation of a comprehensive national drought risk management strategy.

National level response processes

Ideally in any emergency situation, such as the El Niño drought, it is the primary responsibilities of national governments to ensure that they implement programs to respond and secure citizens.

International Cooperating Partners, international and local NGOs as well as donors can come in with additional support. But first the governments would have to be convinced that they are in an emergency situation. During the 2015/16 drought, five countries out of the 15 declared national disasters launched international appeals for support from December 2015 to July 2016, while eight of South Africa's nine provinces declared the drought a disaster by July 2016 (SADC, 2016a). In Lesotho, where more than 500,000 people were in need of emergency food at the onset of the emergency the government had developed a National Multi-Hazard Contingency Plan (2015—18), declared a national emergency by December 2015, and reallocated US$ 10 million in its national budget to address water shortages among its rural population. Likewise, Zimbabwe and Swaziland had developed Contingency Plans and declared national emergencies by February 2016 followed by Malawi and Mozambique in April 2017. The Government of Malawi had also developed a National Multi-Hazard Contingency Plan and a National Food Insecurity Response Plan for the period 2015—16. In addition, the Government of Malawi took out a US$ 30-million risk insurance from ARC to cover potential food production deficits in future. Due to capacity limitations, Madagascar only issued a message of solidarity by August 2016. All these actions were, however, inadequate to meet the scale of humanitarian needs that were caused by the El Niño—induced drought which meant that all countries in the region required urgent international assistance with responding to the drought emergency (SADC, 2016a).

What these trends tell us is that the declaration of disasters among Southern African states is a complex process. In many cases, due to broader consequences related to declaration of drought, such as loan write offs to farmers, associated insurance implications, costs to central Governments, and impacts to the broader economy, appear to govern how and when governments declare disasters. The bottom line is that countries normally do not want to rush into declaring disasters, even when all other nongovernment sectors see it as apparent.

Another important observation is that the 2015—16 drought emergency occurred at a time when the resource-based economies of most of the countries in Southern Africa were in decline. The regional GDP growth rate was declining and forecast at 1.6% per annum in 2016 against a minimum of 6% required to sustain economic growth. Commodity prices were also in decline which put pressure on the currencies of most countries in the region which are commodity exporters. This has left most countries in the region unable to fund contingency plans they may have developed, set aside financial resources for use in responding to weather-related emergencies, or invest in infrastructure and social programs that would enhance the capacities of the region's populations to respond to disasters. The review also established that national governments in the region assumed leadership in responding to the impending drought as soon as it became evident that large numbers of the region's population were under threat from the disaster. This was despite the fact that the scale of the disaster threatened to overwhelm the individual countries' capacities to respond.

The SADC review report also highlights that national efforts at creating resilient communities have also benefited from the services provided by international and continental financing institutions. The World Bank, for example, has provided financial support amounting to US$ 290 million, policy development support as well as technical assistance for post-disaster recovery in a number of countries around the region. The African Development Bank has set up a Feed Africa objective under which it is developing an African Agriculture Transformation Strategy in support of the Comprehensive African Agricultural Development Programme (CADAP). The AfDB also has supported programmes targeting emergency support, budget support, drought, and resilience building and support for long-term disaster risk insurance which is implemented through strengthening the early warning and management systems of regional climate centers in each of the four subregional economic communities across the continent.

Lessons learned—what are the lessons from the responses[4]
Lessons at policy and institutional level

The analysis in the preceding sections has shown that weather-related emergencies are recurring phenomena over most of the Southern African region with the 2015—16 El Niño—induced drought emergency highlighted as the worst such phenomena in 35 years. While the immediate reaction to the emergency by governments and development partners was to secure and distribute food aid to save the lives of affected populations, a number of lessons can be learned at the policy and institutional level especially to enable the affected communities to break out of the repetitive cycles of disasters and building the long-term resilience of these populations over the long term. These lessons include the following:

Government's leadership in planning, coordinating, and allocating resources makes response effective

Government leadership in planning and coordinating emergency response including allocating resources is critical as governments provide the policy and legal frameworks for the implementation of responses to emergencies. Some countries have made some progress, for example, Dzama (2016) notes that South Africa has made good progress in establishing integrated climate change and Disaster Risk Reduction Management (DRR-M) policy frameworks, including a focus on drought planning and management for agriculture; however, this can be improved through adopting a long-term, national drought policy and strategy to improve the country's response to future droughts. Where resources are available, governments should make budgetary provisions for addressing emergencies as happened in Lesotho where government allocated resources from the national budget to address water shortages that were affecting rural farmers in the early stages of the El Niño—induced emergency. According to SADC (2017), the experience from Zimbabwe where disaster response was coordinated from the Office of the President and Cabinet is a case in point as government leadership helped guide the sourcing, transportation, and distribution of relief food to affected communities despite the severity of the economic problems the country was experiencing at the time. In addition, the need for building of coordination and capacity of regional bodies and member states to address cross-border trade issues—customs waivers, import bottlenecks, GMOs—emerged as a major lesson emerging during the response to the 2015/16 drought (USAID, 2017).

Timeous decision-making in response to early warning is paramount

Synthesis of the response process followed by SADC reflects the need to accompany early warning systems with timeous decision-making for emergency response to be effective. The El Niño—induced drought emergency was predicted through the Southern Africa Regional Climate Outlook Forum in August 2015. By February 2016, SADC Secretariat, with support from WFP and FAO, had convened

---

[4]The lessons presented in this section are synthesized from a number of reports including SADC (2017), FAO (2017), UNICEF (2017), USAID (2017)., Dzama (2017)

a consultative meeting to discuss possible response to the impacts of the impending emergency especially given the fact that the 2014—15 agricultural season had also been a poor one. By June 2016, SADC had launched a regional international humanitarian appeal and humanitarian support organizations organized as the Regional Inter-Agency Standing Committee had compiled an action plan detailing the scale of the humanitarian needs and projected budget to meet these needs. This timeline reflects timeous decision-making on the part of SADC and its Member States which facilitated the early mobilization of resources for the procurement of emergency relief supplies for the estimated 13 million people in Southern Africa who were in need of emergency relief (SADC, 2017).

Emergency response measures should include resilience building strategies for long-term sustainability

The analysis of response strategies reflects the importance of integrating emergency response measures with strategies for building long-term resilience. For example, SADC (2017) highlighted that SADC and its Member States need to integrate early recovery strategies into emergency response to ensure sustainability of response strategies over the long term. FAO (2017) points out that emergency response interventions were of short-term nature and were difficult to systematically evaluate, as such, measuring outcomes and impact remained a big challenge. Most of the initial interventions under the regional response to the El Niño—induced drought emergency were focused on providing immediate relief to people who were in need of food and did not therefore include provisions for resilience building. Resilience building has, however, been recognized as a critical component of emergency response, and regional bodies, member states, and development partners have started putting in place measures that will promote the integration of resilience building initiatives into future relief responses to emergencies.

Effective coordination mechanisms at regional, national, and subnational levels are a sine qua non

Emphasis on the four Ws (Who is doing What, Where, and When?) as highlighted by USAID (2017) together with harmonization of modalities, geographic coverage, ration size/composition, and assessments with host governments is necessary for such emergencies. From a regional perspective, the establishment of the SADC Logistics and Coordination Team at the SADC Secretariat facilitated the expeditious movement of imported food into the region to support the immediate humanitarian needs of affected people at national and subnational level.

The use of multisectoral approaches to disaster response helps move responses forward very quickly

Disaster risk response and management can be organized to address multisectoral issues including food security, provision of clean water and sanitation services, and support to treatment of malnutrition of children under the age of five and management of natural resources including water, land, and forestry (SADC, 2017). This approach facilitated the timeous control of a potentially disastrous humanitarian situation across the Southern African region and helped avert human deaths. USAID (2017) highlighted the potential of multiple modalities, agreeing in advance on consumption gaps, transfer value, and viability of various modalities by population/geography and identify common indicators for use by all partners.

Lessons at programmatic and project level

Appropriate packaging and timely dissemination of weather information is necessary in early warning and response to disasters

Lesotho and Zimbabwe have started programs aimed at packaging and disseminating weather information in the local languages. The SADC (2016a,b,c) review proposes that these initiatives could be adopted across the region to facilitate communication of the likelihood of the occurrence of emergencies in languages that the ordinary people understand. Wherever possible, ethno-meteorological knowledge systems should also be incorporated into the interpretation and dissemination of weather information. While the transport plan and the associated policy measures developed by the SADC Logistics cell facilitated the moving of the large volumes of food aid required to feed the large numbers of people affected by the El Niño drought phenomenon, the SADC (2016) review established that their effectiveness was affected by poor flows of information between national authorities responsible for importing food and port authorities and transport operators especially with respect to quantities of food that needed to be transported. What this calls for is that in future, SADC Secretariat and its Member States need to improve on their handling and processing of data relating to potentially affected people. Countries should consider adopting a common reporting system based on the mobile vulnerability assessment system that is used under the IPC assessment approach. Further, a common reporting system seems to be necessary during emergencies that have a regional footprint such as El Nino—induced 2015/16 drought.

It is critical to strengthen capacities of institutions that are involved in coordinating vulnerability assessments, early warning, and national response plans

A number of early warning and response activities require specialized skills and capacities. The profiling of hazards, risks, and disasters which the region and countries are prone to and designing the most appropriate response protocols and escalation systems; stronger and more effective regular (yearly) and ad-hoc/needs-based vulnerability assessments; maintenance of online databases in the early warning and information system; coordinating the vulnerability assessments, early warning information - Systems, implementation response plans require specialized skills. There is evidence of lack of such skills and malalignment of administrative across the board, not just in government, but also in the bilateral and multilateral agencies. For instance, FAO (2016) review of response to El Nino reported that there were late provision of inputs related to procurement delays; inadequate capacities in the FAO representations related to planning, coordination, and understanding procurement procedures and policies; difficulties related to obtaining of technical specifications and clearances particularly for livestock; delays in authorization of bulk procurement handled through FAO headquarters; restrictive national laws governing procurement of certain agricultural inputs. At the country level, most countries had challenges in quantifying agricultural intervention needs due to lack of clear methodologies for assessment and lack of credible data. Targeting of beneficiaries appears to have remained a challenge in many of the programs implemented by different agencies and organizations, mainly due to lack of harmonized criteria for selection of beneficiaries. Consequently, there were some reported duplications and inclusion of some underserving households in some interventions. There was evident lack of capacities among implementation partners, particularly government structures—ranging from absence of extension service delivery structures on the ground (e.g., in Madagascar) to lack of operational resources such as transport facilities to enable them to effectively follow-up interventions (e.g., in Lesotho, Mozambique, and Malawi).

Greater efficiencies are experienced in situations where humanitarian agencies and development partners are organized and coordinated

The coordination of the Humanitarian Country Teams whose membership includes international cooperating partners through the United Nations Resident Coordinator's Office greatly improved the delivery of disaster management support in all the countries that were visited.

The use of standardized indicators for collecting data on vulnerable populations improves targeting of beneficiaries and efficiency of delivery of emergency support

Vulnerability assessment methodologies that are used across the region are not standardized. The results of these assessments cannot therefore be used to identify target beneficiaries for food relief. In some countries, the methods used to identify households to be provided with emergency relief were open to political influence, while in some cases these resulted in duplication and inclusion of undeserving households. In other instances, households that were already on social protection programs were excluded from receiving relief even if their level of vulnerability to the drought emergency was high. The use of common indicators such as those that have been developed under the IPC approach proved to be effective and should be adopted across the region.

Disaster response programs need to ensure that productive assets at community level are protected in times of emergencies

Unless they get early support, communities that are affected by emergency situations usually dispose of their livelihood assets such as livestock as a coping strategy for dealing with disasters such as the El Niño drought phenomenon. While this reaction saves lives in the immediate term, it compromises the ability of these communities to regroup and reestablish themselves in the aftermath of disasters. It is therefore important that responses to emergencies include elements which enhance the capacity of affected communities to recover without having to dispose of those assets that form the foundation of their lives. Local insurance schemes to protect communities against production losses can also contribute to resilience building if they are administered properly. The experience gained from the introduction of the ARC-based insurance system in Malawi needs to be further refined through the engagement of regional and national insurance service providers for them to offer similar services to rural populations.

The use of cash-based transfers as "top up" on already existing social protection systems improves community capacities to respond to shocks from emergencies over the long term

Integrating short-term cash transfers into already existent social protection services will enhance community capacities to respond to shocks and contribute to resilience building. Social protection services which are already implemented as medium- to long-term interventions can be enhanced and expanded by introducing cash transfers as top ups which would make them address resilience building among recipient communities. Having said that, it is important to keep in mind that cash transfers are not a "silver bullet" that will address all the problems associated with responding to emergencies as they can be limited in their applicability due to unfavorable market conditions or limited platform coverage.

More effort should be put toward mechanisms that build consensus among various actors to ensure timely implementation of decisions during emergencies

Significant progress was made toward strengthening of the analysis of food security and nutrition data, national vulnerability assessments and integration of HIV, gender, nutrition, and IPC in vulnerability assessment and analysis. However, lack of data and/or consensus on its reliability delayed the preparation of response plans. Evidence-based information from countries was insufficient for planning. Generally, there was inadequate reliable data at the country level, including questionable quality of food security and nutrition data to inform detailed IPC analysis.

Conclusions

It is important to have profiles and response protocols for various hazards, risks, and disasters which the Southern Africa region and specific countries are prone to including recurring extreme weather events (droughts, snowfall, hailstorms, early frost, and strong winds), pests and diseases which exacerbate food and nutrition insecurity in the region. The early warning and information systems facilitates both regular and needs-based/ad hoc vulnerability assessments to update the profiles of risks, hazards, and disasters which form the basis for emergency responses. For it to be effective, the early warning and information system requires stronger coordination capacity. The disaster management teams at different levels also need to maintain up-to-date databases, profiles, and information on various hazards and risks. The disaster management needs technical capacities to coordinate various actors involved in vulnerability assessments as well as responding to hazards, risks, and disasters.

References

Benson, C., Clay, E., 1994. The Impact of Drought on Sub-Saharan African Economies: A Preliminary Examination: Working Paper 77 ISBN 0 85003 212 1. Overseas Development Institute: Regent's College Inner Circle. Regent's Park London NW1 4NS.

Bruntrup, M., Tsegai, D., 2017. Drought Adaptation and Resilience in Developing Countries. Retrieved from: https://www.die-gdi.de/uploads/media/BP__23.2017.pdf.

Chikoore, H., 2016. Drought in Southern Africa: Structure, Characteristics and Impacts. University of Zululand, p. 203.

Dzama, K., 2016. Is the Livestock Sector in Southern Africa Prepared for Climate Change? Southern African Institute for International Affairs (SAIIA). Policy Briefing 153.

Engelbrecht, F., Engelbrecht, C., 2017. Climate change over South Africa - from the next few decades to the end of the century. CSIR Natural Resources and the Environment Agricultural Research Council - Institute for Soil, Climate and Water, Pretoria: South Africa.

FAO, 2016. FAO in Southern Africa El Niño Response Plan 2016. http://www.fao.org/3/a-i5981e.pdf.

FAO, 2017. Review of FAO Response to 2015/16 El Niño Induced Drought in Southern Africa: October, 2017.

FEWSNET, 2018. FEWS NET El Niño and La Niña Monitoring Resources. 2015/16 El Nino. http://www.fews.net/fews-net-el-ni%C3%B1o-monitoring-resources.

Goussard, H., 2016. Energy diversification key to manage El Niño. ESI Africa: Africa's Power Journal. https://www.esi-africa.com/energy-diversification-key-to-manage-el-nino-says-riscura/.

IOM, 2017. Spaces of Vulnerability and Areas Prone to Natural Disaster and Crisis in Six SADC Countries Disaster Risks and Disaster Risk Management Capacity in Botswana, Malawi, Mozambique, South Africa, Zambia and Zimbabwe. International Organization for Migration, 17 route des Morillons 1211, Geneva, Switzerland.

Jury, M.R., 2013. Climate trends in Southern Africa. South African Journal of Science 109 (1/2). Art. #980, 11 pages. https://doi.org/10.1590/sajs.2013/980.

Laosuthi, T., Selover, D., 2007. Does El Niño affect business cycles? (2007). Eastern Economic Journal 33 (1), 21−42.

Lyon, B., 2009. Southern Africa Summer Drought and Heat Waves: Observations and Coupled Model Behavior: International Research Institute for Climate and Society. The Earth Institute, Columbia University, New York, New York.

Masih, I., Maskey, S., Mussá, F.E.F., Trambaue, P., 2014. A review of droughts on the African continent: a geospatial and long-term perspective. Hydrology and Earth System scinces 18, 3635−3649. https://doi.org/10.5194/hess-18-3635-2014. www.hydrol-earth-syst-sci.net/18/3635/2014/.

Obasi, G.O.P., 2005. The impacts of ENSO in Africa. In: Low, P.S. (Ed.), Climate Change and Africa. Cambridge University Press, Cambridge, pp. 218−230. https://doi.org/10.1017/CBO9780511535864.030.

OCHA, 2016. Overview of El Niño Response in East and Southern Africa. United Nations Office of Coordination of Humanitarian. Briefing note (as of 1 December 2016).

Paek, H., Yu, J.-Y., Qian, C., 2017. Why the 2015/2016 were and 1997/1998 extreme El Niño's different? Geophysical Research Letters 44. https://doi.org/10.1002/2016GL071515.

Relief Web, 2015. Southern Africa Humanitarian Outlook 2015/2016: Special Focus on El Niño - World | Relief Web. Retrieved from: http://reliefweb.int/report/world/southern-africa-humanitarian-outlook-20152016-special-focus-el-ni-o.

SADC, 2015. Southern Africa Development Community Food Security Early Warning System Agro Met Update, 2015/16 Agricultural Season Issue 04 Mid-December Update Season: 2015-2016 21-12-2015. http://www.sadc.int/files/6314/5249/5535/SADC_Agromet_Update_Issue-04_-_2015-2016_Season.pdf.

SADC, 2016a. Regional Humanitarian Appeal 2016. Southern African Development Community. https://www.sadc.int/files/4814/6840/2479/SADC_Regional_Humanitarian_Appeal_June_20160713.pdf.

SADC, 2016b. Southern Africa Development Community Food Security Early Warning System Agro Met Update, 2015/16 Agricultural Season: Issue 06 January/February Update Season: 2015−2016. http://www.sadc.int/files/4214/5591/1070/SADC_Agromet_Update_Issue-06_-_2015-2016_Season.pdf.

SADC, 2016c. SADC Regional Vulnerability Assessment and Analysis Synthesis Report 2016: State of Food Insecurity and Vulnerability in the Southern African Development Community Compiled from the National Vulnerability Assessment Committee (NVAC) Reports Presented at the Regional Vulnerability Assessment and Analysis (RVAA) Annual Dissemination Forum on 6−10 June 2016 in Pretoria, Republic of South Africa.

SADC, 2017. Review of the SADC Response to the El Niño Induced Drought Emergency in Southern Africa. Final Report.

SADC, 2018a. Regional Vulnerability Assessment & Analysis Programme (RVAA): Southern African Development Community Food Agriculture Natural Resources Directorate. https://www.sadc.int/sadc-secretariat/directorates/office-deputy-executive-secretary-regional-integration/food-agriculture-natural-resources/regional-vulnerability-assessment-analysis-programme-rvaa/.

SADC, 2018b. SADC Regional Vulnerability Assessment & Analysis (RVAA) Synthesis Report on the State of Food and Nutrition Security and Vulnerability in Southern Africa 2018.

SADC, 2018c. Stakeholders Call for Closer Collaboration to Reduce Disaster Risks, as SADC Holds Regional DRR Conference. SADC News Portal. https://www.sadc.int/news-events/news/stakeholders-call-closer-collaboration-reduce-disaster-risks-sadc-holds-regional-drr-conference/.

UNICEF, 2017. Rapid Internal Stock-taking of UNICEF Response to the 2015 − 2016 Southern Africa El Niño Induced Drought, February 2017.

USAID, 2017. Location-specific Stakeholder Engagement, Action Plans, and Priorities: Common Action Planning Themes across Locations: Presentation on Evaluation of El Nino.

Wilhite, D.A., 2016. Managing drought risk in a changing climate. Clim Res 70, 99−102. https://doi.org/10.3354/cr01430.

World Meteorological Organization (WMO), 1992. International Meteorological Vocabulary, second ed. Geneva, Switzerland.

World Meteorological organization (WMO), 2017. Climate Breaks Multiple Records in 2016, with Global Impacts. Press Release Number: 04/2017: https://reliefweb.int/report/world/wmo-statement-state-global-climate-2016.

Zoltan, D., 2008. Survey of the Drought Indices in Agrometeorology: Hungarian Meteorological Service H-1675 Budapest POB 39, dunkel.z@met.huSymposium on Climate Change and Variability - Agro Meteorological Monitoring and Coping Strategies for Agriculture, Oscarsborg, Norway, June 4−6, 2008. http://www.wamis.org/agm/meetings/sycram08/S1-Dunkel.pdf.

World Meteorological Organization, 2014. El Nino Southern Oscillation. WMO-No-1145, Geneva, Switzerland, p. 12.

Further reading

FAO, 2018. Integrated Food Security Phase Classification. http://www.ipcinfo.org/.

RIASCO, 2016. Response Plan for the El Niño-Induced Drought in Southern Africa May 2016−April 2017. Regional Inter-Agency Standing Committee (RIASCO), Johannesburg, South Africa. https://reliefweb.int/sites/reliefweb.int/files/resources/RIASCO%20Action%20Plan%20Draft%20Document%20PDF%20version.pdf.

SADC/RVAC, 2016. Southern Africa Development Community Regional Vulnerability Assessment Committee Situation Report; Based on 2015/16 National Vulnerability Assessments.

https://www.sadc.int/news-events/news/stakeholders-call-closer-collaboration-reduce-disaster-risks-sadc-holds-regional-drr-conference/.

Thanarak, L., David, D.S., 2007. Does El Niño affect business cycles? Eastern Economic Journal 33 (1), 21−42.

Drought preparedness and livelihood implications in developing countries: what are the options?—Latin America and Northeast Brazil

Antonio Rocha Magalhães[a], Marilia Castelo Magalhães[b]

[a]*CGEE-Center for Strategic Studies and Management, Brasilia, Brazil,*
[b]*Consultant, Brasília, Brazil*

Introduction

This chapter discusses droughts and drought policies in the Americas and the Caribbean region and presents the case of Northeast Brazil (NEB), which has a long experience in dealing with droughts.

The Americas have been affected by droughts all over: North America, including Canada, the United States, and Mexico; Central America and the Caribbean; and South America. There have been periodic droughts in all regions, dry and humid. However, the same meteorological drought affects differently in humid and dry areas. While, in humid areas, like the Amazon, a fall in precipitation of up to 30% may still result in enough water for its several uses. In climatic marginal dry semi-arid regions the same rainfall reduction may result in serious water deficits with their undesired consequences. Therefore, droughts have more impacts in the dry regions, though the impacts in other regions may not be negligible.

Droughts are not new phenomena, they have always existed. Extreme droughts caused the end of the Maya civilization in Central America (Gill et al., 2007). The Incas in South America developed irrigation systems that permitted them to become more resilient to droughts and other climate variations (Mamani-Pati et al., 2014).

Presently, droughts continue to plague the Americas. From 2012 to 2016, extreme droughts have affected practically all countries in the region: from California to NEB, through most of Mexico, Central America, and other countries of South America. Drought impacts have affected millions of people, their activities, and the landscape. In most of these places, climate in that period was the driest in historical record. Extreme droughts have been associated with the ENSO phenomenon—El Niño Southern Oscillation—both in its positive (El Niño) and negative phases (La Niña).

It is known that droughts are the kind of natural disasters that impose the highest costs to society. Impacts can be economic, social, and environmental, but also political, institutional, and cultural. In the developing countries of the Americas, including the whole of South and Central Americas and the

Drought Challenges. https://doi.org/10.1016/B978-0-12-814820-4.00004-3

Caribbean and the south of North America (Mexico), all these types of impacts can be observed. But the social impacts are the heaviest: the impacts on poor people, who are the most vulnerable population. In fact, what distinguishes developed and developing countries with regard to droughts is the weight of each type of impact, with social impacts being dominant in developing countries. In developed countries, most impacts are of an economic nature.

In each of the countries in the Latin America region, between 20% and 60% of their territory is classified as drylands—hyper-arid, arid, semi-arid, or sub-humid. These are the cases of NEB, Northern Mexico, Northeastern Venezuela, Coastal Pacific, and Central areas of Honduras and Nicaragua (Quijandria et al., 2001). Semi-arid and dry sub-humid regions are usually populated regions. Differently from arid regions, where there is a dearth of rainfall every year, in semi-arid and dry sub-humid regions there is, on average, sufficient rainfall to allow for crops to grow and to support the growth of population and an urbanization process. As a result, these regions can be occupied and maintain a civilization based on the primary products of agriculture, especially crops linked to local food provision, cattle raising, and fibers.

Along the years during the last century, a water infrastructure, composed of dams, aqueducts, and wells, was built that increased water security and allowed for a larger population to live there. In developing Latin America, from Argentina and Brazil to Mexico, this was the scenario for complex results on overall vulnerability to droughts. On one hand, the social and economic vulnerability have decreased due to drought policies that aimed at protecting the affected population, such as social protection programs and crop insurance programs. While there is some improvement, vulnerability is still high. On the other hand, environmental vulnerability has increased, due to land degradation and desertification. The situation gets worse in case of multiyear droughts such as in 2010−17, when many reservoirs dried up.

During the recent drought periods in this century, there was a sharp reduction in rain-fed agriculture, which is the realm of smallholder farmers and rural workers. A fall in crop production means loss of employment and occupation and this exacerbates social crises, as it affects the livelihoods of the poor and requires immediate relief assistance which is not always effective. In all countries in the region, the capacity to respond has improved over time. In larger countries, like Brazil and Mexico, the response has been provided mainly with national resources. In smaller countries, like in Central America and the Caribbean, international donors have played a more important role in relief assistance.

Water supply has been affected in the recent prolonged droughts. Usually, droughts hit more sharply the rural and urban semi-arid regions. However, in extreme droughts like the ones observed since 2010, even large cities in humid places like São Paulo, the most industrialized city of the Latin America and the Caribbean (LAC) region, with its 22 million inhabitants, were affected. In 2014 and 2015, São Paulo faced the most serious water supply crisis in its history, with the reservoirs that serve the city becoming almost dry. The crisis provides lessons, but there is always the risk of the so-called hydro-illogical cycle (Wilhite, 2012), with society and governments becoming unconcerned with droughts once the rainy season comes back. In California, for instance, during the recent 2012−16 drought, water consumers reduced consumption by 25%. However, in 2018, once the drought was over, roughly 80% of water utilities reduced their conservation goals to 0% (Underwood, 2018).

Drought policy responses in LAC have so far been mostly reactive. Being reactive does not necessarily mean that permanent works are not done to reduce future vulnerability. The construction of hydraulic infrastructure has been done in several countries, but they have been decided or implemented in reaction to a drought. However, most of the responses are relief and emergency intervention measures such as distribution of water, food, and medical assistance; provision of emergency employment; or, in some cases, social protection in the form of conditional or unconditional cash transfers.

In most countries of LAC, progress has been made to implement disaster risk reduction systems aligned with the Sendai Framework and the United Nations International Strategy for Disaster Reduction (UNISDR). The recognition of the state of emergency or public calamity is processed through a Civil Defense System, which allows states and communities to receive federal or international resources to finance relief actions. A proactive drought policy also comprises relief responses and permanent responses, but planned and implemented in a proactive way, not as a reaction to a given drought period. It needs to be a permanent activity. The most comprehensive drought policy in the region so far was developed by Mexico, with the adoption of the PRONACOSE—National Program Against the Droughts (Arreguín-Cortés, 2015).

The case of Northeast Brazil

The characteristics of droughts are different in time and in space. There are, however, similarities that can be inferred. The case of NEB region is a good example: a large, poor region that in some way represents what occurs in other large poor regions of LAC when the drought hits. The NEB is a region that comprises nine of the 27 states of Brazil, with 1.5 million km^2, of which 900,000 km^2 are semi-arid, with annual precipitation of up to 800 mm, high evapotranspiration rates of up to 3000 mm per annum, and, normally, eight dry months per year (Map 4.1). It is a densely populated region, with 53 million inhabitants in 2010, most of them concentrated in metropolitan areas along the coast: Salvador, Recife, and Fortaleza, with 3 million inhabitants each. The carrying capacity of the NEB is about 22 million people (including small cities), which makes this region one of the most densely populated semi-arid landscapes in the world.

Droughts have been present in the history of the semi-arid NEB. Table 4.1 shows the years of drought in the Northeast. In the beginning, during the 16th and the 17th centuries, one may argue that records were not accurate, as the interior of the Northeast was not yet populated.[1] So, it is possible that there were more droughts than was recorded, based on voyagers' reports of the time. The 18th century registered a more massive occupation of the backlands by the Portuguese, and the quantity and quality of records improved and continued improving on the 19th and the 20th centuries. According to the data, there were droughts in 20% of the years in the 18th and 19th centuries. In the 20th century, there was one drought for each 3.3 years, and in the beginning of the 21st century, from 2000 to 2017, one in each 1.3 year period was dry. Table 4.1 also shows that back-to-back droughts are common, when the droughts have a duration of two years or more.

Can we conclude that there was an increase in dryness? And that, due to climate change, this increase is being more evident in the 21st century and will probably be even more serious in the future as the IPCC says (IPCC, 2014)? This is one possibility that cannot be discarded, but there are also other complementary hypotheses. One is population increase. From the first census in 1872 to the last census in 2010, the population of the Northeast increased from 4.6 million to 53.1 million, a 13-fold increase. Rural population reached 14.8 million in 1960 and 14.3 million in 2010. So, the population that depends

---

[1]The region was populated by the Amerindian peoples. Population densities, however, were very low.

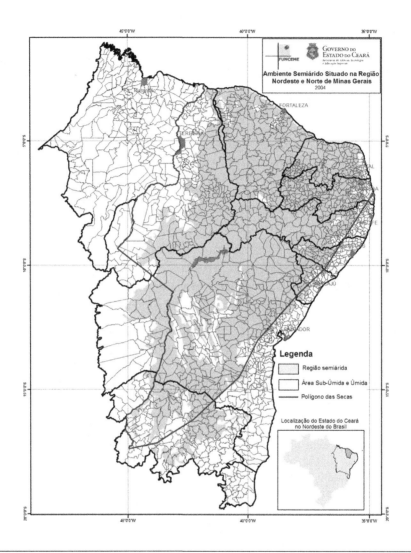

MAP 4.1

Semi-arid region of Brazil.

Source: FUNCEME, 2005.

directly on agriculture has decreased a little in the last 50 years, while urban population has increased at higher rates (though the rate of total population growth has also decreased). Most of the urban population also depends, directly or indirectly, on income and employment based on the use of natural resources and, above all, has its water supply source in the dry semi-arid areas (Table 4.2).

Another hypothesis is land degradation and desertification which renders the region more vulnerable, given unsustainable land-use practices. Nature is less resilient now, so that the same meteorological drought for which nature was possibly adapted to in the past is now reflected in lack of water for

Table 4.1 Record of droughts in the Northeast Brazil.

Century	Drought years	Number of droughts	Number of years of drought
16	1583–85	1	3
17	1603, 1624, 1692	3	3
18	1711, 1723–24, 1744–46, 1754, 1760, 1766–67, 1772, 1777–80, 1784, 1790–94	10	21
19	1804, 1809–10, 1816–17, 1824–25, 1827, 1830–33, 1845, 1877–79, 1888–89, 1891, 1898	11	20
20	1900, 1902–03, 1907, 1915, 1919, 1932–33, 1936, 1941–44, 1951–53, 1958, 1966, 1970, 1976, 1979–83, 1987, 1992–93, 1997–98	17	30
21 (2000–2017)	2001–02, 2005, 2007, 2010, 2012–16	5	12

Source: Marengo, J.A, Torres, R.R., Muniz Alves, L., 2016. Drought in Northeast Brazil: past, present, and future. Theoretical and Applied Climatology 124(3–4) Springer; Magalhaes, A.R., et al., 1988. The effects of climate variations on agriculture in Northeast Brazil. In: Parry, M., Carter, T., Konijn, N. (Eds.), The Impact of Climate Variations on Agriculture. In: Assessments in Semi-arid Regions, vol. 2. Kluwer Academic Press, Amsterdam.

Table 4.2 Population increase in the Northeast—1872–2010.

Year	Number of people		
	Rural	Urban	Total
1872	–	~	4.636.6
1960	14.748.2	7.680.7	22.428.9
2010	14.260.7	38.821.2	53.081.9

Source: IBGE – Brazilian Institute of Geography and Statistics, 2010. Demographic Censuses. Rio de Janeiro.

the population and reduction in agricultural yields. A recent study done by Center for Strategic Studies and Management (CGEE) and Foundation of Meteorology and Water Resources of Ceará (FUN-CEME) identified a total of 70,000 km^2 of land that is already desertified (CGEE, 2016a). More than that, most of the land in the semi-arid of NEB continues to suffer from land degradation and desertification. Vegetation has been deforested, water resources have vanished, soil erosion has increased, environmental resilience to droughts has decreased.

At the same time, government response has contributed to increase in societal resilience. Campos (2014) for instance, argues that overall resilience has increased, so that drought impacts are now reduced when compared to previous droughts. In deed social impacts used to be catastrophic in the past. In the big drought of 1877−79, which is frequently cited, besides the heavy impacts on agriculture and cattle raising that ceased all activities, hundreds of thousands of people died. According to one voyager, an American naturalist who visited the state of Ceará in 1878, about 500,000 people died in that state alone and other 300,000 died in the rest of the Northeast (Smith, 1879). Human deaths continued to be caused by droughts, in a lesser degree, in the rest of the 19th and in the 20th centuries and have been absent in the droughts of the 21st century.

Government's response to drought

Government response has occurred in two ways, basically. One is to provide relief actions to the affected population; the other is to reduce future vulnerability.

Relief actions

The first part, regarding relief actions, is coordinated through a system which is reasonably well organized, the Civil Defense System, which follows the lines of the Sendai Framework for Disaster Risk Reduction and the UNISDR. There is a long list of relief activities that tackled mainly two problems: (i) access to water (which becomes even scarcer during droughts); and (ii) access to income. In the drought of c.1979−83, 3 million workers were employed in work fronts in NEB. This income allowed workers to buy food. In the 21st century, the government gave up the idea of work fronts and adopted a cash transfer system to benefit the poor population in the whole country, including poor people in the drought-stricken region. This makes sense, because most of the poor in Brazil live in the rural Northeast.

This system of social protection has been able to provide a basic income to all affected families, so there was no need, even in extreme droughts like the one from 2012 to 2016, to provide emergency jobs via work fronts. This system, which is called "Bolsa Família," a conditional cash transfer program subject to children attending school and to family attending the health-care system, is complemented in some special cases with a crop insurance system, in case of loss of crops above 50%.

Water distribution is an action that is still largely used. In 2016, the Federal Government alone supported the works of 8000 water tanker vehicles, in the whole semi-arid region. They started distributing water to rural communities and isolated households, but lately they have also benefited whole cities that run out of drinking water. All these actions involve a combined effort of the three levels of government—federal, state, municipal—and several federal and state agencies. In fact, distribution of water through water tankers has been present even in years that were not considered drought years, showing the persistence of a water deficit in the rural semi-arid.

Reduce future vulnerability

Concerning the second way, aiming at reducing future vulnerability, the main line of action has been what is called "transportation of water in time" through the construction of dams to accumulate water in the region's river basins. Since 1906, when the first big dam was inaugurated, thousands of dams have been built. In 1909, the federal government created what came to be the National Department of Works Against Droughts (DNOCS, for its acronym in Portuguese). DNOCS was key in the process of reducing vulnerability in the water sector and in creating conditions for population to grow in the semi-arid rural and urban areas.

Most of the small dams were built by private owners or municipalities, basically with the goal of assuring water during the dry months of the year (even in what we could call a "normal" rainy year, there are 7–8 months without rains). There is another big number of medium dams, built by the federal government and, more recently, by states governments, which envisage the continuity of water supply during 1 or 2 years of drought. Some of these dams have been constructed in collaboration between the federal government and landowners. And finally, there is a smaller number of big and very big dams, mostly constructed by DNOCS, of the Federal Government, aiming at protecting the water sector during longer-term droughts.

The biggest and most recent dam, called Castanhão, in Ceará state, accumulates approximately 6.5 billion m^3 of water and is able to face a long-term drought of up to 5 years. In 2017, however, the total accumulated capacity of this dam fell to below 3%, causing big concerns to water managers and water users.

Besides these actions which were implemented as direct responses to droughts, there were other actions that reduced vulnerability. First, still in the water sector, there was increasing intensity in exploring underground water, with the construction of thousands of wells and the use of desalinators to improve the quality of the saline water. Aqueducts and inter-basin water transfer systems were built and continue to be built to improve distribution of water in the region. The biggest of the water transfer projects is the Program for Basin Integration of the São Francisco River with the rivers of the north of the Northeast (PISF).[2] The PISF project is 470 km long and aims at providing water security for 12 million people in the states of Ceará, Pernambuco, Rio Grande do Norte, and Paraíba. The source of water is the São Francisco river, the only permanent river that crosses the NEB.

Second, another action that reflected in reducing vulnerability was overall growth of infrastructure, especially transportation. In 1877, 1900, 1915, and 1932, many people died during migration from the interior to the coast. People would go months almost without food or water, and death associated with starvation, thirst, and malnutrition occurred. Now people can move from the dry backlands to the coastal areas in one day, and societal and government assistance can reach affected people much more easily.

Third, economic growth, as in other places, reduced the dependency of the economy and of the people on agriculture. In the Northeast, in the 1960s, about 30% of the GDP (Gross Domestic Product) came from rain-fed agriculture. Presently, this ratio is about 7%, and part of it is irrigated agriculture which is resistant to short-term droughts (but not resistant to long-term droughts). This means that there is less economic vulnerability now. However, social vulnerability continues because most of the poor still depend on rain-fed agriculture.

Fourth, there was a strengthening in the capacity of government (federal, state, and municipal) which allowed the government to allocate more resources to drought responses. And fifth, there was a process of migration and urbanization that resulted in a society that is less dependent on rains.

---

[2]PISF—Projeto de Integração do São Francisco (http://www.integracao.gov.br/web/projeto-sao-francisco/entenda-os-detalhes).

In summary, social and economic vulnerability decreased, but environmental vulnerability increased. Land degradation and desertification have increased, and for those whose economic activities depend on natural resources, vulnerability has also increased. The recent multiyear drought has demonstrated that water security, is not assured in such cases. An analysis of the last big drought, between 2010 and 2017, illustrates well this statement. In this period, only 2011 was not a drought year, but not much water had been accumulated. From 2012 to 2016, there were five consecutive years of drought that brought many reservoirs to zero or close to zero water accumulation and caused crises of water supply in several cities, including large metropolitan areas. Finally, 2017 was not a meteorological or agricultural drought but was a hydrological one because there was no accumulation of water and the soils soon became dry. Were it not for the support of the government with cash transfers and social assistance, a calamity would have happened. In fact, it came close to a calamity in certain small and medium cities where taps ran out of water and people living in cities had to fetch water in designated points for their own consumption.

Looking into the future, we do not have good news from the climate side. With the expected climate change, the number of extreme droughts and floods may increase, and more response capacity will be needed. Currently, response to present droughts is still not adequate because notwithstanding the progress the impacts of the last drought were still serious. In November 2017, a group of experts and policy makers from all nine states of the Northeast met in Fortaleza, Ceará, to assess the impacts and responses to the last drought. Impacts were serious in all states, mainly on rain-fed agriculture and water supply, but also in other sectors. During the drought, resources had to be redirected to fighting the drought impacts instead of their originally intended purposes. Clearly, existing response is not yet enough. With climate change, the need for new capacity is even greater (CGEE, 2016b).

Towards a proactive drought policy in NEB

In the beginning of the 21st century, the NEB still faces the challenge to reach sustainable development and to adapt to recurrent droughts. The challenge is twofold: first, under present climate variability, droughts still cause a huge impact which is magnified by land and water degradation and desertification; second, under climate change, droughts are becoming more extreme and more frequent. More than ever, society and government need to make progress in the improvement and consolidation of a drought policy to face future hazards. How to live in the future with a larger population, higher economic growth, and less water?

Sustainable development is the path. In 1992, after a successful international conference on climate variations and sustainable development of semi-arid regions (ICID), the NEB governments followed up on its recommendations and took an initiative of integrating sustainable development in government planning (Ribot et al., 1996). This resulted in the Aridas Project that recommended the development of trend and sustainable scenarios and a sustainability strategy that included reducing vulnerability to climate crises (Projeto Aridas, 1995). However, this is much complex and requires changes in mindsets and in decision-making processes. The implementation of sustainable development will take time as it requires overcoming short-term economic interests versus the rights of future generations.

More recently, an international initiative led by World Meteorological Organization (WMO); United Nations Convention to Combat Desertification (UNCCD); and the Food and Agriculture

Organization (FAO) of the United Nations convened a High-Level Meeting on National Drought Policies (HMNDP) and invited country partners to develop and implement national drought policies to further reduce vulnerability to droughts and improve the quality of assistance to drought-stricken populations. The recommended national drought policies aim at removing obstacles and difficulties faced by countries and regions in dealing with drought and in better preparing for future events. Several international organizations committed themselves to support countries' efforts in designing and implementing national drought policies.

National drought policies should be designed around three pillars: (1) monitoring, prediction, and early warning; (2) impact and vulnerability studies; and (3) drought mitigation and response. The first and the second pillars provide past, current, and future information on the climate and on its economic, social, and environmental impacts, and feed into the third pillar, which comprises all programs and projects that aim at reducing vulnerability to drought and at responding adequately to ongoing droughts. Of course, many countries have experiences in drought policies and will not start from scratch. There are several shortcomings which have to be dealt with: lack of coordination and adequate institutional framework; lack of financial resources; and lack of capacity.

Following on the recommendations of the HMNDP, the Ministry of National Integration of Brazil, with the support of the World Bank, started an initiative to adapt drought policies to the new guidelines. Several workshops were held in different states of the NEB, with the participation of stakeholders from the federal and states governments and from civil society. A major outcome of this process was the launching, by the National Water Agency (ANA), with the collaboration of the states' meteorological and water services, of a new drought monitoring system which is communicated to the public every month in the webpage of the ANA.[3] This experience has been documented by the World Bank and the CGEE in a book called *Droughts in Brazil: Proactive Management and Policy* (De Nys et al., 2016).

The process of organization and implementation of the Northeast Drought Monitor took two years, from 2014 to 2016, and involved not only national governments and civil society organizations but also international cooperation from the United States, Mexico, and Spain, besides the World Bank. The Monitor is based on the experiences of the United States and Mexico, with study tours and training of Brazilian teams to these countries. The Monitor produces a map of the Northeast which is published monthly by ANA showing the conditions of drought in the region, one color for each type of drought: no relative drought, moderate drought, severe, extreme, and exceptional drought. It also indicates, for each type of drought, if the impacts are short term or long term. In 2016, for instance, most of the Northeast was under extreme drought, with long-term impacts due to low levels of reservoirs. Each month, the drought monitor is prepared by a different state, through its water and meteorological organization, based on indicators, such as the Standardized Precipitation Index (SPI), the Standardized Precipitation and Evaporation Index (SPEI), and the Standardized Runoff and Dry Spell Indicators, and is validated by local stakeholders. Only after this validation can the map be delivered to users, especially stakeholders in the government, municipal water companies, and farmers. Besides the map, municipalities and states send other information to the National Civil Defense System with local data on the state of drought. This is the condition for the declaration of emergency in municipalities and the possibility for them to participate in the federal programs that address the effects of the droughts, in particular in regard to water distribution.

---

[3]Monitordesecas.ana.org.br.

A second component of the first pillar is climate prediction, or climate forecast. Developments in the climate science and in global circulation models, during the last decades, allowed for improvements in climate prediction, usually 3−4 months in advance, in certain parts of the world. In the Northeast, droughts are associated with El Niño, a warming of the waters of the Pacific Ocean, and with a cooling of the sea surface temperature of the equatorial south Atlantic. These are the two main causes of drought in the region, but there are instances when there is an El Niño and no drought in the Northeast. Climate scientists meet generally in December each year to evaluate the situation and conclude, using all available information, about the trends of the next rain season. The group meets again in January and incorporates the latest information in the prediction. In January 2018, for instance, prediction informed that there was a higher probability of raining above average (O Povo, 2018). This information is useful for policy makers and farmers to support their decisions.

Communication, or early warning, which is the third component of pillar 1, needs to be done carefully. In fact, decisions imply economic and social costs, such as in the case of planting or not planting. If the prediction is wrong, there is an associated cost of unemployment and fall in agricultural production. The credibility of the meteorological services will collapse. However, society and government need to be prepared to act when needed. Communication between experts and policy makers and the public is an issue where some progress has been made but where much needs to be done on an ongoing basis.

There is, historically, an extensive literature on droughts and its impacts in the Northeast, which is the object of pillar 2. The droughts of 1877−79, 1900, 1915, 1932, 1958, 1979−83, and the recent period of 2010−17 are well documented. Narratives tell about the impacts of droughts, especially human suffering, and responses. There are fictions based on the realities of these droughts, now part of the classic Brazilian literature. There are many technical and scientific reports, especially coming from the Academia, where numerous dissertations have been written. And there are the results of scientific, academic and policy meetings, involving the collaboration of governments, civil society, and academic organizations from Brazil and elsewhere. However, the knowledge is scattered and not always available. There is no single place where the information can be found. Under pillar 2, gaps of information should be identified, updated, and brought to the attention of policy makers and researchers.

On pillar 3, drought mitigation and response, there is also a history of government policies and a consolidated experience that includes different ways of short-term relief as well as long-term policies aiming at reducing vulnerability to droughts in the future. Based on the NEB experience, policies are identified: water policies to distribute water during climate crises and to build resilience in the future, through an ambitious water infrastructure program and integrated water resources management; income generating policies aiming at providing the poor affected population with means to acquire food and meet basic needs; and policies to feed animals. These policies involve the cooperation among the three levels of government, with the federal government providing the bulk of financial resources through national programs. Such policies have evolved over time. Income generating policies, for instance, most of the time were based on different kinds of emergency works, have been substituted by conditional cash transfers to all poor population. Relief responses are integrated under the Civil Defense System, which involves institutions at the federal, states, and municipalities levels. However, there is the need for more integration between the three pillars, more coordination, and, especially, more stability in terms of financial resources.

Conclusion

Though droughts occur in all regions of Latin America and Brazil, they are more frequent and severe in the drylands. The NEB contains a large semi-arid region and has a long history of dealing with droughts. As in other poor dry landscapes, social impacts have been heavy, with loss of lives, cattle, and economic assets. As a large middle-income country, Brazil has been able to finance its drought mitigation and response programs. However, there is much space for international, bilateral, and multilateral cooperation, along with the guidelines of the HMNDP.

Drought policies in the NEB need to consolidate the three pillars suggested since the HMNDP and be incorporated into national, regional, state, and local development policies. In the future, with droughts more frequent and severe, only a comprehensive policy of sustainable development that considers sensitiveness to climate variability and change will result in reduced vulnerability and increased resilience. This requires changes not only in the policies and the role of the government but also in mindsets.

References

Arreguin-Cortés, F.I., Pérez, M.L., Ibañez, O.F., 2015. National drought policy in Mexico, a paradigm change: from reaction to prevention. CGEE, Brasilia 20 (41). Available at: www.cgee.org.br.

Campos, J.N.B., 2014. Secas e Políticas Públicas no Semiárido: ideias, pensadores e períodos" (Droughts and public policies in the semi-arid: ideas, thinkers, and periods). Estudos Avançados, USP, São Paulo 28 (82).

CGEE, 2016a. Centro de Gestão e Estudos Estratégicos. Desertificação, degradação da terra e secas no Brasil. CGEE, Brasília.

CGEE-Center for Strategic Studies and Management, 2016b. Seminar of Evaluation of the 2010—2016 Drought in the Brazilian Semi-arid. Fortaleza.

De Nys, E., Engle, N.L., Rocha Magalhães, A., 2016. Droughts in Brazil: Proactive Management and Policy. CRC Press, NY.

Gill, R.B., Mayewski, P.A., Nyberg, J., Haug, G.H., Peterson, L.C., 2007. Drought and the Maya collapse. Ancient Mesoamerica 18 (2), 283—302. Cambridge University Press.

IBGE — Brazilian Institute of Geography and Statistics, 2010. Demographic Censuses. Rio de Janeiro.

AR5 Report IPCC-Intergovernmental Panel on Climate Change, 2014.

Magalhaes, A.R., et al., 1988. The effects of climate variations on agriculture in Northeast Brazil. In: Parry, M., Carter, T., Konijn, N. (Eds.), The Impact of Climate Variations on Agriculture, Assessments in Semi-arid Regions, vol. 2. Kluwer Academic Press, Amsterdam.

Mamani-Pati, F., Clay, D.E., Smeltekop, H., 2014. Modern landscape management using Andean technology developed by the Inca Empire. Chapter 19. In: Jock, C.,G., Landa, E.R. (Eds.), The Soil Underfood. Infinite Possibilities for a Finite Resource. CRC Press, Boca Raton.

Marengo, J.A., Torres, R.R., Muniz Alves, L., 2016. Drought in Northeast Brazil: past, present, and future. Theoretical and Applied Climatology 124 (3—4). Springer.

O Povo, F., 2018. Prognóstico para a Quadra Chuvosa de 2018 (Funceme Forecast for the Rainy Period 2018). www.opovo.com.br.

Projeto Aridas, 1995. A Strategy for Sustainable Development in Brazil's Northeast. Ministry of Planning and the Budget, Brasilia.

Quijandria, B., Monares, A., de Peña Montenegro, R.U., 2001. Assessment of Rural Poverty, Latin America and the Caribbean. International Fund for Agricultural Development, Santiago, Chile.

Ribot, J.C., Magalhães, A.R., Panagides, S.S. (Eds.), 1996. Climate Variability, Climate Change and Social Vulnerability in the Semi-arid Tropics. Cambridge University Press, Cambridge, UK.

Smith, H.H., 1879. Brazil, the Amazons and the Coast. Charles Scribner Sons, New York. Available at: www.forgottenbooks.org.

SUDENE, 2017. Delimitação Do Semiárido. Available at: http://sudene.gov.br/planejamento-regional/delimitacao-do-semiarido.

Underwood, E., 2018. California Water Savings Dwindle when Drought Fears Subside. Preventionweb.net.

Wilhite, D.A., 2012. Breaking the hydro-illogical cycle: changing the paradigm for drought management, American Geosciences Institute. Earth Magazine 57 (7). Lincoln, Nebraska.

Drought in the Yucatan: Maya perspectives on tradition, change, and adaptation

Miguel Paul Sastaretsi Sioui

Geography and Environmental Studies, Wilfrid Laurier University, Waterloo, ON, Canada

Introduction

The impacts of climate change on Indigenous peoples' livelihoods are often overlooked in academic research about this topic. In Yucatan (Mexico), Maya communities, which continue to be centered on millennia-old milpa-based agricultural systems, are particularly vulnerable to climate change—related impacts such as drought. The Yucatan Peninsula is located between the Gulf of Mexico (to the North) and the Caribbean Sea (to the South), covering an area of nearly 44,000 km². Geopolitically, the Peninsula incorporates parts of the Mexican states of Yucatan, Campeche, and Quintana Roo, in addition to northern Belize and Guatemala's El Petén Department. Scientific evidence shows that the Yucatan's climate has been getting hotter and drier, with more frequent unpredictable weather patterns. It is also clear that the effects and impacts of climate change in the Yucatan, such as drought, will continue to increase in severity in coming years and decades. Thus, climate change will increasingly pose a challenge to the Yucatec Maya in their milpa-based land-use patterns and activities. However, what is less clear is how exactly climate change is influencing and reshaping Maya land-use patterns and practices. In this chapter, I present new empirical data about how climate change is affecting the relationships between the Maya and the land, particularly in their traditional milpa farming practices, as well as some possible policy implications of community-based adaptations to climate change in the Yucatan.

Background: Yucatec Maya land-based livelihoods

Archaeological evidence indicates that agriculture in the Yucatan Peninsula has been practiced for at least 5000 years (Anderson, 2005). The *milpa* is the traditional Maya system of rotational agriculture, in which lands are cropped for a few seasons and then temporarily abandoned for new clearings to ensure long-term soil regeneration and overall ecosystem vitality. Agriculture (and the milpa), to the Maya, is "not just a source of the staple food; it is a religious act and a moral duty" (ibid., p. 44). The Yucatan has rich biodiversity, and many plant and wild and domesticated animal species are still managed by a large percentage of the Maya population. This is also a result of a warm, tropical climate with an abundance of rainfall (Dahlin, 1990).

Drought Challenges. https://doi.org/10.1016/B978-0-12-814820-4.00005-5

The Yucatan Peninsula's vegetation is dominated by both short and tall dry tropical forests that are home to thousands of species of plants and animals, a large number of which have for centuries been actively managed by the Yucatec Maya (Gomez-Pompa et al., 2003). From northwest to southeast, the Yucatan has a vegetation gradient that transitions from mangrove and dry deciduous shrub (on the northern coast on the Gulf of Mexico), semi-deciduous low to medium to high forest (western interior, to moister medium to high broadleaf evergreen forest (nearer eastern coast on the Caribbean Sea) (Rico-Gray and Garcia-Franco, 1992). Land cover in the region has been greatly modified due to human use even prior to the arrival of the Spanish in the early 16th century. Yucatan's forests have been greatly modified and cleared by agriculture and, more recently, cattle ranching. Much of the dry forest has been converted into secondary growth communities as a result of henequén (agave) plantations during Spanish colonial time, and from cattle grazing (Gonzalez-Iturbe et al., 2002). The region's climate is characterized as tropical (average yearly temperature of 25−27°C) and humid year-round due to its proximity to the warm waters of the Caribbean Sea and the Gulf of Mexico (Curtis et al., 1996). The annual average rainfall exceeds 100 cm, the majority of which occurs from May to August (the Yucatec Maya agricultural season).

At present, many Yucatec Maya still observe a number of agricultural rituals and ceremonies throughout the year, some of which were practiced millennia ago. The Maya agricultural cycle in the Yucatan begins between April and May, depending on weather conditions, when religious offerings are made in the milpas before clearing the fields of brush and organic debris. As the crops begin to sprout in late July to August, offerings of corn and beans are made to the earth. The cycle concludes during the harvest in September, when thanks are given to the crops (de Frece and Pool, 2008). Present-day Yucatec Maya ejidal communities are still organized according to the milpa lifestyle, and agriculture remains central to community life in Xuilub, as well as social organization. As in the past, the crops most commonly grown in the milpa are corn, beans, and squash. In addition to the main milpa crops, Rico-Gray et al. (1990) described the important plant and vegetable species such as tomatoes and chillies that were grown in small family gardens looked after by women. de la Cerda and Mukul (2008) more recently observed that some Maya look after more than dozens of species of vegetables and plants in their home gardens.

The changing yucatan climate

Climate models for the Yucatan Peninsula have predicted a warming of as much as 4°C by 2100, and this region is predicted to be one of the worst off in terms of mitigation and adaptation potential (Orellana et al., 2009). A 4°C rise in the mean continental temperature by 2100 could potentially be devastating for the Yucatec Maya, whose culture and identity is centered on milpa agriculture. The projected consequences of such accelerated warming in the Yucatan region include rising oceans, more frequent and intense tropical storms, and more frequent droughts (ibid.; Sandoval et al., 2014).

Case study: drought and yucatec livelihoods in xuilub

With an area of approximately 32 ha, Xuilub is located in the eastern part of Yucatan state approximately 40 km southeast of Valladolid (see Figs. 5.1 and 5.2), roughly 2 to 3 km northwest of the Yucatan−Quintana Roo state border. According to the most recent census of the community done by the

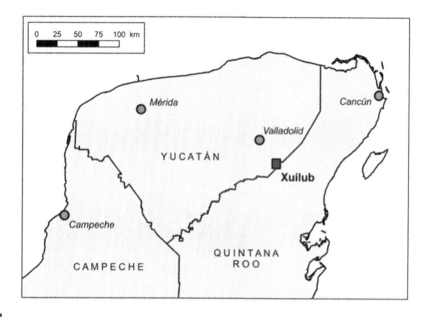

FIGURE 5.1

Map of the location of Xuilub in relation to the Yucatan.

Source: Author and D. Smith.

local medical clinic in 2015, Xuilub counted 707 residents, of which 362 were male and 345 were female. Youths make up a large portion of Xuilub residents, evidenced by the community's median age of 23 years, which is 4 years younger than that of the population of Mexico.

Field data were collected during five trips to Xuilub between 2014 and 2016. Data collection consisted of participant observation and semi-structured interviews. Throughout the field research, I participated in work on seven different milpas (usually 1- or 2-day outings) that were looked after by seven different milperos in the community. I participated in all the different types of milpa work (at different times in the milpa cycle): fallowing, planting, and harvesting. In all, 42 milperos were interviewed. They shared with me invaluable Indigenous knowledges (IKs) and information relating to the changing climate, all facets of milpa farming and its spiritual dimensions, and stories about other land-based knowledges and practices.

Yucatan's changing climate through the eyes of Xuilub's milperos

The question of how climate change is influencing and reshaping Maya land-use patterns and practices has been largely unexplored, even though it could potentially have a significant impact on Yucatec Maya livelihoods. As the effects and impacts of climate change in the Yucatan continue to become more severe, this factor will increasingly pose a challenge to the people of Xuilub in their milpa-based land-use patterns and activities. Milperos' views of current climate trends in the region are

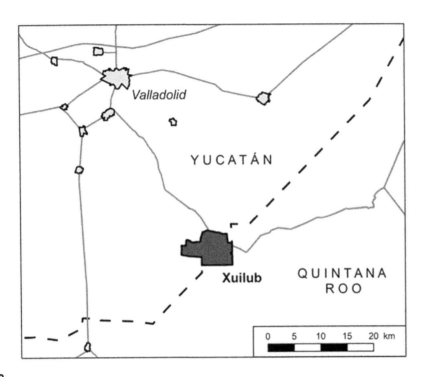

FIGURE 5.2

Map of the boundaries of Xuilub and Xuilub's position in relation to Valladolid.

Source: Author and D. Smith.

consistent with those of the literature generally and with the opinions of local climate scientists based in Valladolid. Generally, the climate has been getting hotter and drier, with more unpredictable weather patterns. For example, one milpero lamented that "dry seasons have been particularly dry and sometimes it rains when it's not supposed to rain. It's also getting hotter and drier earlier in the summer. As a result, harvests have been getting smaller."

Another milpero commented on the excessively hot and dry conditions that led to a poor harvest in 2016, saying:

The harvest was very small because of the weather, the sun. A bad drought for the last four months. In September it rained a tiny bit, so there was a small harvest. But no squash. The soil around here is now red. There's barely any black soil. Using fertilizer helps a little bit, but it's no guarantee. Sometimes it works, sometimes it doesn't. The weather around here, the environment… all there is, is drought. It's not favorable. Too much heat. You suffocate. In the last few years there has been much loss and bad harvests, but even a few years ago it wasn't as bad. When I was growing up, the work in the milpa was not like it is today. Today when you plant a milpa, it doesn't yield anything. When I was growing up, we never used fertilizers. If you carried out the usual tasks, like fallowing, burning, and harvesting, the harvest was always bountiful. But now with the same amount of work, some years you get something, some years you don't. They say that all this is because of climate change. Most of the time now there is drought. It doesn't rain, the earth doesn't give you anything anymore.

Commenting on the increasingly common periods of drought that are longer than in past decades, and the resulting scarcity of water resources, another milpero stated:

When the rain is falling, the earth is wet, compact. But today we're feeling heat that is more intense, stronger. More drought. But when there is rain, it's more humid and fresher. In the past, the earth would bear many fruits. There was a lot of maize, squash, and beans. That's because there was rain. The earth was blessed in that she could produce more. But today this has changed a great deal and there is much more heat. Even if you carry out my milpa work dutifully, the earth doesn't bear like it used to. My crops always dry out or die. There is some harvest, but it's very little. When I was younger and planted maize, fertilizers were not needed to make the earth more fertile. It was always natural. No chemicals were required for the earth to bear fruits. But today, even with chemicals, nothing comes out.

Informal discussions with two climatologists in the nearby city of Valladolid (see Figs. 5.1 and 5.2) gave further credence to the stories and anecdotes of the milperos. They mentioned that they had noticed significant changes in the weather patterns since 2009. However, more marked changes began around 2011, and that the climate has ever since been becoming hotter and drier, with more unpredictable precipitation patterns.

The increasing use of chemical fertilizers

In the past, the soil in the milpas of Xuilub is said to have been very fertile and to produce bountiful harvests every year, almost without fail, according to the community's milperos. Since the early 2010s, fertilizers are increasingly being used by the milperos of Xuilub. Various statements by milperos point to a shared feeling of regret and sadness about the recent and growing need to use fertilizers and, thus, failing to do things the "way of the Old People." One of them commented on this trend: "Sadly, now I have no choice but use fertilizers to create favorable conditions for my crops to grow." In the last 5 years, as I was informed by six interviewees, milpas have been becoming less productive due to a drying and warming climate in the Yucatan region. Reportedly, as recently as 10 years ago, the use of fertilizers was almost inexistent and no synthetic chemicals were required for the milpa to yield abundant harvests. "We did things very naturally back then," explained one interviewee. He continued, "and now we have to use these chemical products that are not from the earth. This is not the way we were intended to carry out our work and responsibilities to the earth."

A number of interviewees lamented that at present harvests are usually very small, even with the use of fertilizers. One young milpero commented on this:

When the climate or weather is good, when there's rain, and there's balance on the land, my harvest turns out very nicely. But when there are periods of great heat and great drought, my crops die. Last year, that's what happened. Many of my maize plants began to dry and die, little by little. They say that when this happens, when there is extreme heat or drought, it's a form of punishment from the gods. So when it happens, we perform a special ceremony to beg Cha-Chak [the Maya god of rain] for rain.

Agrochemicals, most notably chemical fertilizers, are now being used by virtually all milperos in the community. Several interviewees reflected on the changing climatic conditions that have been intensifying over the last decade. Many milperos in the community have been reluctant to use chemical fertilizers on their milpas, but have increasingly felt the pressure to do so in the last few years. More frequent bouts of extreme weather (tropical storms) and a generally reported hotter and drier climate during the growing months (usually April to September) have left Xuilub's milperos with little choice but to implement the use of chemical fertilizers, much to their lament. One middle-aged milpero describes his (and a very common) dilemma:

> Whereas my father and my grandfather were able to work the land naturally and sow a bountiful harvest—more than enough to feed their families very well—the climate isn't like it used to be in those days. Even 15 years ago! I remember being able to work the soil naturally and getting rewarded handsomely. In the last decade, all that has changed. We're getting more and more periods of overwhelming heat and dryness. There's not enough rain. I need to use fertilizers or else my milpa would produce nothing. Only tiny cornhusks. With that I wouldn't be able to feed anyone.

Although chemical fertilizers are now commonly used by Xuilub's milperos, there is a common belief that it is best to avoid their overuse, but this is proving to be increasingly difficult for them. A number of milperos expressed their distrust in "unnatural" agricultural products. "We should only use the bare minimum," said one milpero. When I ask him to explain why this should be, he told me the reason is "because they don't belong in the soil. I think that using too much of them can poison the land, and maybe ruin it for a long time. It's better to do things as naturally as possible, like our forefathers." "The milpa is very beautiful," said another milpero, "when you know how to rotate the crops, it produces the essential Mayan foods that sustain our people. We use fertilizers not because we want to. We use them because they are now necessary." There is concern among several milperos that the chemical fertilizers will eventually parch the soil and render it infertile, although none of them reported any evidence to that effect during the field research.

"I Will never leave my milpa": the responsibility to carry out traditional Maya work

Despite threats to subsistence and the Maya land-based culture in the Yucatan, none of the interviewees seemed to suggest that drought will eventually drive them out of their community and force them to break their relationship with the land. Indeed, as put by one milpero, it is inconceivable for a Maya person to abandon her/his land, and it is very much cultural and spiritual duty to keep a milpa:

> Well, to me, the earth means a lot. In the past, when I used to work on my milpa, the harvest was plentiful. The earth provided my family and me with food to sustain us. But things have changed too much. Now, when it's the rainy season, and we're expecting rain, it doesn't rain, nothing. Many corn plants that I recently planted are dying because of too much heat and sun. But even though I am old, I have a great deal of respect for the land, and I continue to carry out my Maya work. I don't foresee a time when I'll stop working on my milpa, because I'm so used to the work and I will always carry the work deep within me, so maybe I'll die on the field, working.

Another middle-aged milpero echoed the above thoughts and added that he much prefers being in the community and working the land, however unproductive it may be becoming, over city life:

> It's something beautiful to me, because the land protects us from the heat and the wind and the warming that's occurring. Even with increased drought, at least we have land here that we can grow our food on. When you live in the city, there is no room to have land. Everything is paved, so there is too much heat and you can't do anything about it. Here, at least we have our land.

What remains clear is that the milpa will remain the linchpin of the community and the essence of the Maya identity for many years to come. The land-based relationships that the Maya have entertained for millennia with their homeland of the Yucatan have been maintained through changing environmental, political, and sociocultural conditions, not least of which were the ones brought about by centuries of colonialism.

Future outlooks: incorporating Maya knowledges and perspectives into drought mitigation policies

In order for developing regions like Yucatan, Mexico, to more effectively adapt to climate change, it is important for governments to engage with Indigenous communities. The Yucatec Maya face many challenges in adapting to the projected impacts of an increasingly warmer and drier regional climate. Sandoval et al., 2014 believe that most pressing of those is the building of strong social capital networks to compensate for lack of support from weak and fickle state institutions in Yucatan. Indigenous groups such as the Maya hold the key to adaptation and mitigation in their traditional knowledges and everyday land-use practices. Indeed, Kronik and Verner (2010) suggest that locally developed initiatives anchored in traditional land knowledges are at least as—if not more—important as larger-scale institutional initiatives to help rural areas cope with climate change—related impacts. This leads the authors to argue that bilaterally (i.e., government—Indigenous) designing successful adaptation measures rests on government recognition of Indigenous autonomy and sovereignty. The following excerpt sums up this key argument:

> It is important to be thinking in terms of Indigenous peoples' rights when designing and negotiating the new mitigation instruments [...] In Indigenous communities, cultural institutions play a significant role in the adaptation to climate change and variability [...] Indigenous peoples have an important role to play in adaptation—by promoting the incorporation of experience with new and previously unknown phenomena (climate change, and thus flora and fauna) [...] (p. 161).

The inclusion of local IKs is perhaps the most important factor with regard to drafting practicable adaptation strategies for Indigenous communities. The authors' conclusions are in fact in agreement with recent research on local community adaptation to climate change. For example, studies by Brklacich and Bohle (2006), Sander-Reiger et al. (2009), and Head et al. (2011) on small, resource-dependent Canadian and Australian communities propose that locally developed, community-based, climate adaptation measures seem to be more effective than top-down approaches. According to these authors, it is likely that, ultimately, locally conceived and implemented adaptation plans—in both Indigenous and non-Indigenous contexts—have the best potential to prove successful in long-term community survival in the face of climate-related (and a host of other) challenges. In the context of Latin America, it is also becoming increasingly evident that local IKs should underpin any climate change adaptation and mitigation strategies. This inclusion would not only increase the likelihood of Indigenous cultural survival but also oblige governments to recognize the territorial sovereignty of its Indigenous groups. It all begins with a commitment to valorizing and recognizing Indigenous land-based values and practices on the land and seeking to understand more about them by meaningfully engaging with these groups and communities. In the case of the Yucatan, what is needed is a firm commitment from policymakers to understand the needs of its Maya communities. Such a committed effort is the first step in ensuring that the Maya will be able to carry out their milpa work and other land-based relationships and responsibilities for generations to come.

The current degradation of milpa-based practices in the Yucatan presents a threat to the cultural survival of the Maya. Xuilub's milperos are being faced with emerging challenges and must find ways of adapting to drought and other climate change—related impacts. However, the people of Xuilub, and all other Yucatec Maya communities, are still carrying out the same land-based practices as their ancestors thousands of years ago. The Maya people have demonstrated time and again, and through drastically changing conditions and contexts over the centuries, remarkable resilience and a remarkable ability to deal with such changes. In the words of Miguel Leon Portilla, an eminent Mayanist scholar, "[…] Mesoamerica has not disappeared. Some will say it is a miracle of history that its languages, peoples, and cultural heritage are alive. Mesoamericans [including the Yucatec Maya] believe it is their destiny for them to be as long as this cosmic age will continue to exist and this sun will shine" (1992, p. 175).

References

Anderson, E.N., 2005. Political Ecology in a Yucatec Maya Community. University of Arizona Press, Tucson.

Brklacich, M., Bohle, H.G., 2006. Assessing human vulnerability to global climatic change. In: Ehlers, E., Kraft, T. (Eds.), Earth System Science in the Anthropocene. Springer, Berlin/Heidelberg, pp. 51−61.

Curtis, J.H., Hodell, D.A., Brenner, M., 1996. Climate variability on the Yucatan Peninsula (Mexico) during the past 3500 years, and implications for Maya cultural evolution. Quaternary Research 46 (1), 37−47.

Dahlin, B.H., 1990. Climate and prehistory on the Yucatan Peninsula. Climatic Change 5 (3), 245−263.

de Frece, A., Pool, N., 2008. Constructing livelihoods in rural Mexico: milpa in Mayan culture. The Journal of Peasant Studies 35 (2), 335−352.

de la Cerda, H.E.C., Mukul, R.R.G., 2008. Homegarden production and productivity in a Mayan community of Yucatan. Human Ecology 36 (3), 423−433.

Gomez-Pompa, A., Allen, M.F., Fedick, S.L., Jimenez-Osornio, J.J., 2003. The Lowland Maya Area: Three Millennia at the Human-Wildland Interface. Routledge, London.

González-Iturbe, J.A., Olmsted, I., Tun-Dzul, F., 2002. Tropical dry forest recovery after long term Henequen (sisal, Agave fourcroydes Lem.) plantation in Northern Yucatan, Mexico. Forest Ecology and Management 167 (1−3), 67−82.

Head, L., Atchison, J., Gates, A., Muir, P., 2011. A fine-grained study of the experience of drought, risk and climate change among Australian wheat farming households. Annals of the Association of American Geographers 101 (5), 1089−1108.

Kronik, J., Verner, D., 2010. Indigenous Peoples and Climate Change in Latin America and the Caribbean. World Bank Publications, Washington.

Orellana, R., Espadas, C., Conde, C., Gay, C., 2009. Atlas escenarios de cambio climático en la Península de Yucatán. Mérida: Centro de Investigación Científica de Yucatán (CICY).

Rico-Gray, V., Garcia-Franco, J.G., Chemas, A., Puch, A., Sima, P., 1990. Species composition, similarity, and structure of Mayan homegardens in Tixpeual and Tixcacaltuyub, Yucatan, Mexico. Economic Botany 44 (4), 470–487.

Rico-Gray, V., García-Franco, J.G., 1992. Vegetation and soil seed bank of successional stages in tropical lowland deciduous forest. Journal of Vegetation Science 3 (5), 617–624.

Sander-Regier, R., McLeman, R., Brklacich, M., Woodrow, M., 2009. Planning for climate change in Canadian rural and resource-based communities. Environments 37 (1), 35–57.

Sandoval, C., Soares, D., Munguía, M.T., 2014. Vulnerabilidad social y percepciones asociadas al cambio climático: Una aproximación desde la localidad de Ixil, Yucatán. Sociedad y Ambiente 5.

The shifting limits of drought adaptation in rural Colombia

Brianna Castro
Department of Sociology, Harvard University, Cambridge, MA, United States

Introduction

Two landmark reports released in 2018—the Intergovernmental Panel on Climate Change special report on the impacts of global warming of 1.5°C above pre-industrial levels (IPCC, 2018) and the World Bank's Groundswell Report on climate-related migration (Rigaud et al., 2018)—projected that immediate and long-term climate change impacts, including sea level rise, drought, and extreme weather events, are causing and will continue to cause internal migration, especially in developing nations. In the case of Colombia, a country with one of the highest rates of internal displacement due to conflict in the world, internal migration related to environmental change is understudied and often overshadowed by the historically high levels of conflict displacement. However, environmental influences on migration are becoming observable in Northeastern Colombia, where erratic rainfall patterns, increasing temperatures, and more frequent and severe droughts are expected (IPCC, 2018).

From May 2014 through late 2016 the Caribbean coast of Colombia experienced an El Niño drought unlike any in recent memory. In the rainfed agricultural region of El Carmen de Bolívar, Colombia, the breakdown of traditional wet and dry seasons over multiple years created a slow-onset environmental crisis that resulted in multiple internal migration responses. Better-off families migrated temporarily in the early months of drought with intentions to return to their rural homes with the rains. Remaining families sent adaptive migrants out of their households to other agricultural areas or to urban centers to seek work and remit income. By the third year of the drought, repeated failed harvests and completely dried-up water stores displaced capital-depleted families who left their agricultural homes permanently and in desperate conditions.

These internal migration responses to protracted drought reflect research to date on the topic. Migration as a response to climatic hazards takes many forms in the case of slow-onset stressors such as drought. This is due to the presence of multiple tipping points in the nexus of adaptation in place and migration (McLeman, 2018). Often in the case of slow-onset environmental stressors, migration is a strategy of last resort after in situ adaptations fail (Findlay, 2011; McLeman, 2011). Migration becomes a more reasonable adaptation only after households deplete their resources by adapting in place (Dow et al., 2013; McLeman, 2011). Though we recognize how these tipping points occur at the broad level, the nuance of individual and household level decisions to adapt in place or migrate during climate stress is less understood. This ethnographic case showcases when and how families reach tipping points and migrate in response to drought.

Drought Challenges. https://doi.org/10.1016/B978-0-12-814820-4.00006-7

Drought is all too often a crisis of circumstances, and the Colombian case was no different. Vulnerability to drought today is a product of not just the environmental stressor, but rather environmental stress in the sociohistorical context of a marginalized, impoverished region bearing scars of a lengthy and complicated conflict history. Slow-onset environmental stressors disproportionately impact rural, natural resource—dependent households (Obokata et al., 2014). In El Carmen de Bolívar, subsistence farmers relying on rainfall-dependent agriculture lost their livelihoods, while Colombians in municipal centers experienced the same drought but no crisis as they had access to water infrastructure and their employment was not dependent on natural resources. This chapter discusses how the combination of socioeconomic vulnerability and families' perceptions of their own risks and resources during an El Niño drought drove the timing and permanence of their migrations.[1] This grounded view reveals reasons why migration as an adaptive strategy is not universally available to everyone at the same moment, and this varying ability to adapt comes with enormous livelihood consequences.

Vulnerability and adaptation during drought

An environmental stressor does not affect everyone in the same way, and for that reason we expect individual responses to vary (Black et al., 2011; Hunter et al., 2015; Raleigh, 2011). The El Niño drought considered here demonstrates households' adaptive capacities, including how they cope and recover their livelihoods, in an atypical drought. These adaptive capacities depend on the depth and breadth of households' "toolboxes" when adapting to increasingly severe and frequent climate change droughts.[2] Families' vulnerability to drought is produced by the combination of both social context and environmental stress, including socioeconomic and institutional dynamics which limit adaptive capacity during slow-onset environmental crisis (Adger et al., 2006, 2009; Black et al., 2011; Freudenburg et al., 1995).

Social, cultural, and financial capital facilitate the mobilization of resources, making those with more resources less vulnerable to livelihood disruptions (Oliver-Smith, 2009, 2012). More socially and naturally resourced, less politically and socially marginalized, families are less vulnerable and thus better able to weather environmental stress (Cutter et al., 2003; Kelly and Adger, 2000). Perceptions of vulnerability rooted in environmental uncertainty further complicate households' socio-ecological vulnerability. Drought, like many slow-onset climate phenomena, is riddled with such uncertainty regarding its length and severity. Households think through the stakes of waiting out the drought alongside perceived losses from migration.

Migration is a high-stakes, often costly, adaptation to drought that is more or less likely depending on the sensitivity of local socioeconomic systems (McLeman, 2018; Hunter et al., 2015). How households and communities adapt to environmental stress, through in situ means or through migration, depends on context. Regions like El Carmen de Bolívar, dependent on subsistence agriculture, are more vulnerable to climate changes due to limited alternate livelihood options and possibilities for adapting in place (McGregor, 1994; Meze-Hausken, 2000; McLeman and Smit, 2006; Mortimore and Adams, 2001). In the Colombian case, the array of factors shaping families' adaptation to drought posits migration as one among a host of adaptations not equally available to every household.

---

[1]This chapter is based on qualitative fieldwork conducted by the author in 2016 and 2017 in El Carmen de Bolívar, Colombia. Data include 80 in-depth interviews with farming families, local experts, and government officials, as well as copious field notes from ethnographic observation.
[2]For detailed climate change predictions, see the IPCC: http://www.ipcc.ch/.

Drought in the Colombian context

When talking about migration in Colombia, the discourse tends to focus on conflict. Yet, Colombia is extraordinarily biodiverse, unevenly developed, and faces severe environmental challenges. Colombia's *Third National Communication on Climate Change* (2017) estimates average temperatures in 2070 to increase by 2−4°C and average annual rainfall to decrease throughout the country. El Carmen de Bolívar specifically is predicted to have 15% less precipitation by 2040 and a greater than 40% reduction in average rainfall by 2070.[3]

El Carmen is the municipal hub of the southern El Carmen de Bolívar region's 128 surrounding farming villages. The primary industry is rainfed agriculture, complemented by an informal labor market in domestic service, construction, and selling prepared foods or goods on the streets. Formerly known as the Zone of Gold, El Carmen de Bolívar has been historically recognized as a particularly fertile, productive agricultural zone known for cultivating tobacco, avocado, sesame seeds, and cotton. Tobacco companies from the United States and Germany had offices in El Carmen for decades where they processed and exported black tobacco. In the context of Colombia's civil conflict, the fertile hills of El Carmen became the seat of the conflict's most intense territorial fighting between the national army, paramilitary groups, and guerrilla fighters in the mid-1990s, which continued off and on for nearly 15 years. Trapped in the midst of the fighting, farming families and urban dwellers alike were massively displaced, and international tobacco and cotton companies left the region.

After intense fighting subsided, the majority of farming families reestablished homes and returned to working the land in the region between 2010 and 2012. Land ownership had been precarious prior to the conflict, and during the conflict years much land was bought and sold, both legally and illicitly. As a result, the 2012 Colombian Victims and Land Restitution Law (Law 1448) met with mixed success in this region. In addition to small-scale land grabbing, multinational agriculture corporations purchased land in the region and introduced nonnative, large-scale cattle and palm oil operations. Some families returned to villages around El Carmen de Bolívar with formal claims to land restitution and state support, while others settled on land without state support and began farming it with the legal right to establish ownership of land after 10 consecutive years of farming the property.

Farmers in the region understand the rhythm of rain and drought, and have historically planted according to the annual cycle of two wet and dry seasons. The El Niño weather phenomena that began in 2014, however, wrought havoc on the seasonal cycles in Central America and along the northern Colombian Coast. Each year, harvests progressively decreased as rainfall remained far lower than usual, and in the fall of 2015, community water reservoirs were drying up. By February 2016, over 426,000 people were affected by drought on the Caribbean coast of Colombia, more than 98,000 animals had died, and 64 municipalities were in formal states of emergency. Economic losses in the region totaled over 138 billion Colombian pesos, with the drought being particularly damaging in El Carmen de Bolívar due to sparse infrastructure (El Carmen acoge, 2016).

Because of its complicated conflict history, El Carmen de Bolívar and the Montes de María region more broadly is one of the poorest regions in Colombia with the highest level of conflict displacement. Rural farming villages typically have no water, natural gas, or electricity infrastructure. Unlike most residents of the Caribbean coast of Colombia, farming families in El Carmen de Bolívar cook over

---

[3]The executive summary of Colombia's *Third National Communication on Climate Change* can be reviewed here: http://documentacion.ideam.gov.co/openbiblio/bvirtual/023732/RESUMEN_EJECUTIVO_TCNCC_COLOMBIA.pdf.

wood fires rather than gas and use stored rainwater for crops, livestock, and household use. By February 2016, news reports cited the drought-induced displacement of 93 farming families from El Carmen de Bolívar to the nearby municipal center (El Carmen acoge, 2016). During conflict, the environmentally displaced could potentially pass as victims of conflict and receive support based on such circumstances. However, after the national peace accords were signed in late 2016, registering as conflict displaced person was no longer valid. Formal protections for environmental migrants in Colombia are nonexistent. Those who migrate internally during a protracted drought currently have no option but to do so by their own financial means, without formal support from government or non-government actors.

Migration decision-making during drought

When thinking about environmental stressors and migration, different tipping points occur where thresholds are reached and individuals or families migrate (McLeman, 2018). Thresholds for decision-making can be considered in various ways. One concept of thresholds is a point at which a significant change in social behavior occurs—in drought migration, for example, a threshold is when in situ adaption fails and individuals adapt by migrating in a way that alters previous migration patterns (Bardsley and Hugo, 2010). To understand individual and family migration decisions at the microlevel in the context of drought, a broader conception of thresholds is most useful. Here a "threshold is simply that point where the perceived benefits to an individual of doing the thing in question exceed the perceived costs" (Granovetter, 1978).

In the midst of protracted El Niño drought in El Carmen de Bolívar, farming families faced the difficult decision of when and how to migrate without any idea of when the drought would end. In situ adaptation, primarily through relying on familiar strategies for short, biannual dry seasons, were costly to enact for months on end. For example, families reduced their food and water consumption, shared the food they had among neighbors, and sold some of the household's livestock for quick cash. Adaptation through migration, though, is also especially financially and socially expensive for most families. As a result, the decision to migrate in El Carmen de Bolívar was both stratified by pre-existing inequality and came at high livelihood costs.

This balance of opportunities and costs of resilience and migration is something I captured through on the ground observations during drought crisis. The debate of whether to stay or go was often a point of family conflict. As one wife explained, "somehow he [her husband] can look around and swallow all the suffering. This is too hard for me. I cannot continue watching my children go hungry, becoming malnourished and so that's why I'm going. And he'll stay, but I'll do cleaning and bring back money… maybe on Saturdays… Just until the rain."[4] Yet male heads of household faced different opportunities and costs in their migration calculus, "My family is in a critical situation because, to be honest, the drought is whipping us hard—we don't have food, we don't have harvest, we don't have water, we don't have work, we don't have anything… more than 30 years I've lived on this land and I don't have a land title. Can you believe I was born here and I don't have title? Without it I don't have the right to go anywhere. There would be no coming back. And if we go, how will we do it? Even

---

[4]Interview with a farming wife, aged 37 years.

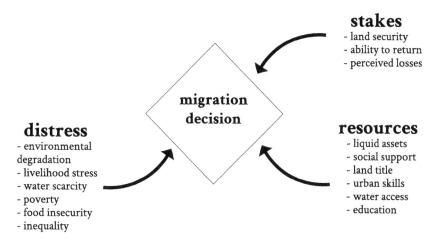

FIGURE 6.1

Migration decision-making during drought.

if the bank loaned us money it would be based on our harvest and there is no harvest to pay. And then, how? We move to town with no money, no job, no food, nothing. Migrate to town, with nothing? Who would do that?"[5]

Environmental migration occurs in a contextual web including an array of factors (Black et al., 2011), and in this case, three elements (see Fig. 6.1) were most salient to migration decisions during drought: perceived distress, perceived resources, and perceived stakes. Perception is key because it is subjective understandings of distress, resources, and stakes rather than the objective measures of those elements that guide daily decisions to stay or go. In this rural, subsistence agricultural context, migration decisions were made at the household and family level rather than at the individual level.

In terms of the severity of their distress, families considered when the drought might end, how much environmental damage would occur as a result, the progressive state of water scarcity, their current poverty level, food insecurity, and how migration might affect their social vulnerability. Families faced high uncertainty in terms of knowing how long the drought and its resulting impacts would last, and drew upon their own recent memories as the best evidence for the environmental distress they could anticipate. In addition to distress, families also considered the resources they perceived to be available to them either to adapt in situ or to facilitate a migratory adaptation to the environmental crisis. These resources included financial, social, and cultural capital that would serve them in staying put or migrating. Ultimately, critical to the migration decision were families own evaluations of the stakes of migrating, or what they stood to lose by staying or going. This included financial capital, property, or the perceived temporality of the potential migration. If a family believed they would be able to return to their homes after the drought then they perceived the migration to be a temporary one and weighed the option in terms of how much longer the drought might last and the migration's

[5]Interview with a farming husband, aged 32 years.

financial feasibility. If, however, a family thought they would lose their rural home by migrating or would be so financially desperate after migrating they could not afford to return, then they considered the potential migration a permanent decision they avoided for as long as possible.

The first threshold

Though reports of drought displacement emerged in February 2016, the first group of farming families to migrate during the El Niño drought did so in June 2015, after the anticipated wet season failed to arrive. The first wave of adaptive migrants in the El Niño drought was the earliest to hit a tipping point in the face of high levels of distress due to drought. These families had greater resources to support what they expected would be temporary migrations to the nearest urban center of El Carmen, where they would wait for the drought to end. By early 2016, these families, who were able to migrate with lower perceived stakes, could be found staying in homes they owned or with family who had room to house them temporarily in El Carmen. Their cost–benefit analysis of the move involved little risk of losing their rural homes, as these families had secure land tenure for their farms as well as someone from their local community to care for their farms while they were away. Rather than endure life without water and limited access to food in the countryside, they lived from other income sources, such as a family member's meager pensions, social security of an elderly family member, or remittance support from extended family.

When explaining their decisions to adapt by migrating rather than staying in place, this first wave of migrants described high levels of distress due to lost income, water scarcity, and food security from the drought, as compared with relatively low stakes associated with migrating as a result of having secure tenure for their farming parcels and thus a high perceived ability to return after El Niño ended. They also described their resources for migrating including housing alternatives in the urban center, secure rural land title, and social support from extended family.

Some families that decided to stay in place rather than migrate did so based on their perceptions of their resources and stakes that may or may not have been accurate. For example, families explained they had no friends or relatives to temporarily house them in the urban center. Yet, the families made this assumption without asking their friends or relatives if they would be willing to take them in for the remainder of the drought. Families also unanimously believed that if they had an active land restitution claim after the conflict they would lose it if they did not physically remain on their land. The Office of Land in Colombia stated this was incorrect, and that families could leave the land but must register their new location with the Office of Land. This affirmation from the Office of Land was confirmed by the Office of Conflict Victims; however, such affirmations were not enough to assuage the fears of formerly conflict-displaced families in a region with a long history of land grabs.

The shifting limits of in situ adaptation

Of all the migrant families during the drought in El Carmen de Bolívar, half were adaptive migrants in the first wave.[6] These strategic migrants moved to their own homes or family homes in the urban center, usually made from concrete with either dirt or tile flooring. The other half migrated not at one

---

[6]Sample of interviews with migrant and nonmigrant families in El Carmen de Bolívar.

specific tipping point but rather as a slow drip from the countryside as failed harvests piled up. For each family in the second migrant wave, the tipping point to migration came at distinct moments that shared similar characteristics. Each occurred at the moment when that family's in situ adaptation efforts failed because their resources to continue adapting in place ran out and migration was the only remaining option. These desperate migrants started from scratch in the municipal center by constructing homes from debris at the periphery of town without running water or electricity access.

For the most socioeconomically advantaged residents in the region, whose livelihoods were not based on subsistence agriculture, the drought was not a crisis nor did it provoke migration. For the large majority, whose livelihoods depended on agriculture, irrigation would have been necessary to successfully make it through a period of over 2 years without rain. In this region, however, there was no proximate source of water infrastructure to irrigate from, leaving that adaptation unavailable. Families' in situ adaptations were reduced to the extension of strategies they used in typical wet and dry seasons. From 2014 through mid-2015, these strategies included efforts such as planting only the least water intensive crops (yucca and ñame), planting less total acreage, sending one migrant out of the household to work on farms in nearby regions and remit income, and procuring loans for seeds to plant after failed harvests emptied seed stores.

By mid-2015, however, in situ agricultural adaptions failed, and adaptation in place took the form of typical dry season resilience strategies. These included reducing communal purchasing of limited water for subsistence, reducing household consumption of food and water, selling livestock and pieces of land (when a family had legal land title), removing children from school, and sending female migrants to urban centers to work in domestic service. Male household heads could not earn income by migrating because their only marketable skills were in the agricultural labor market (which, by mid-2015, had been completely eliminated by drought throughout the entire region). These strategies were a gamble for families who slowly emptied their savings and assets hoping they could stay put until the drought ended. The additional length of time combinations of strategies could buy varied from one household to another, depending on the value of assets they were able to sell off and the amount of resources available to be shared between households.

Each family's tipping point occurred when their resources for managing in place ran out. Half of El Carmen de Bolívar's farming families had sufficient resources to outlast the drought and adapt without migrating. Families who did have to adapt through migration in the second wave of this drought did so in desperate circumstances that were far different from the first wave in 2015. These families left their farming homes permanently without the necessary resources to establish secure homes in the municipal center, nor were they able to return to the countryside after the drought ended. Without secure title to leave in 2015, families perceived leaving under distress without title in 2016 as a permanent migration. In considering their decision to ultimately leave, families prioritized their resources in their migration decision over the perceived distress and stakes. This decision was made based on the complete lack of resources to stay (no water access, no harvest, no remaining resources to purchase food, no further assets to liquidate). Though families acknowledged the high stakes of losing their farming homes by migrating permanently and their continued distress facing the uncertain drought length and abject poverty in the countryside, their utter lack of remaining resources to cope took center stage in the decision to migrate. One year after the drought no migrants had managed to return to their farming homes.

Conclusion

The 2014−16 El Niño drought's impacts on the livelihoods of farmers in El Carmen de Bolívar are consistent with slow-onset environmental migration research. Socioeconomic vulnerability determined when and how families migrated during the El Niño drought in Colombia, either strategically or in desperate circumstances, though all families whose livelihoods depended on rainfed agriculture suffered extreme losses. In both the first and second threshold of migration, resources were critical to migration decisions—either the resources needed to migrate temporarily, migrate permanently, or adapt in place. The nature of a strategic or desperate migration also determined the destination of the migrant family in the nearest municipal center.

Though temporary, adaptive migration during drought in this rainfed agricultural region would have been preferable to depleting resources through in situ adaptation, this research signals that, even for families with the same livelihood, migration as an adaptive strategy is not universally available at the same moment. Furthermore, when faced with the uncertainty of the length and severity of drought families make guesses about the costs of different adaptations and then gamble with the limited resources they have in their efforts to adapt in place.

In cases such as El Carmen de Bolívar where the most vulnerable families respond to drought through resource-depleting in situ adaptations, each successive period of drought leaves families more vulnerable. The livelihood implications of coping with drought raise cause for concern—in this case depleted savings, accrued debts, and scattered livelihoods. For example, when this drought ended in December 2016 with a moderate rainy season, neither temporary nor permanent migrants had returned to their farming homes 1 year later. The long-term effects of this unprecedented, protracted drought remain to be seen. The first two harvests following the drought were poor, with low crop yields due to insect infestations and fungi—two phenomena that are common after droughts.

Perhaps such extreme consequences of droughts in regions like northern Colombia with highly uneven development could be averted through early warning systems such as those described in other chapters of this volume. One clear finding from this case was that in areas with reliable water infrastructure, no one migrated. This finding implies that providing water infrastructure in increasingly drought-prone regions is critical to protecting the sustainability of farming livelihoods and preventing droughts from becoming crises of circumstance.

Acknowledgments

The author is grateful to Robert McLeman for his helpful editorial comments. The author benefitted from the support of the National Science Foundation Graduate Research Fellowship, the David Rockefeller Center for Latin American Studies, and the Graduate Student Research Council at Harvard University.

References

Adger, W.N., Paavola, J., Huq, S., Jane Mace, M., 2006. Fairness in Adaptation to Climate Change. MIT Press, Cambridge, MA.

Adger, W.N., Eakin, H., Winkels, A., 2009. Nested and teleconnected vulnerabilities to environmental change. Frontiers in Ecology and the Environment 7 (3), 150−157.

Bardsley, D.K., Hugo, G.J., 2010. Migration and climate change: examining thresholds of change to guide effective adaptation decision-making. Population and Environment 32 (2−3), 238−262.

Black, R., Adger, W.N., Arnell, N.W., Dercon, S., Geddes, A., Thomas, D.S.G., 2011. The effect of environmental change on human migration. Global Environmental Change 21, S3−S11.

Cutter, S.L., Boruff, B.J., Shirley, W.L., 2003. Social vulnerability to environmental hazards. Social Science Quarterly 84, 242−261.

Dow, K., Berkhout, F., Preston, B.L., Klein, R.J.T., Midgley, G., Shaw, M.R., 2013. Commentary: limits to adaptation. Nature Climate Change 3 (4), 305−307.

El Carmen acoge el primer caso de desplazados por la sequía, 2016. El Heraldo. Retrieved from: https://www.elheraldo.co/nacional/defensoria-denuncio-desplazamiento-de-313-personas-en-bolivar-por-sequia-244841.

Findlay, A.M., 2011. Migrant destinations in an era of environmental change. Global Environmental Change 21, S50−S58.

Freudenburg, W.R., Frickel, S., Gramling, R., 1995. Beyond the nature/society divide: learning to think about a mountain. Sociological Forum 10, 361−392.

Granovetter, M., 1978. Threshold models of collective behavior. American Journal of Sociology 83 (6), 1420−1443.

Hunter, L.M., Luna, J.K., Norton, R.M., 2015. Environmental dimensions of migration. Annual Review of Sociology 41 (6), 1−21.

IPCC, 2018. Summary for policymakers. An IPCC Special Report on the impacts of global warming of 1.5°C above pre-industrial levels and related global greenhouse gas emission pathways, in the context of strengthening the global response to the threat of climate change, sustainable development, and efforts to eradicate poverty. In: Masson-Delmotte, V., Zhai, P., Pörtner, H.O., Roberts, D., Skea, J., Shukla, P.R., Pirani, A., Moufouma-Okia, W., Péan, C., Pidcock, R., Connors, S., Matthews, J.B.R., Chen, Y., Zhou, X., Gomis, M.I., Lonnoy, E., Maycock, T., Tignor, M., Waterfield, T. (Eds.), Global Warming of 1.5°C. World Meteorological Organization, Geneva, Switzerland, 32 pp.

Kelly, P.M., Adger, W.N., 2000. Theory and practice in assessing vulnerability to climate change and facilitating adaptation. Climatic Change 47, 325−352.

McLeman, R.A., 2011. Settlement abandonment in the context of global environmental change. Global Environmental Change 21, S108−S120.

McLeman, R.A., 2018. Thresholds in climate migration. Population and Environment 39 (4), 319−338.

Mcleman, R.A., Smit, B., 2006. Migration as an adaptation to climate change. Climatic Change 76, 31−53.

McGregor, J., 1994. Climate-change and involuntary migration: implications for food security. Food Policy 19 (2), 120−132.

Meze-Hausken, E., 2000. Migration caused by climate change: how vulnerable are people in dryland areas? A case-study in Northern Ethiopia. Mitigation and Adaptation Strategies for Global Change 5 (4), 379−406.

Mortimore, M.J., Adams, W.M., 2001. Farmer adaptation, change and 'crisis' in the Sahel Global. Environmental Change 11, 49−57.

Obokata, R., Veronis, L., McLeman, R.A., 2014. Empirical research on international environmental migration: a systematic review. Population and Environment 36, 111−135.

Oliver-Smith, A., 2009. Climate change and population displacement: disasters and diasporas in the twenty-first century [Chapter 4]. In: Crate, S., Nuttall, M. (Eds.), Anthropology and Climate Change. Left Coast Press, Walnut Creek, CA.

Oliver-Smith, A., 2012. Debating environmental migration: society, nature and population displacement in climate change. Journal of International Development 24 (8), 1058−1070.

Raleigh, C.A., 2011. The search for safety: the effects of conflict, poverty and ecological influences on migration in the developing world. Global Environmental Change 21, S82−S93.

Rigaud, K.K., de Sherbinin, A., Jones, B., Bergmann, J., Clement, V., Ober, K., Schewe, J., Adamo, S., McCusker, B., Heuser, S., Midgley, A., 2018. Groundswell: Preparing for Internal Climate Migration. World Bank, Washington DC. https://openknowledge.worldbank.org/handle/10986/29461.

Further reading

Hunter, L.M., 2005. Migration and environmental hazards. Population and Environment 26 (4), 273−302.

Drought and the gendered livelihoods implications for smallholder farmers in the Southern Africa Development Community region

7

Everisto Mapedza[a],*, Giriraj Amarnath[b], Karthikeyan Matheswaran[c], Luxon Nhamo[d]

[a]*International Water Management Institute (IWMI), Accra, Ghana,*
[b]*International Water Management Institute (IWMI), Colombo, Sri Lanka,*
[c]*Stockholm Environment Institute (SEI), Bangkok, Thailand,*
[d]*International Water Management Institute (IWMI), Pretoria, South Africa*
**Corresponding author*

Introduction

Droughts continue to disrupt human activities and in a number of developing country contexts resulting in disastrous consequences on livelihoods. According to Ban Ki-moon, the former Secretary General of the United Nations, "Over the past quarter-century, the world has become more drought-prone, and droughts are projected to become more widespread, intense, and frequent as a result of climate change."[1] The Centre for Research on the Epidemiology of Disasters (CRED) summarizes the drought events from 1900 to 2016 and shows that in Africa 867,131 people lost their lives over the same period with a total damage of US$ 5,238,593,000 as shown in Table 7.1.

Fig. 7.1 represents the frequency of drought events that occurred in seven Southern Africa Development Community (SADC) countries between 1960 and 2016. In most drought-prone countries, drought occurs once in every 3 years (Guha-Sapir et al., 2017).

This section provides an overview of the key global gender developments. The global gender developments can be summed up by the 1979—United Nations Convention on the Elimination of All Forms of Discrimination Against Women (CEDAW); the 1995—Beijing Fourth World Conference on Women; the 2000—United Nations Millennium Development Goals (Goal 3); 2002—Johannesburg World Summit on Sustainable Development; the 2012—Rio +20 (Rio −20) and the Sustainable Development Goals with Goal 5 out of the 17 Goals focusing on gender equality. In 2017, at COP 13 the United Nations Convention to Combat Desertification (UNCCD) adopted a Gender Action Plan and in 2019 United Nations Environment Programme (UNEP) published a report which looked

---

[1]Ban Ki-moon, the former United Nations Secretary General, Statement of June 17, 2013, on the World Day to Combat Desertification. http://www.unccd.int/Lists/SiteDocumentLibrary/WDCD/WDCD%202013/desertification%20day%202013%20SG%20message%20final%20cleared.pdf.

Drought Challenges. https://doi.org/10.1016/B978-0-12-814820-4.00007-9

Table 7.1 Impact of drought events across continents (1900–2016).

Continent	Number of			Total Damage ('000 USD)
	Events	People killed	Affected people (million)	
Africa	313	867,131	415.9	5,238,593
Americas	152	77	109.4	60,205,139
Asia	170	9,663,400	2107.8	56,555,264
Europe	42	1,200,002	15.5	25,481,309
Oceania	29	684	10.7	11,586,000
Total	**706**	**11,731,294**	**2659.2**	**159,066,305**

Source: From EM-DAT- The International Disaster Database. Centre for Research on the Epidemiology of Disasters – CRED, EM-DAT, CRED / UCLouvain, Brussels, Belgium – www.emdat.be (D. Guha-Sapir).

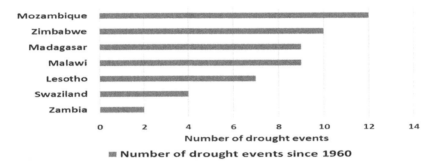

FIGURE 7.1

Average number of drought events in selected Southern Africa Development Community countries from 1960 to 2016.

Source: From Guha-Sapir, D., Below, R., and Hoyois, P. (2017) - EM-DAT-The International Disaster Database: Centre for Research on the Epidemiology of Disasters (CRED): The CRED/OFDA International Disaster Database, Université Catholique de Louvain, Brussels, Belgium. Available at: http://www.emdat.be/database (Accessed on: 13 April 2017). EM-DAT, CRED / UCLouvain, Brussels, Belgium — www.emdat.be (D. Guha-Sapir).

at how gender and environment statistics could help unlock information for action and measuring progress toward the Sustainable Development Goals (SDGs) (UNEP, 2019). There seems to be consented effort to address gender at global level. Most African countries, including within the SADC region, are part of the global initiatives. The key is how such global commitments are translated into national and subnational action within individual countries.

The literature on development including drought tends to focus on the aggregate impacts of drought on humanity in the developing world. Table 7.2 shows figures from the SADC case study countries.

The gendered statistics are important for all SDGs including SDG13, which focuses on climate impacts and strengthening resilience to disasters such as droughts. Addressing gender is described as helping to accelerate the achievements of the SDGs (UN Women and UNDP, 2018). This chapter seeks to explore gendered nuances on how drought affects women and men differently. For purposes of

Table 7.2 Impact of drought events across seven SADC countries (1900–2016).

Country	Number of			
	Events	People killed (number)	Affected people (million)	Damage (million USD)
Lesotho	7		3.7	1.0
Madagascar	8	200	4.7	
Malawi	8	500	28.3	
Mozambique	13	100,068	20.1	50.0
Swaziland	6	500	2.1	
Zambia	5		4.2	
Zimbabwe	8		21.3	551.0
Total	55		84.3	603.7

Source: From EM-DAT- The International Disaster Database. Centre for Research on the Epidemiology of Disasters – CRED, EM-DAT, CRED / UCLouvain, Brussels, Belgium – www.emdat.be (D. Guha-Sapir).

this chapter, the UN Women definition is applied which defines gender as referring to "the roles, behaviors, activities, and attributes that a given society at a given time considers appropriate for men and women. In addition to the social attributes and opportunities associated with being male and female and the relationships between women and men and girls and boys, gender also refers to the relations between women and those between men. These attributes, opportunities and relationships are socially constructed and are learned through socialization processes. They are context/time-specific and changeable. Gender determines what is expected, allowed and valued in a woman or a man in a given context. In most societies, there are differences and inequalities between women and men in responsibilities assigned, activities undertaken, access to and control over resources, as well as decision-making opportunities. Gender is part of the broader socio-cultural context, as are other important criteria for socio-cultural analysis including class, race, poverty level, ethnic group, sexual orientation, age and others" (UN Women).[2]

Gender determines the roles, power, resources for males and females, access to services, opportunities, and information. It is important to note that these roles are ascribed to men and women not because of what they can do, but it is based on what society deems appropriate for men and women. Gender is a central organizing principle of societies and often governs the processes of production and reproduction, consumption, and distribution. Gender roles are the "social definition" of women and men and vary among different societies and cultures, classes and ages, and during different periods in history (Mapedza et al, n.d). Gender research in rural development is therefore essential for poverty reduction and the sustainability of development interventions (van Hoeve, n.d; van Hoeve and van Koppen, 2006). Most gender analysts tend to approach gender in two ways based on gender needs. This division was largely developed by Mayoux and widely extended by Moser (1989) see (Wieringa, 1994). First, there is the Practical Gender Needs Approach. The Practical Gender Needs Approach

---

[2]https://trainingcentre.unwomen.org/mod/glossary/view.php?id=36&mode=letter&hook=g&sortkey= Accessed on 19 April 2019.

tries to offer gender-specific solutions, which address women's concerns within their current roles. For instance, looking at how women could better work within the existing patriarchal system. Second, the other approach is the Gender Transformative Approach[3] (Cole et al., 2014; Kabeer, 2001; Mukhopadhyay, 2004). This approach, also viewed as a feminist perspective, argues that for meaningful gender changes there has to be a change in the norms and values that define gender roles (Cole et al., 2014; Kabeer, 2001; Mukhopadhyay, 2004). This approach argues that as long as the social structure promoting patriarchy is in place, gender inequalities will continue.

The El Nino—induced drought in Southern Africa negatively reduced the amount of water available for livelihood options in a gender-specific manner. Coping mechanisms toward drought depends on the access to assets and social capital, which are a reflection of gender relationships. Women and children were more vulnerable to the negative impacts of drought than men, especially in the case studies of Madagascar, Malawi, Zambia and Zimbabwe. Hence, gender, needs to take a central role in the planning for drought preparedness and drought response initiatives. Drought affects men, women, children, and the elderly in a different manner (Stehlik et al., 2000), and it is important to understand structural vulnerabilities such as gender and equity in general (Oxfam, 2016).

Climatic shocks such as droughts are increasingly threatening agricultural and economic development in sub-Saharan Africa (SSA). An estimated 70% of economic losses in SSA are attributed to droughts and floods (Bhavnani et al., 2008). Studies have indicated a fall in the annual growth in gross domestic product (GDP) in SSA countries by 2%—4% (Brown et al., 2011); a substantial decline of about 22% in yields of maize, 17% each for sorghum and millet, and 18% for groundnuts by 2030 (Schlenker and Lobell, 2011); and a reduction in food security (Parry et al., 2005; Lybbert and Sumner, 2012; Wheeler and von Braun, 2013) as a consequence of climatic shocks. Identifying and promoting options that will help address the problems posed by climatic shocks is, therefore, a formidable challenge for policy in SSA. Since most sub-Saharan countries depend on agriculture, most of their GDP graphs clearly demonstrate the co-relationship of a decline of GDP whenever there is a drought.

Methodology

The broader objectives of the water assessment initiative within SADC were to

i. develop frameworks and systems for monitoring availability of water for livestock and agriculture and capacitate national institutions to implement the frameworks and systems;

ii. develop a predictive disaster risk reduction early warning tool to provide early warning information on the potential impact of droughts on agricultural water resources in the affected areas;

iii. establish a baseline of water sources and develop a framework to facilitate mapping of water bodies for agricultural livelihood activities in drought-affected areas;

iv. evaluate water resources information with reference to the social, institutional, gender, and livelihoods arena of the affected communities.

This chapter is based on the part of the last objective, which looked at gender and drought.

---

[3]This is also referred to as Strategic Gender Needs Approach.

The study is based upon field research in Madagascar, Malawi, Zambia, and Zimbabwe. As part of the research we looked at the effects of the 2015/16 El Nino on the water levels and availability and the impacts on livelihoods in the SADC region. Key informant interviews were conducted with government, nongovernment, and international organizations dealing with drought. Focus group discussions (FGDs) with men and women were also conducted in Madagascar and Malawi. The country field data collection was facilitated by the Food and Agriculture Organization (FAO) of the United Nations. In addition, the study looked at Lesotho, Mozambique, and Swaziland where remote sensing data were used to better understand how the El Nino drought progressed within the SADC region and how this could be used to contribute toward an early warning system (EWS). Fig. 7.2 shows the case study countries as well as the three additional countries where only remote sensing studies were conducted. This chapter further draws from literature review on gender and droughts.

Literature review was conducted building upon the drought EWS research that had been conducted in South Asia. Both published and gray literature were utilized as some of the available information

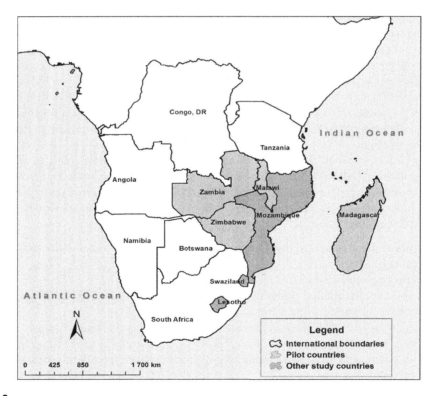

FIGURE 7.2

Map of Southern Africa Development Community showing the scoping pilot countries as well as the other study countries.

Source: Luxon Nhamo, International Water Management Institute (IWMI).

was based on past reports. Drought conference presentations and their respective reports, especially the United Nations Convention to Combat Desertification (UNCCD) Africa Drought Conference organized in Windhoek, Namibia, August 15—19, 2016, were good opportunities to interact with drought experts from different parts of the world.

Research findings on gender and drought

This section will provide findings from the study in the SADC region and how they have implications for the SSA at large. The findings are organized according to different themes with the first section looking at drought and gender sex-disaggregated statistics.

Drought and gender: sex-disaggregated statistics

Gender experts understand that droughts affect men, women, children, and the elderly in a different manner. Policy makers and most civil servants, on the other hand, still perceive droughts as affecting all people equally. Despite gender understanding by gender experts, most of the data collected does not disaggregate information by sex or age. The Zimbabwe Vulnerability Assessment Committee (Zim-VAC), for instance, does not provide sex-disaggregated information. In one of the key informant interviews, it was pointed out that the raw data distinguish between men and women. Therefore, it was possible to disaggregate the analysis, but the assumption was that it was more expensive to carry out the sex-disaggregated data analysis. The challenge starts at the global level (see Table 7.1 earlier) to the regional bodies such as the SADC Regional Vulnerability Assessment Committee (RVAC) and to the National Vulnerability Assessment Committees (NVACs). These committees coordinate drought coordination, with the support of agriculture, disaster units, meteorological departments based on local institutional architecture.

According to the European Union,[4] "sex-disaggregated statistics are data collected and tabulated separately for women and men. They allow for the measurement of differences between women and men on various social and economic dimensions and are one of the requirements in obtaining gender statistics." The European Union[5] further goes on to clarify that gender statistics are defined by the sum of the following characteristics:

a. Data are collected and presented disaggregated by sex as a primary and overall classification, that is sex-disaggregated data;
b. data are reflecting gender issues;
c. data are based on concepts and definitions that adequately reflect the diversity of women and men and capture all aspects of their lives; and
d. data collection methods take into account stereotypes and social and cultural factors that may induce gender biases (some examples of sex bias in data collected are underreporting of women's economic activity, undercounting of girls, their births or their deaths, or underreporting of violence against women).

---

[4]https://eige.europa.eu/gender-mainstreaming/concepts-and-definitions accessed on 20 April 2019.
[5]https://eige.europa.eu/gender-mainstreaming/methods-tools/gender-statistics accessed on 20 April 2019.

All the study countries collected data that only went as far as stage "a" above. It is important to note that most key informants highlighted the need for gender-disaggregated data, but this was not being conducted. In Zimbabwe, ZIMVAC further proposed that our study could help with an initial pilot on sex-disaggregated dataset. Doss (2013) clearly argues that sex-disaggregated data should not be seen as an extra expense but helps to design response and programs which meet the requirements of both men and women (Upton et al., 2016). Drought data are often given in aggregate terms, which will most likely result in gender blind responses, as there is neither sex nor age disaggregation. It is a missed opportunity as, in some instances, the disaggregated data is collected, but most of the key informants never made the extra effort to sex-disaggregate the data. Some of the key informants did not understand the importance of gender in the results they produced. The often-highlighted cost assumption during key informant interviews was not backed by any real cost—benefit analysis evidence. It was simply assumed that it was more expensive to do a gender analysis.

Gender and access to assets

While drought will affect both men and women, how they cope with drought impacts depends on gender. Most study countries studied are patriarchal societies where inheritance of assets such as land was through the male lineage. According to the World Bank (2019), women in Africa tend to have fewer assets than men. Women, therefore, tend to be more vulnerable to droughts due to their limited assets and control of natural resources (Neumayer and Plümper, 2007; UNDESA, 2016; UNDP, 2017). In few exceptions in some parts of Malawi, for instance, where it was matrilineal society's assets such as land are inherited through the female line. In such circumstances, the uncles played key decision-making roles in deciding how the assets were inherited (Mapedza et al., 2017). After a drought, assets that a man or woman has access to will determine how well they can cope with the impact of the drought. This is one of the missing elements in the drought vulnerability assessments databases. A better understanding of who has been affected and the status of their assets will inform a better response mechanism for the affected men and women. The World Bank (2019) report notes that women only own 25% of the land. Yet, if ownership of land is one of the key assets for resilience, then women already begin from a disadvantaged point. With limited assets in both Madagascar and Malawi, girls were forced to drop out of school first before boys in order to engage in economic activities as one of the coping mechanisms with the drought. In Malawi, in some FGDs, it was pointed out that with the limited economic opportunities some girls were also forced into early marriages so that the parents would get grain as part of the bride price. The differential impact of the same drought on boys and girls also meant that the future opportunities for girls were already circumscribed by dropping out of school. One of the communities had organized a group of concerned mothers, with the support on an NGO, to try to identify cases of food shortages and help families before the girl child is withdrawn from school or getting married off.

Gender and education

The patriarchal nature of most of the society in the developing world prioritizes the education of boys than girls. While a number of governments are investing in the education of girls, women are still less educated than men. Mechanisms such as Climate Smart Agriculture to cope with increased droughts are usually disseminated through agricultural extension. Most extension staff are still biased toward men as the "real" farmers. In Malawi and Zimbabwe, it was mentioned that they are promoting female extension staff to help engage with female farmers who, due to migration, are the default farmers. It might seem like a simple issue, but addressing the requirements of women as farmers is often poorly done. Deciding where and when to have an extension meeting will encourage or make it impossible for women to attend an agricultural extension meeting (Mapedza et al., 2017). Doss and Morris (2001) noted how gender influenced the adoption of agricultural innovations such as improved maize varieties in Ghana. In Zimbabwe, irrigation that is meant to minimize drought and rainfall variability challenges saw men making most of the decisions despite men being the minority within the irrigated scheme (Tagutanazvo et al., 2015). Doss (2001) further notes that the risk of innovation is itself gendered and cannot be simply understood at the household level. This further reinforces the importance of nuanced gender understanding of droughts, which goes beyond the aggregated, usually household level data.

Drought insurance schemes have recently been on the rise as a mechanism for reducing the impact of drought on men and women in Africa. At the African Union level, the Africa Risk Capacity (ARC) is an initiative established to enable African governments to better plan and respond to weather-related disasters such as droughts. At the national level, countries like Malawi have received a payment for the El Nino—induced drought experienced in the 2015/16 farming season. However, there are other local level insurance schemes, which might not be better understood by women who tend to be less informed and less educated. With limited knowledge, they might be more risk-averse and not contribute to drought and other weather-indexed insurance schemes. According to (FAO, 2011), female farmers harvest 20%—30% less yield due to lower use of inputs.

Yet, women are potential reservoirs of drought knowledge and innovation. Knowledge of drought and how to respond to drought through use of drought-tolerant crops are gendered. Gender researchers Kabeer and Natali argue and recognize that women are not inherently vulnerable but are agents of transformative change (Kabeer and Natali, 2013; OSAGI, 2001). According to the UN Women and UNDP (2018:29), women's knowledge, skills, and experiences regarding the environment and management of food, water, and other natural resources are central for designing solutions to cope with climate change and disasters such as drought. However, in most of the SADC region and sub-Saharan region, men tend to be, in general, better educated than women. This trend seems to apply to the whole of Africa (World Bank, 2019) and is an opportunity lost for the whole continent.

Gender and social capital

Social capital is a key asset in coping with droughts. Social capital is defined as the glue that keeps society together. Rydin and Falleth (2006) further define social capital into bonding social capital, which unites people together, and bridging capital that enables people to connect or build bridges with others. Social capital studies have also noted that there are differences in levels of social capital between men and women. The World Bank (2019) uses the term network and points out that women tend to be less networked due to gendered restrictions and limited opportunities. A better grasp on gender and social capital will enhance the response mechanisms for drought in SSA and the developing world at large. Social networks can help not only in terms of exchange of information, but the networks are also important for access to food resources during drought. Failing to provide sex-disaggregated data will blind analysis of how social capital could be leveraged to minimize the impacts of drought. Assuming that both men and women have the same social networks will most likely result in gender blind solutions that do not address the needs of women. Women tend to cope under very difficult circumstances of drought. A gendered analysis will help track what are the women's experiences with providing food during a drought. In Madagascar's southern part where there was a greater impact on drought, women organized themselves to harvest cactus fruit that became a major source of food for households that had run out of food. Such innovations toward solutions need to be highlighted, and how such local knowledge could help inform policy and practices during future droughts needs to be noted.

Gendered drought solutions and innovations

Understanding of gender has often been viewed in terms of seeing women as victims and suffering in the hands of drought and other climate-related disasters. This chapter further engages with literature, which argues that a better understanding of gender will not only highlight specific vulnerabilities for women but also will offer solutions that women have innovated over time (Aguilar et al., 2015; Carvajal Escobar et al., 2008; McKinney and Fulkerson, 2015). Gendered understanding of droughts will help bring out the innovative solutions that women are developing to help cope and ameliorate the impact of droughts within the developing countries. Women need to be viewed as sources of solutions as they have to deal with the consequences of drought.

Gender equality is defined by OSAGI (2001) as "the equal rights, opportunities and outcomes for girls and boys and women and men. It does not mean that women and men are the same, but their rights, responsibilities and opportunities do not depend on whether they are born female or male." This is central for better disaster management cycle from mitigation, preparedness, response, and recovery (WEDO, 2007). Beyond the gender stereotypes (Doss et al., 2018) of victimhood for women, it is important to note that it is women who have to go through the disasters such as drought (Charan et al., 2016).

Gender equality is also an issue of human rights. According to UNDESA (2017), women comprised 49.6% of the global population in 2017. By not specifically focusing on women's circumstances pertaining to drought, policy solutions will not specifically address the requirements of nearly half of its global citizens. The United Nations Women (UN, 2016) argues that what is required is a sustainable and inclusive growth that takes into account women—hence leaving no one behind. The World Bank (2019) notes that it must also encompass the financial inclusion of women in Africa.

Policy implications on gender and drought

Most of the countries in Africa are in the process of preparing their reporting mechanisms for the SDGs. This could be a good opportunity to make sure that the data collected across all SDGs are sex and age disaggregated. Organizations such as the African Union and regional bodies which include SADC could develop a common template for national data collection. The guideline would be more important if it addresses the four elements on collecting gendered statistics that has already been developed for the European Union. Agencies of the United Nations for drought, specifically UNCCD, could collaborate with case study countries in all the SSA subregions. This could be linked to the UNCCD Gender Action Plan which intends to establish gender baselines. Such good practices on collecting gendered statistics will then be shared with all the countries in SSA. Most countries have accepted that SDGs are now preparing the local reporting mechanisms. This offers an opportune moment for collecting and analyzing sex- and age-disaggregated data. It is also encouraging that most of the

SSA countries are signatories to the gender conventions such as CEDAW, Hyogo Framework, the 2005 World Conference on Disaster Reduction, and Gender Policy at African Union level and have been actively developing gender policies at national level.

UNCCD in conjunction with other UN organizations such as UN Women could also set up longitudinal case studies to better understand how norms and cultural values change over time. With a better understanding of how gender relationships are transformed over time, this will offer a better understanding of entry points to help trigger gender transformation within SSA. Such studies could also explore opportunities for enhancing both bonding and bridging social capital among women.

Funding was often highlighted in key informant interviews as a reason for not collecting gender-disaggregated statistics. The research team felt that while funding can be an issue the real issue was lack of commitment and good gender ideas. None of the respondents had developed a gendered statistics methodology that could not get funded. Lack of good gender ideas and commitment was one of the perceived challenges. It is therefore important that the gender profile is elevated by developing implementation plans and allocation resources, especially in partnership with development partners who also bring current gender ideas and experiences.

While few countries such as Namibia are in the process of developing drought policies, this could be also an opportunity to develop new drought policies in SSA that address gender as a key consideration. Incorporating gender in drought policies will ensure that the needs of women are addressed as well as tapping from the knowledge that the women have in ensuring that droughts will not necessarily result in disasters through better preparedness.

The holistic nature of gender would entail that planning for gender within drought needs to collaborate with other sector ministries in order to promote positive gender changes over time. While drought offers an arena where gender strides can be achieved, more gender transformation will be realized if there is cooperation across the different silos or ministries. Gender needs action within the drought arena while taking cognizance of the importance of collaboration for more win-win gender outcomes.

One of the areas, which need to be promoted, is women's agency and decision-making (UN Women and UNDP, 2018). After analyzing sex-disaggregated data, it is paramount that women have a say in what that data mean and mapping the way forward. Decision-making and women's control of what happens in drought policies and responses need to take into account the decisions that women make. This entails women's empowerment investments to increase their decision-making powers over drought and other household, community, and national decisions. One such area is to recognize women's rights to land. In Ethiopia, both husband and wife are being coregistered on the rural land titles.

Conclusion

Gender is central in understanding how men and women cope with drought and how they develop their resilience over time. This chapter sought to argue that it is important to know the circumstances of both men and women toward drought. It is important that gender statistics will help highlight the fact that women are more than drought victims and contribute toward drought solutions. The ongoing SDG reporting mechanisms offer a good opportunity to collect sex-disaggregated gender statistics. In other words, this chapter sought to argue that an informed understanding on gender would contribute toward solutions for resolving the drought challenges which increasingly seem to impact on the developing countries especially in the context of climate change. Ignoring the specific needs and requirements of women would entail leaving out nearly half of the global citizens, which would be a great gender injustice.

While this chapter has looked at gender using the lens of drought, it is important to note that addressing gender issues calls for a more holistic approach to address gender inequality in the developing world. Some of the approaches, as also reinforced by the SDGs and UN Women, include changing the social norms and values. It is important to reconfigure the socialization process that helps reduce gender stereotypes from an early age. From the way children are brought up, they already start defining boundaries on what is expected from a boy or a girl. Such norms and values can be reshaped over time. Transformative education for both boys and girls has the potential to reconfigure the gender relationships over time.

Livelihoods and gender need to be better integrated into vulnerability assessments. At the moment, most figures are aggregated, and they need to be sex disaggregated to best inform coping mechanisms for women, men, youths, and other gendered categories. Such discussions need to begin from SADC level to national, subnational and local levels.

Regional bodies such as SADC must link with ongoing global initiatives such as the UNCCD efforts as outlined in the Africa Drought Conference held on August 15−19, 2016, which proposed a Drought Resilient and Prepared Africa (DRAPA). Such global efforts will enhance drought planning, management, monitoring, and preparedness within the SSA.

References

Aguilar, L., Granat, M., Owren, C., 2015. Roots for the Future: The Landscape and Way Forward on Gender and Climate Change. International Union for Conservation of Nature and Global Gender and Climate Alliance, Washington, D.C.

Bhavnani, R., Vordzorgbe, S., Owor, M., Bousquet, F., 2008. Report on the Status of Disaster Risk Reduction I the Sub-Saharan Africa Region. Commission of the African Union, United Nations and The World Bank. Available from: http://www.unisdr.org/files/2229DRRinSUBSaharan AfricaRegion.pdf.

Brown, C., Meeks, R., Hunu, K., Yu, W., 2011. Hydroclimate risk to economic growth in Sub-Saharan Africa. Climatic Change 106, 621−647.

Carvajal-Escobar, Y., Quintero-Angel, M., Garcıa-Vargas, M., 2008. Women's role in adapting to climate change and variability. Advances in Geosciences 14, 277−280.

Charan, D., Kaur, M., Singh, P., 2016. Indigenous Fijian women's role in disaster risk management and climate change adaptation. Pacific Asia Inquiry 7 (1), 106−122.

Cole, S.M., van Koppen, B., Puskur, R., Estrada, N., DeClerck, F., Baidu-Forson, J.J., Remans, R., Mapedza, E., Longley, C., Muyaule, C., Zulu, F., 2014. Collaborative effort to operationalize the gender transformative approach in the Barotse Floodplain. Penang, Malaysia: CGIAR Research Program on Aquatic Agricultural Systems. Program Brief: AAS-2014-38.

Doss, C., 2013. Data Needs for Gender Analysis in Agriculture. IFPRI Discussion Paper 01261. IFPRI, Washington DC, USA.

Doss, C., 2001. Is risk fully pooled within the household? Evidence from Ghana. Economic Development and Cultural Change 50 (1), 101−130.

Doss, C., Meinzen-Dick, R.S., Quisumbing, A.R., Theis, S., 2018. Women in Agriculture: Four Myths. IFIPRI, Washington DC.

Doss, C., Morris, M., 2001. How does gender affect the adoption of agricultural innovations? The case of improved maize technology in Ghana. Agricultural Economics 25 (1), 27−39.

FAO, 2011. The State of Food and Agriculture 2010−11: Women in Agriculture: Closing the Gender Gap for Development. Food and Agriculture Organization of the United Nations, Rome.

Guha-Sapir, D., Below, R., Hoyois, P., 2017. EM-DAT-The International Disaster Database: Centre for Research on the Epidemiology of Disasters (CRED): The CRED/OFDA International Disaster Database. Université Catholique de Louvain, Brussels, Belgium. Available at: http://www.emdat.be/database.

Kabeer, N., Natali, L., 2013. Gender Equality and Economic Growth: Is There a Win-Win?. In: IDS Working Paper, vol. 417 Institute of Development Studies, Brighton.

Kabeer, N., 2001. 'Reflections on the measurement of women's empowerment. In: Discussing Women's Empowerment: Theory and Practice. Novum Grafiska AB, Stockholm. Sida Studies No. 3.

Lybbert, T.J., Sumner, D.A., 2012. Agricultural technologies for climate change in developing countries: policy options for innovation and technology diffusion. Food Policy 37, 114−123.

Mapedza, E., Tagutanazvo, E., van Koppen, B., Manyamba, C., 2017. Agricultural water management in matrilineal societies in Malawi: landownership and implications for collective action. In: Suhardiman, D., Nicol, A., Mapedza, E. (Eds.), Water Governance and Collective Action − Multi Scale Challenges. Routledge, Taylor and Francis Group, London and New York, pp. 82−95.

Mapedza, E., Tilahun A., Geheb, K., Peden, D. Boelee, E., Tegegne S. D, van Hoeve, van Koppen, B., n.d. Why Gender Matters: Reflections from the Livestock-Water Productivity Research Project, IWMI and ILRI, Addis Ababa, Ethiopia.

McKinney, L., Fulkerson, G., 2015. Gender equality and climate justice: a cross-national analysis. Social Justice Research 28 (3), 293−317.

Moser, C.O.N., 1989. Gender planning in the third world: meeting practical and strategic gender needs. World Development 17 (11), 1799−1825.

Mukhopadhyay, M., 2004. Mainstreaming gender or 'streaming' gender away: feminists marooned in the development business. Institute of Development Studies (IDS) Bulletin, Repositioning feminism in Development 35 (4), 95−103.

Neumayer, E., Plümper, T., 2007. The gendered nature of natural disasters: the impact of catastrophic events on the gender gap in life expectancy, 1981−2002. Annals of the Association of American Geographers 97 (3), 551−566.

OSAGI (Office of the Special Advisor on Gender Issues and Advancement of Women), 2001. Gender Mainstreaming: Strategy for Promoting Gender Equality. Factsheet. http://www.un.org/womenwatch/osagi/pdf/factsheet1.pdf.

Oxfam, 2016. Harvest of Dysfunction: Rethinking the Approach to Drought, its Causes and Impacts in South Africa. Report for Oxfam South Africa, Johannesburg.

Parry, M., Rosenzweig, C., Livermore, M., 2005. Climate change, global food supply and risk of hunger. Philosophical Transactions of the Royal Society B 360 (1463), 2125−2138.

Rydin, Y., Falleth, E., 2006. Networks, Institutions and Natural Resource Management. Edward Elgar, London.

Schlenker, W., Lobell, D.B., 2011. Robust negative impacts of climate change on African agriculture. Environmental Research Letters 5 (1), 1−8.

Stehlik, D., Lawrence, G., Gray, I., 2000. Gender and drought: experiences of Australian women in the drought of the 1990s. Disasters 24 (1), 38−53.

Tagutanazvo, E., Dzingirai, V., Mapedza, E., Van Koppen, B., 2015. Gender dynamics in water governance institutions: the case of Gwanda's Guyu-Chelesa irrigation scheme in Zimbabwe. wH$_2$O Journal of Gender and Water 4 (1), 55−64.

UNDESA (United Nations Department of Economic and Social Affairs), 2016. World Economic and Social Survey 2016. Climate Change Resilience: An Opportunity for Reducing Inequalities. ST/ESA/363. United Nations, New York.

UNDESA (United Nations Department of Economic and Social Affairs), 2017. World Population Prospects: The 2017 Revision. United Nations, New York.

UNDP (United Nations Development Programme), 2017. Overview of Linkages between Gender and Climate Change. Policy brief. UNDP, New York.

UNEP, 2019. Gender and Environment Statistics: Unlocking Information for Action and Measuring the SDGs. UN Environment, Nairobi, Kenya.

UN, 2016. UN Secretary-General's High-Level Panel on Women's Economic Empowerment 2016. United Nations, New York.

UN Women and UNDP, 2018. Gender Equality as an Accelerator for Achieving the Sustainable Development Goals. UN Women and UNDP, New York, USA.

Upton, J.B., Cisse, J.D., Barrett, C.B., 2016. Food security as resilience: reconciling definition and measurement. Agricultural Economics 47, 135−147.

van Hoeve, E., n. d.undated. Gender Research. ILRI.

van Hoeve, E., van Koppen, B., 2006. Beyond Fetching Water for Livestock: A Gendered Sustainable Livelihood Framework to Assess Livestock-Water Productivity. ILRI Working Paper 1. ILRI, Nairobi.

WEDO (Women's Environment and Development Organization), 2007. Changing the Climate: Why Women's Perspectives Matter. http://wedo.org/changing-the-climate-why-womensperspectives-matter/.

Wheeler, T., von Braun, J., 2013. Climate change impacts on Global Food Security. Science 341, 508−513.

Wieringa, S., 1994. Women's interests and empowerment: gender planning reconsidered. Development and Change 25 (4), 829−848.

World Bank, 2019. Profiting from Parity: Unlocking the Potential of Women's Business in Africa. World Bank, Washington, DC. © World Bank. https://openknowledge.worldbank.org/handle/10986/31421 License: CC BY 3.0 IGO.

Further reading

Mapedza, E., 2013. Why gender matters for farming within the Limpopo River. AGRI-DEAL Magazine 2, 83−85.

Moser, C., 1993. Gender Planning and Development: Theory Practice and Training. Routledge, London and New York.

Integrating regional climate and drought characteristics for effective assessment and mitigation of droughts in India

Rajendra Prasad Pandey[a],[*], Rakesh Kumar[a], Daniel Tsegai[b]

[a]*National Institute of Hydrology, Roorkee, India,*
[b]*United Nations Convention to Combat Desertification (UNCCD), Bonn, Germany*
[*]*Corresponding author*

Introduction

In India, the occurrence of drought is linked with the amount, distribution, time of onset, and withdrawal of monsoon rainfall. Since monsoon rainfall is highly erratic and unevenly distributed, drought conditions may prevail in almost any year in one part of the country or another. Droughts have substantial economic, environmental, and social impacts, especially in arid and semi-arid regions of India where rainfed agriculture predominates. For example, persistent drought conditions between 2001 and 2003 resulted in serious crop losses and contributed to mass migration of people and livestock in the states of Rajasthan and Gujarat.

Most notably, the drought from 2006—2010 in Bundelkhand region of India resulted in serious crop losses and drinking water supply shortages in the region. Yet, past government responses to drought have largely been in the form of small financial assistance to farmers, creation of temporary employment opportunities for affected people, as well as the suspension of loan recovery payments. Generally, the current drought management system in India seems to be very much steered towards relief and crisis management rather than risk mitigation. There is thus a pressing need to manage drought with a proactive mitigation approach in addition to emergency relief actions.

In some developed countries like the United States and Australia, drought preparedness planning has emerged as a widely accepted tool for governments at all levels seeking to reduce the risk of future drought events (Wilhite et al., 2000). For example, 47 of the 50 states in the United States have developed formal drought preparedness plans, of which 11 are more proactive with a focus on drought preparedness (WMO and GWP, 2014). By contrast, India does not yet have a comprehensive drought mitigation policy despite the recurrent drought experiences and hardship in some parts of the country. India responds to each new drought event on an ad hoc basis.

There is thus a pressing need to shift to a proactive risk management approach, which requires area-specific drought mitigation planning. Indeed, a drought management plan for India should consist of

Drought Challenges. https://doi.org/10.1016/B978-0-12-814820-4.00008-0

three basic components: (1) near real-time monitoring and early warning, (2) vulnerability and risk assessment, and (3) region-specific mitigation planning/response actions. In the remainder of this chapter, we present an overview of drought characteristics in different climatic regions of India and describe the challenges and a way forward for region-specific drought monitoring, assessment, and mitigation planning to manage drought effectively.

Primary causes and characteristics of droughts in different climatic regions of India

India encompasses a wide range of climatic regions (Fig. 8.1). The occurrence of droughts at any location in India is linked with the erratic distribution and limited amount of rainfall during southwest monsoon season (June–September). In the case of the states of Tamil Nadu and Karnataka, the northeast monsoon season (October–November) is also important. Drought may occur due to any one or more of the following causes: (1) less than usual amounts of rainfall at a given place and time, (2) late arrival (or onset) of monsoon, (3) occurrence of prolonged dry spells during the monsoon period, and (4) early withdrawal of monsoon. The characteristics of drought in terms of frequency, severity, and persistence in successive years vary across the climatic regions, with mean annual precipitation,

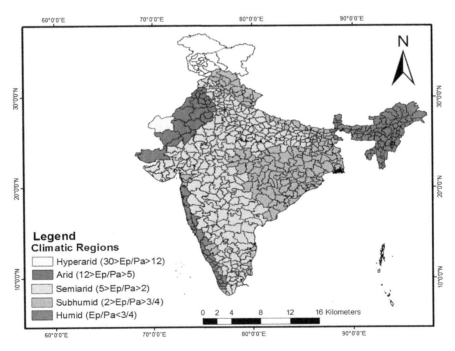

FIGURE 8.1

Spatial distribution of climatic regions in India.

Source: Pandey, R.P., Daradur, M., Jain, V.K., Jain, M.K., 2016. Assessment of vulnerability to drought towards effective mitigation planning: the case of Ken river basin in India. Journal of Indian Water Resources Society 36 (4), 36–48.

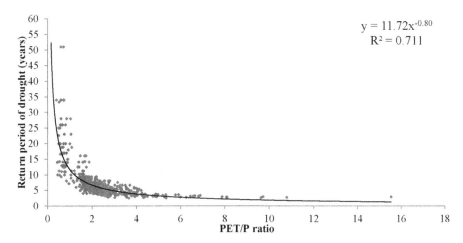

FIGURE 8.2

Relationship of drought return period with Ep/Pa ratio.

Source: Pandey, R.P., Daradur, M., Jain, V.K., Jain, M.K., 2016. Assessment of vulnerability to drought towards effective mitigation planning: the case of Ken river basin in India. Journal of Indian Water Resources Society 36 (4), 36–48.

evapotranspiration, length of wet season, and annual temperature variations further influencing governing regional drought characteristics.

A relationship between average return period of drought and ratio of mean annual potential evapotranspiration (Ep) and precipitation (Pa) in India is presented in Fig. 8.2. Based on 113 years of monthly rainfall data (1901–2013) for 536 stations, it is clearly shown that drought return period decreases with increasing wetness. Whereas the frequency of drought in the dry-sub-humid regions varies once in every 5–6 years, the variation for sub-humid and humid regions is between 7 and 8 years and more than 9 years, respectively.

The drought frequency map (in terms of return period vs. Ep/Pa) in various parts of the country is presented in Fig. 8.3. The arid and semi-arid regions of Rajasthan, Gujarat, Haryana, Punjab, Uttar Pradesh, Karnataka, and Tamil Nadu experience droughts with an average frequency of once in 3 years to once in 4 years (see Fig. 8.3). On the other hand, the relationship between drought return period and normal annual temperature variation (derived from monthly temperature data for 268 stations as shown in Fig. 8.4) indicates that the average drought return period decreases when the normal annual temperature variation (T_d) increases.

As illustrated in Fig. 8.5, the magnitude of drought severity also varies across different climatic regions. The maximum rainfall departures from the normal correspond with the normal annual temperature variation (T_d) (Fig. 8.6). The severity increases with the increase in the T_d.

The occurrence of persistent drought events that last for two or more consecutive years is most common in arid, semi-arid, and dry subhumid regions. However, persistent drought events in the sub-humid regions (i.e., those with mean annual rainfall of 1200–1600 mm) and humid regions (with annual rainfall >1600 mm) are relatively rare. As such, drought mitigation plans devised for one climatic region may not be effective or applicable to another.

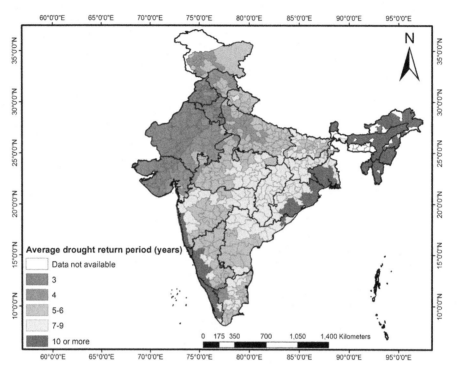

FIGURE 8.3

Average drought return period map of India.

Source: Pandey, R.P., Daradur, M., Jain, V.K., Jain, M.K., 2016. Assessment of vulnerability to drought towards effective mitigation planning: the case of Ken river basin in India. Journal of Indian Water Resources Society 36 (4), 36–48.

Current approach to drought declaration and management

As noted in the Government of India Manual for Drought Management (2016), an official declaration of drought depends on data from multiple indicators, including weekly information of rainfall deficit, crop area sown, satellite-based vegetation index (normalized difference vegetation index (NDVI) or vegetation condition index (VCI)), stream flow, groundwater level, lake and reservoir storage (Table 8.1). Other indicators that may also be considered include distress sales and movements of cattle, human migration, fodder availability, drinking water, animal health, as well as employment opportunities in the agriculture sector.

Each state has to establish a Drought Monitoring Cell (DMCs) to monitor data regularly on the above indicators, which is responsible for apprising the State Executive Committee (SEC)/Disaster Management Department in the State Government on the spread and severity of drought. The three steps for triggering a drought declaration are as follows:

Step 1:

Mandatory indicators such as rainfall deviation, SPI, or dry spell must be considered as per the matrix given in Table 8.2. This is to determine if the first drought warning trigger is met.

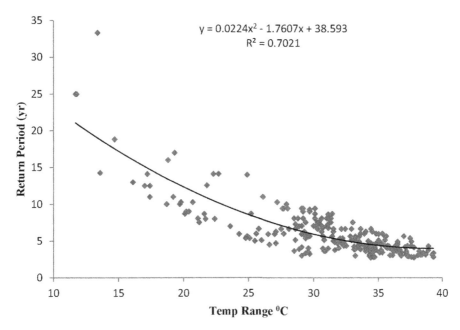

FIGURE 8.4

Average drought return period corresponding to temperature variation.

Source: Pandey, R.P., Daradur, M., Jain, V.K., Jain, M.K., 2016. Assessment of vulnerability to drought towards effective mitigation planning; the case of Ken river basin in India. Journal of Indian Water Resources Society 36 (4), 36–48.

Step 2:

If the first drought trigger is set off in Step 1, the impact indicators will be examined as stipulated in the matrix outlined in Table 8.3.

Explanation: The severity of the drought depends upon the values of at least two out of four categories of impact indicators. Drought severity is assessed as follows:

- "Severe" drought: if two of the selected three impact indicators are in "severe" class
- "Moderate" drought: if two of the selected three impact indicators are in "moderate" or "severe'" class.
- "Normal conditions": if none of the above conditions are met.

If a finding of "severe" or "moderate" drought is determined, the second drought trigger is released. It is important to note that the state has the option of reducing the drought category from "severe" to "moderate" (or moderate to normal) if more than 75% of the administrative region (district/taluk/block/mandal), for which drought is being declared, is irrigated. However, the State Government will still be required to conduct field verification as prescribed in Step 3 below.

Step 3: If the second drought trigger is set off, states conduct a field verification exercise to make a final determination of severity. Once a formal declaration of drought is made and its

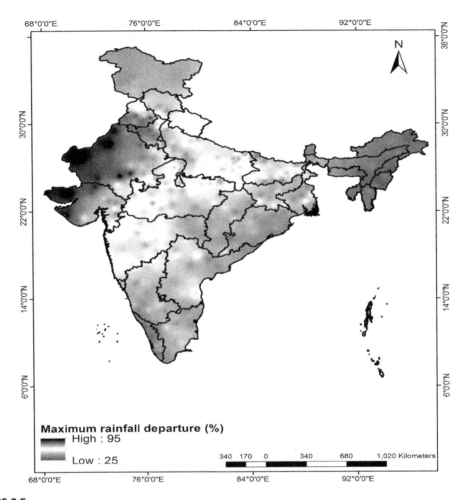

FIGURE 8.5

Distribution of observed drought severity (maximum rainfall deficiency) in India.

Source: Pandey, R.P., Daradur, M., Jain, V.K., Jain, M.K., 2016. Assessment of vulnerability to drought towards effective mitigation planning: the case of Ken river basin in India. Journal of Indian Water Resources Society 36 (4), 36—48.

severity determined, the State Government may then seek assistance from the Central Government for implementation of mitigation and response actions.

Indicators for early season drought declaration based on rainfall deviation or SPI and the dry spell as shown in Table 8.2 continue to be mandatory for declaration.

Need for suitable drought indices

The lack of universally accepted definitions and widely acceptable indices are major constraints to the timely identification of onset and assessment of drought severity level. For hydrologists and

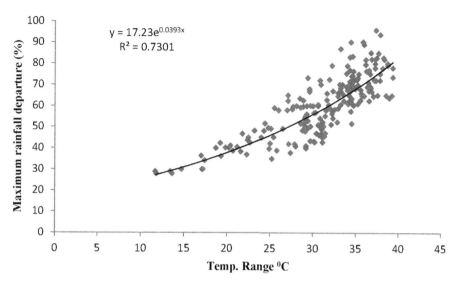

FIGURE 8.6

Maximum rainfall departures corresponding to temperature variation.

Source: Pandey, R.P., Daradur, M., Jain, V.K., Jain, M.K., 2016. Assessment of vulnerability to drought towards effective mitigation planning: the case of Ken river basin in India. Journal of Indian Water Resources Society 36 (4), 36–48.

Table 8.1 Key variables, indicators, and sources of data for drought monitoring.

S. No.	Key variable	Indicator/Index(ices)	Sources of data
1	Rainfall	Rainfall deviation/SPI/Dry spell	IMD, State Government
2	Crop area sown	Deviation from normal	State Government (Department of Agriculture)
3	Satellite-based crop condition	NDVI, NDWI deviation from normal VCI form of NDVI/NDWI	MNCFC, NRSC, ISRO and State Remote Sensing Centers
4	Stream flow	SFDI	CWC/India-WRIS
5	Groundwater level	Groundwater levels	SGWD CGWB
6	Reservoir storage	Reservoir level/storage	CWC, Irrigation Department, Water Resources Department

SPI, standardized precipitation index; NDVI, normalized difference vegetation index; VCI, vegetation condition index.

Source: MoA & FW, 2016. Manual for Drought Management. Departmenttof Agriculture, Cooperation and Farmers Welfare, Ministry of Agriculture and Farmers Welfare Govt. of India. New Delhi. pp. 1–155.

Table 8.2 Matrix for rainfall deviations and dry spells (Trigger-1).

RF Dev./SPI	Dry spell	Drought trigger
Deficit or scanty RF/SPI < −1	Yes	Yes
Deficit or scanty RF/SPI < −1	No	Yes if rainfall is scanty or SP < −1.5, else No
Normal RF/SPI > −1	Yes	Yes
Normal RF/SPI > −1	No	No

SPI, *standardized precipitation index; RF Dev., rainfall deviation.*

Source: MoA & FW, 2016. Manual for Drought Management. Departmenttof Agriculture, Cooperation and Farmers Welfare, Ministry of Agriculture and Farmers Welfare Govt. of India. New Delhi. pp. 1–155.

Table 8.3 Matrix for impact indicators (Trigger-2) (Trigger-1).

Mandatory indicators		Impact indicators				
Rainfall indices		**Agriculture**	**Remote sensing**	**Soil moisture**	**Hydrology**	**Category of drought**
Rainfall deviation or SPI	Dry spell	Crop area sown	VCI or NDVI deviations	PASM/MAI	SFI/RSI/ SGW	

The states may consider any three of the four types of the Impact Indicators (one from each) for assessment of drought severity.
SPI, *standardized precipitation index; NDVI, normalized difference vegetation index; VCI, vegetation condition index; PASM, percentage average soil moisture; MAI, moisture adequacy index; SFI, stream flow index; RSI, reservoir storage index; SGW, shallow ground water.*

Source: MoA & FW, 2016. Manual for Drought Management. Departmenttof Agriculture, Cooperation and Farmers Welfare, Ministry of Agriculture and Farmers Welfare Govt. of India. New Delhi. pp. 1–155.

researchers, it is one of the most challenging jobs to develop suitable and acceptable hydrometeorological drought indices to correctly describe the regional drought attributes. This notwithstanding, a variety of drought indices have been developed in other parts of the world. Some of these indices include the "effective drought index" (EDI) (Byun and Wilhite, 1999) and the "Palmer drought severity index" (PDSI) developed by Palmer in 1965.

Alongside, the Food and Agriculture Organization of the United Nations (FAO) has developed an algorithm for the generation of time series of moisture adequacy index for agricultural droughts. However, none of these indices are fully accepted by researchers due to the limitations associated with these indices. McKee et al. (1993) introduced standardized precipitation index (SPI), which has relatively better acceptability in some regions.

The SPI is used as a primary indicator for drought monitoring in India. Like the SPI, the applicability of the EDI has been evaluated in dry-subhumid and semi-arid climatic regions of India. It is found that the EDI provides more precise assessment of drought conditions. Nonetheless, it requires further evaluation for other regions to ascertain its broad applicability and acceptability. In this light, there is a need for long-term empirical focus in attempting to devise regional thresholds/indices. This will provide for near real-time determination of onset of drought, quantification of event severity and vulnerability to drought in order to suitably utilize them to describe regional drought problems.

Amid the varying drought identification indices and levels of applicability, various states in India are developing a regular drought monitoring system to provide critical information for decision-making on a continuous basis to enable near real-time recognition of occurrence of drought, its severity, and management actions. To achieve the above goals of near real-time drought monitoring and mitigation, improvement of drought vulnerability indices (DVIs) is needed. Further, development of techniques for formulation of area-specific water management plan is needed to deal with drought conditions in different climatic regions of the country.

Demarcation of areas vulnerable to drought

Drought risk at a given place is a function of hazard and vulnerability. Hazard refers to climatological factors, while vulnerability includes physiographic, social, environmental, and other factors like coping ability, awareness, and regional economic activities that influence risk (Hamouda et al., 2009). It is understood that geographical location, physical characteristics of a watershed, surface and groundwater availability, variability of rainfall, regional climatic factors, and socioeconomic factors contribute to water deficits and crop losses from time to time (Pandey et al., 2008, 2010). In line with this, the National Institute of Hydrology (NIH) has developed a method for integrated assessment of vulnerability to drought using hydrometeorological factors, topography, land use, soil type, spatial and temporal water demand, surface and groundwater availability, irrigation support, and socioeconomic aspects expressing coping ability of the region (Pandey et al., 2010) as categorized in Table 8.4.

Using this method, data on the regional topography, land use, soil, regional climate, water demand, water availability, irrigation support (if any), socioeconomic conditions, coping ability, and so forth are weighted and entered into a geographic information system (GIS). The integrated values of weights of various indicators are organized according to a spatial grid, allowing for the generation of a DVI value for each cell, which can then be mapped. The DVI in this instance is defined as the ratio of sum of the weights of factors to the sum of their maximum weight values. This method has been used to demarcate drought vulnerability zones in the Tel sub-basin of Mahanadi river system and the Ken basin in central India (Fig. 8.7).

Planning drought mitigation measures and way forward

There is a pressing need in India to systematically demarcate zones vulnerable to drought. This will help prioritize the planning and implementation of mitigation actions, as well as facilitate the development of region-specific drought mitigation plans as follows:

i. **Arid areas**: Having average frequency of drought as once in every 2—3 years, the chances of severe and extreme droughts are higher in comparison with other regions. In these areas, droughts are likely to persist for two or more consecutive years and hence, the need for decision makers to prioritize. This is necessary because mitigation options for these areas are often very limited. For such areas, mitigation options may include the following:
   - Intensive rain water harvesting/water conservation—this can moderate adverse impacts to some extent only.

Table 8.4 Different category of factors used to demarcate areas vulnerable to drought.

Static factor of vulnerability (physiographic factors which vary w.r.t. space)	Semi-static factors of vulnerability (which vary with the phase of development w.r.t. space and long-term temporal variability)	Variable factors of vulnerability (which vary w.r.t. space and time)
1. Topographic factors (general slope, drainage, etc.) 2. Soil (soil water holding capacity) 3. Climatic components (precipitation, etc.)	1. Water resources development (irrigation supports) and in situ water conservation 2. Status of surface water storage availability 3. Status groundwater availability 4. Population density (population concentration, industrial/commercial activities) 5. Land use 6. Regional cropping system/crop water demand 7. Region-specific activities (like cattle farming/wildlife preservation, etc.)	1. Rainfall (monthly/seasonal/annual) 2. Stream flow 3. Storages in tanks/reservoirs (if any)

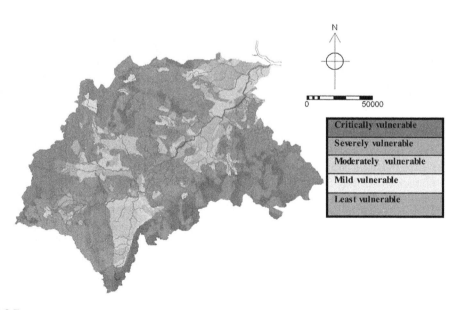

N

0   50000

Critically vulnerable

Severely vulnerable

Moderately vulnerable

Mild vulnerable

Least vulnerable

FIGURE 8.7

An example of demarcation of areas vulnerable to drought in tel sub-basin in Odisha.

- Recycling of wastewater particularly for supplemental irrigation.
- Promotion of microirrigation systems (sprinkler/drop systems).
- Measures for evaporation reduction from tanks/reservoirs.
- Need to pursue interbasin water transfer as an effective alternative—to mitigate impacts of drought and bring prosperity in the region

ii. **The semi-arid areas**: With an average frequency of drought of once in every 4 years, and often also encountering severe drought events, the mitigation option for these areas may include the following:
- Creation of water storage in tanks and reservoirs, or in situ water retention through gully checks and stop dams
- Provision of within-stream storage and water diversions
- Intensive water conservation/rain water harvesting
- Increase water use efficiency
- Provision of extensive groundwater recharge
- Recycling of wastewater for nondrinking purposes/irrigation
- Interbasin water transfer—remains a better mitigation option for bringing prosperity in semi-arid areas.

iii. **The subhumid areas**: These areas face drought with an average frequency of once in every 6–8 years and the occurrence of severe drought events in about once in 10 or more years. Persistent drought events are not very common. The mitigation option for these areas may be as follows:
- Increase water storage in tanks and reservoirs
- Water conservation/in situ water retention through gully checks and stop dams
- Intensive groundwater recharge may be a better option to enhance groundwater availability
- Within-stream water storage and water diversions may be used to enhance both surface and groundwater availability in such areas.
- Rejuvenation of tanks/ponds/springs, particularly in case of hilly regions

iv. **The humid areas**: These areas face drought with an average frequency of once in every 10 or more years; occurrence of severe drought is rare. There exist multiple options to manage drought, such as measures for rejuvenation of springs/ponds/lakes and in situ water conservation to enhance water availability in relatively dry periods.

Summary and conclusions

Overall, this chapter has outlined how the frequency, severity, and persistence of droughts are governed by regional climatic factors that vary significantly across different climatic regions in India. The current approach to drought management in India is still oriented toward a reactive response or relief from crisis, rather than proactive approach. As may be seen in many Indian states, there seems to be no system for regular, real-time monitoring of data needed to assess droughts.

In view of this, it is important to develop a comprehensive drought monitoring system for regular collection and timely dissemination of relevant data in a useable manner. Area-specific planning of mitigation measures in the different climatic regions is also necessary for effective drought management. The integrated assessment of vulnerability to drought using physical, climatic, and social factors in the ways outlined above, along with a more precise demarcation of areas with distinct level of vulnerability, will help both state and central government agencies identify and prioritize appropriate drought mitigation actions.

References

Byun, H.R., Wilhite, D.A., 1999. Objective quantification of drought severity and duration. Journal of Climate 12, 2747–2756.

Hamouda, M.A., Nour El-Din, M., Moursy, F.I., 2009. Vulnerability assessment of water resources systems in Eastern Nile Basin. Water Resources Management 23, 2697–2925.

McKee, T.B., Doesken, N.J., Kleist, J., 1993. The relationship of drought frequency and duration to time scales. Preprints. In: 8th Conference on Applied Climatology, 179–184. January 17–22, Anaheim, California.

MoA & FW, 2016. Manual for Drought Management. Department of Agriculture, Cooperation and Farmers Welfare, Ministry of Agriculture and Farmers Welfare Govt. of India, New Delhi, pp. 1–155.

Pandey, R.P., Singh, S.K.M.R., Ramasastri, K.S., 2008. Streamflow drought severity analysis of Betwa river system (India). Water Resources Management Journal 22 (8), 1127–1141.

Pandey, R.P., Pandey, A., Galkate, R., Byun, H.-R., Mal, B.C., 2010. Integrating hydro-meteorological and physiographic factors for assessment of vulnerability to drought. Water Resources Management 24 (15), 4199–4217.

Pandey, R.P., Daradur, M., Jain, V.K., Jain, M.K., 2016. Assessment of vulnerability to drought towards effective mitigation planning: the case of Ken river basin in India. Journal of Indian Water Resources Society 36 (4), 36–48.

Wilhite, D.A., Knutson, H.C., Smith, K.H., 2000. Planning for drought: moving from crisis to risk management. Journal of the American Water Resources Association 36 (4), 687–710.

World Meteorological Organization (WMO), Global Water Partnership (GWP), 2014. National Drought Management Policy Guidelines: A Template for Action. In: Integrated Drought Management Programme (IDMP) Tools and Guidelines Series, vol. 1. WMO, Geneva, Switzerland and GWP, Stockholm, Sweden.

Further reading

Dai, A.G., 2013. Increasing drought under global warming in observations and models. Nature Climate Change 3 (1), 52–58.

Dracup, J.A., Lee, K.S., Paulson Jr., E.G., 1980a. On the definition of droughts. Water Resources Research 16 (2), 297–302.

IPCC, 2007. Climate Change–2007. Final Report on Climate Change by Intergovernmental Panel. Published for the Intergovernmental Panel on Climate Change. Cambridge University Press.

Liu, N., Liu, Y., Bao, G., Bao, M., Wang, Y., Zhang, L., Ge, Y., Bao, W., Tian, H., 2016. Drought reconstruction in Eastern Hulun Buir Steppe, China and its linkages to the sea surface temperatures in the Pacific Ocean. Journal of Asian Earth Sciences 115, 298–307.

Palmer, W.C., 1965. Meteorological Drought. Research Paper No. 45. U.S. Department of Commerce Weather Bureau, Washington, D.C.

Pandey, R.P., Ramasastri, K.S., 2001. Relationship between the common climatic parameters and average drought frequency. Hydrological Processes 15 (6), 1019–1032.

Ponce, V.M., Pandey, R.P., Sezan, E., 2000. Characterization of drought across climatic spectrum. Journal of Hydrologic Engineering 5 (2), 222–224.

Sharma, K.D., Ramasastri, K.S., 2005. Drought Management. Allied Publishers, New Delhi.

Trenberth, K.E., Dai, A.G., van der Schrier, G., Jones, P.D., Barichivich, J., Briffa, K.R., Sheffield, J., 2014. Global warming and changes in drought. Nature Climate Change 4 (1), 17−22.

Vinit, J., Pandey, R.P., Jain, M.K., 2015. Spatio-temporal assessment of vulnerability to drought. Natural Hazards 76 (1), 443−469.

Wilhelmi, O.V., Wilhite, D.A., 2002. Assessing vulnerability to agricultural drought: a Nebraska case study. Natural Hazards 25, 37−58.

Assessment, monitoring, and early warning of droughts: the potential for satellite remote sensing and beyond

Valerie Graw[a,b,*], **Olena Dubovyk**[a,b], **Moses Duguru**[c], **Paul Heid**[a], **Gohar Ghazaryan**[a], **Juan Carlos Villagrán de León**[c], **Joachim Post**[d], **Jörg Szarzynski**[e,f,h], **Daniel Tsegai**[g], **Yvonne Walz**[e]

[a]*Center for Remote Sensing of Land Surfaces (ZFL), University of Bonn, Bonn, Germany,*
[b]*Remote Sensing Research Group (RSRG), University of Bonn, Bonn, Germany,*
[c]*United Nations Office for Outer Space Affairs (UNOOSA), United Nations Platform for Space-based Information for Disaster and Emergency Response (UN-SPIDER), Bonn, Germany,*
[d]*German Aerospace Center (DLR), Cologne, Germany,*
[e]*United Nations University, Institute for Environment and Human Security (UNU-EHS), Bonn, Germany,*
[f]*Eurac Research, Bolzano, Italy,*
[g]*United Nations Convention to Combat Desertification (UNCCD), Bonn, Germany,*
[h]*Disaster Management Training and Education Centre for Africa (DiMTEC), University of the Free State (UFS), Bloemfontein, Republic of South Africa*
[*]*Corresponding author. Email: valerie.graw@uni-bonn.de*

Introduction

Satellite remote sensing (RS) is an outstanding tool to detect and monitor changing conditions on land in an objective manner. In this regard, RS has a long history in measuring and monitoring drought characteristics and their impacts on land, specifically by developing targeted indices (AghaKouchak et al., 2015). An increasing number of publications underline the need and potential which RS contributes to, e.g., support the implementation and monitoring of the Sustainable Development Goals (SDGs) (Paganini et al., 2018). Also, on the national scale, RS-based drought monitoring has an impact considering the observation of vegetation or precipitation via satellite imagery. In addition to the SDGs (UN, 2015), the Sendai Framework for Disaster Risk Reduction (SFDRR) (UNISDR, 2015) has also triggered a monitoring process of loss and damage due to disasters as basis for understanding underlying drivers and to develop targeted disaster risk reduction (DRR) strategies, specifically for the most exposed and vulnerable livelihoods.

Drought events trigger significant water and food security concerns leading to risks for the economy, environment, and the population itself (AghaKouchak et al., 2015; Kogan, 2019; Schwalm et al., 2017; Wilhite et al., 2007). The conceptual definition of drought describes four different types of drought: meteorological drought, agricultural drought, hydrological drought, and socioeconomic drought (Wilhite and Glantz, 1985; Loon et al., 2016). The original concept of these four drought types along a temporal gradient induced by the lack of rainfall (Wilhite and Glantz, 1985) has recently been

Drought Challenges. https://doi.org/10.1016/B978-0-12-814820-4.00009-2

revised by van Loon et al. (2016) taking into account human activities as additional drivers of droughts, such as irrigation, water abstraction, or dam building, resulting in both ecological and socioeconomic impacts.

Today, we are facing what is called the "golden age" of RS (IAC, 2018; Sara Aparício, 2017). The availability and accessibility of RS data with different sensors and sensor systems for targeted monitoring of land surface characteristics and dynamics has never been better than now. Modern RS enables assessment and monitoring at multiple scales and at different levels of detail. The spatial resolution of sensors ranges from centimeters to kilometers and can be selected in a targeted way depending on the application context, ranging from detailed analysis of small areas to global applications. Freely available sensor data which are mainly considered for land cover assessment include imagery from Landsat with 30 m resolution; from the Moderate Resolution Imaging Spectroradiometer (MODIS) with 250, 500, and 1000 m resolution; or from the National Oceanic and Atmospheric Administration Advanced Very High-Resolution Radiometer (NOAA-AVHRR) with 1 km resolution. One core mission with regard to these golden times is the Copernicus Sentinel Mission of the European Union (EU). In total, five Sentinel satellites will be in space in 2020.

Besides the spatial resolution, also the temporal resolution, describing the revisit period of a sensor covering the same area, is relevant and contributes to accurate monitoring of sustainable development. For the Sentinels the recovering period can be up to 5 days which is a high temporal resolution. Nevertheless, imagery from these sensors is only available since 2015. To understand the characteristics of drought events and their dynamics as well as monitoring land stress and successful sustainable land management practices, longer time series are needed (Skidmore et al., 1997). Here, ongoing long-term satellite missions are still relevant, such as the Landsat program, operating since 1972, the NOAA-AVHRR, which captures land surface information since 1982, or the MODIS sensors providing information since 2000.

With regard to droughts, a huge variety of RS products have been developed using different satellite RS data and methods (AghaKouchak et al., 2015). The overall aim of this chapter is to synthesize the role and contribution of satellite RS data and methods for the assessment and monitoring of droughts and their impacts, its potential to provide early warning of future drought events, and highlight remaining challenges. Therefore, we (1) review established and recent developments of satellite RS—based drought products and elaborate the role and contribution of RS along existing applications of drought assessment and early warning; (2) refer to attempts where RS contributes to the 2030 Agenda for Sustainable Development Global Development Agenda in the context of drought monitoring and action, and (3) include insights in how RS products can be used as input data for accurate drought risk monitoring. The strengths and limitations of RS data for assessing, monitoring, and early warning of droughts are discussed, and the opportunity of integrating RS products into comprehensive drought risk assessments is highlighted with the overall aim to develop targeted drought risk reduction strategies for sustainable development.

RS-based drought assessment and monitoring

Satellite RS provides a synoptic view of the land and a spatial context for measuring drought impacts which have demonstrated to be a valuable source of timely, spatially continuous data with improved information on monitoring vegetation dynamics over large areas (Mishra and Singh, 2010). It allows the monitoring of different climatological and biophysical variables playing a key role in the context of

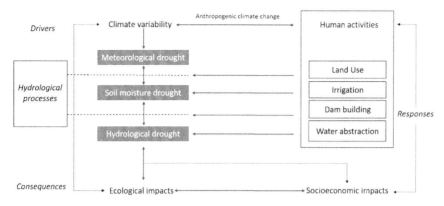

FIGURE 9.1

Drought types.

Based on Van Loon, A. F., Gleeson, T., Clark, J., Van Dijk, A.I.J.M., Stahl, K., Hannaford, J., Di Baldassarre, G., et al., 2016. Drought in the anthropocene. Nature Geoscience 2016 9:2.

droughts. Different key variables or combinations of them are considered appropriate for drought monitoring using RS imagery and geospatial data analysis in general. Besides a various number of single drought indices also a number of combined drought indices have been developed. Monitoring with RS ranges from assessment of the hazard itself—e.g., a shortage in precipitation—to the impacts on environmental conditions on the ground—e.g., diminishing crop growth and reduced yields or land degradation. Fig. 9.1 shows an overview of different drought types and processes where climate has an impact on meteorological, soil moisture, and hydrological droughts and, at the same time, are also driven by human activities as land use and land management impact soil moisture and hydrological drought again. All drought types further lead to ecological and socioeconomic impacts.

Climate variables for drought monitoring

The meteorological drought definition relates to a reduction in precipitation with regard to amount, intensity, and timing (Wilhite, 2011). For meteorological drought monitoring, the standardized precipitation index (SPI) has been widely used, and its forecasting is recommended by the World Meteorological Organization (WMO) (WMO, 2005). The SPI takes into account the long-term precipitation information for a certain time period, fits it into a probability distribution which is then transformed into a normal distribution so that values below 0 indicate less than median precipitation and above 0 respectively higher (WMO, 2005). Depending on the observation period, the SPI is calculated for different time spans and thereby, e.g., adjusted to drought monitoring of agricultural areas with SPI 3 and 6 covering the growing season with 3 or 6 months averages, short-time variability with SPI 1, or long-term observations with SPI 12 and higher (WMO, 2012). This index can be calculated using RS-based precipitation information, but unlike land cover, precipitation is harder to monitor from space due to a high degree of spatiotemporal variability (Liu et al., 2012). As rainfall usually occurs with cloud cover, there is no direct way of measuring precipitation from space, so it is often estimated by measuring the thickness of clouds or the temperature of the top cloud layer (Ceccato and Dinku, 2010). There are a number of RS precipitation products. Most commonly used are the Tropical Rainfall Measurement Mission (TRMM), its successor the Global Precipitation Measurement (GPM), the

Global Precipitation Climatology Project (GPCP), Climate Prediction Centers (CPC), Merged Analysis of Precipitation (CMAP), CPC MORPHing technique (CMORPH), Climate Hazards Group Infra-Red Precipitation with Station data (CHIRPS), Precipitation Estimation from Remotely Sensed Information using Artificial Neural Networks (PERSIANN), and the African Rainfall Climatology (ARC) (Tian et al., 2009; Ceccato and Dinku, 2010; Funk et al., 2015; Poméon et al., 2017). Although, they are based on different satellites or gauge inputs, they share a common methodology as well as potentials and limitations. Calculations are either based on a combination of microwave and (thermal-)infrared satellite data or satellite images merged with gauge data to correct atmospheric errors (Du et al., 2013; Ceccato and Dinku, 2010; AghaKouchak et al., 2015). Accuracy is depending on the number, distribution, and quality of the gauges (Ceccato and Dinku, 2010). In comparison to other products such as vegetation information, the spatial resolution is rather coarse (2.5−5 degrees, at best 0.5 degrees) while the temporal resolution is rather high up to multiple times a day. However, these precipitation products are often not as accurate as in situ measurements as they are prone to over- or underestimation of actual precipitation by up to 50% (Ceccato and Dinku, 2010). Intensive precipitation rates and amounts are well detected, sparse, and small-scale precipitation rates are rather missed out (Du et al., 2013; Ceccato and Dinku, 2010). For these reasons, so far satellite-based precipitation data have not been fully integrated into drought monitoring (AghaKouchak et al., 2015).

With regard to the definition and characteristics of the meteorological drought also high temperatures or low relative humidity play a role. The thermal conditions can be monitored using surface brightness temperature (BT) derived from thermal infrared (TIR) bands from multiple sensors such as AVHRR, MODIS, and Landsat. In general, BT and derived land surface temperature (LST) are used as proxy for moisture availability and evapotranspiration (ET) conditions such that water depletion in the plant root zone leads to stomatal closure, reduced transpiration, and subsequently increased canopy temperatures (Petropoulos and Islam, 2017). Same as precipitation, soil moisture is mainly detected using infrared and passive microwave sensors. The main sensors are Landsat, AVHRR, MODIS, the Advanced Spaceborne Thermal Emission and Reflection Radiometer (ASTER), and Envisat (Cheng and Ren, 2012). Landsat provides the longest thermal data record starting from 1984, and data with spatial resolution of 60−120 m are available in the TIR spectrum from 10.5 to 12.5 µm (Cristóbal et al., 2018). Commonly used LST products are available from NOAA-AVHRR and MODIS (Aqua and Terra) with spatial resolution from 1000 to 8000 m and 1-day temporal resolution (Ceccato and Dinku, 2010). The MODIS LST has been found to be the most reliable product, with only 1 K deviation for homogeneous surfaces. At 1.000 m resolution, surfaces are rarely homogeneous, so the retrieval precision is likely far lower over land than over water (Cheng and Ren, 2012). Similar to precipitation data, the operational use of LST or land surface emissivity (LSE) for drought monitoring is not spread widely as they both fall short on achieving reliable results (Cheng and Ren, 2012).

Based on thermal data, the temperature condition index (TCI) was developed in order to determine temperature-related impact on crops. The TCI is derived based on $TCI = 100*(T_{max}-T)/(T_{max}-T_{min})$, where T denotes the temperature which is scaled between multi-year maximum and minimum (T_{max} and T_{min}, respectively) (AghaKouchak et al., 2015; Zhang et al., 2017). Several studies integrated LST with vegetation indices, assuming that these indicators have inverse relationships. In fact, a higher LST can have an adverse impact on vegetation and further implies drought stress due to increasing ET (Wardlow et al., 2012).

ET further describes an important part of the hydrological cycle representing processes in the soil−water−vegetation system. As it is difficult to measure ET accurately, especially at larger scales, a number of modeling approaches have been developed. In principal ET, modeling algorithms can be categorized into two groups: surface energy balance (e.g., Simplified Surface Energy Balance, MET-RIC) and water balance (e.g., Vegetation ET)—based models (Senay et al., 2015; Senay et al., 2012; Wang et al., 2012). The Food and Agriculture Organization (FAO) of the United Nations suggests estimating vegetation transpiration or soil evaporation for drought monitoring. An operational global dataset was introduced by Vicente-Serrano et al., 2010.

With regard to climate variables and combinations, modeling approaches tend to be more successful and represent higher correlations with ground station data than considering one input variable only. These include integration of cloud cover and temperature analysis (Stephens and Kummerow, 2007), soil moisture information using free available microwave data (Bolten et al., 2010; Urban et al., 2018), but also analysis of ET by integrating LST information (Karnieli et al., 2010; AghaKouchak et al., 2015; Anyamba and Tucker, 2012).

Vegetation and land cover dynamics monitoring in the drought context

Satellite RS allows monitoring of land and vegetation conditions across large areas. In locations with a limited number of sampling gauges, RS data are often the only available information to monitor vegetation response as a result of droughts (Rhee et al., 2010). Several drought indices for vegetation condition monitoring have been established, based on various RS data and methods and from a range of disciplines. These indices utilize unique methods to quantify drought and have differing data requirements. The simplest way to monitor drought using the RS-based normalized difference vegetation index (NDVI) is to calculate NDVI anomalies or vegetation condition, using the vegetation condition index (VCI), over time. The NDVI is predominantly used to estimate biomass (Deering and Haas, 1980) with a direct link to drought. Similar to the NDVI is the enhanced vegetation index (EVI) (Zhangyan et al., 2008), which integrates measurements with less atmospheric distortion and is considered most useful in high biomass areas with higher water saturation (Huete et al., 2002). Studies showed that negative NDVI anomalies can be used to identify drought and also, there are strong links of the NDVI on interannual time scale to the El Nino/La Nina Southern Oscillation (ENSO) phenomenon (Anyamba and Tucker, 2012; Liu and Negrón Juárez, 2001), by understanding the links between ENSO events and drought occurrence, the dynamics of NDVI anomalies can be used to predict an oncoming drought (Liu and Negron Juarez, 2001). An indicator based on NDVI anomalies is the standardized vegetation index (SVI) by Peters et al. (2002). It describes the probability of variation from normal NDVI over multiple years as a Z-score deviation from the mean and is calculated for each pixel location. Due to weekly time steps also fast changing drought patterns can be detected. The VCI developed by Kogan (1995) is scaling NDVI values from 0 to 1, using the formula $VCI = (NDVI - NDVI_{min})/(NDVI_{max} - NDVI_{min})$. Kogan (2002) later also introduced the vegetation health index (VHI) which combines VCI and TCI, all based on similar formula and referred to as the condition indices (Zhang et al., 2017). The addition of the TCI should report more accurately about vegetation health during drought conditions.

The VHI is also used by the Agricultural Stress Index Systems (ASIS), an RS-based tool which was developed by the FAO as a way to address food insecurity due to droughts. The aim of ASIS was to support the detection of hot spots globally where crops could be affected by droughts (Van Hoolst et al., 2016). The VHI is used to monitor vegetation during the crop cycle and assess the degree of dryness to estimate the intensity and duration of dry periods during the crop cycle applying five drought severity classes as well as one category indicating no drought conditions (Table 9.1).

Table 9.1 Classification of vegetation health index (VHI) as used for Agricultural Stress Index Systems.

Magnitude of VHI	Drought category
<35	Exceptional drought
35–45	Extreme drought
46–55	Severe drought
56–65	Moderate drought
66–75	Abnormal dry
>75	No drought

Source: https://www.unccd.int/sites/default/files/relevant-links/2017-09/OSCAR%20ROJAS%20PRESENTACION_0.pdf.

Also, the percentage of the agricultural area affected by drought is calculated and fitted to the administrative boundaries which then can be compared with national agricultural statistics.

ASIS estimates the severity of droughts from 1984 to present every 10 days at 1 km resolution using the combined data from the Advanced Very High-Resolution Radiometer (AVHRR) and Meteorological Operational satellite program (METOP). The large temporal span allows ASIS to compare drought events between each other, also with droughts triggered by, e.g., strong El Niño ENSO events, as in 1997/98 and 2014/16. The global version of the ASIS does not distinguish between crop types, and hence its water requirements. Another limitation relates to the incapacity to carry out long temporal comparisons if crop rotations changed during the observation period, and in irrigated areas. Recently, the country-level ASIS was published including explicit information on crop types to improve the characterization of drought impacts (FAO, 2018; FAO, 2015). Additionally, information on crop phenology is integrated which supports the detection of crop sensitivity to the manifestation of droughts at various phenological stages. It incorporates spatially specific crop areas to tailor the needs of specific countries. ASIS offers the capacity to estimate the drought affectedness of a particular region during stages of the crop cycle, also takes into account actual crop yield information using multiple regression analysis and yield prediction (FAO, 2018).

While vegetation already gives insights into changes of land cover also other land cover dynamics can be monitored using RS. Expansion and contraction of lakes may occur as a result of disturbances (Giardino et al., 2013). These may scale from rapid events operating at small-scale to large-scale events of uncertain duration. Droughts steadily build up as water availability declines and may steadily subside as precipitation increases. Due to the interaction and relationship between rainfall, watershed, catchment, surface water, groundwater, and droughts, it is relevant to consider the above phenomena with lake dynamics (expansion and contraction of water bodies). Also, climate warming, increase in drainage and potential ET play key roles in lake dynamics.

Several RS data sources have been applied extensively for monitoring dynamics of lake surfaces such as Landsat TM and ETM + for monitoring the shrinking of lakes due to droughts (Riordan et al., 2006; Deus and Gloaguen, 2013) but also radar imagery using Sentinel-1 (Xing et al., 2018; Kaplan et al., 2019). UN-SPIDER recommends the use of Sentinel-1 data for water inundation mapping, meaning the extent and depth of flooding. Data fusion of Sentinel-1A and 1B could improve the temporal resolution to 5—6 days and thereby provide more detailed information on changes.

For the majority of cases, not one single indicator or index can represent the complexity of drought conditions across the affected spatial and temporal scales (Hayes et al., 2012; Zhang et al., 2017). In most drought monitoring scenarios, the best approach is to use several indicators for drought analysis. Studies by Liu and Kogan (1996) found that oversaturated soil moisture leads to depressed NDVI and low VCI values, wrongly indicating drought where the TCI would perform better. Thereby, a tool was created to monitor both drought and excessive wetness (Anyamba and Tucker, 2012): The scaled drought condition index (SDCI), which combines temperature, vegetation, and precipitation condition, was identified as an optimum RS-based index that can be used for agricultural drought monitoring in humid/subhumid regions as well as arid/semi-arid regions, after testing several RS variables and their combinations for Arizona, New Mexico, North Carolina, and South Carolina (Rhee et al., 2010). The vegetation drought response index (VegDRI) integrates RS vegetation indices with other climate-based drought indices and biophysical information such as land use/-cover types, soil types, elevation, and ecological condition (Anyamba and Tucker, 2012). Other drought indices include the temperature vegetation dryness index (TVDI), the microwave integrated drought index (MIDI), the synthesized drought index (SDI), the optimized meteorological drought index (OMDI), and the optimized vegetation drought index (OVDI). Zhang et al. (2017) provided a detailed review on the mentioned indices application for the United States.

Assessment and monitoring of drought impacts

The consequences or impacts of droughts are not only driven by drought hazard severity as, e.g., a lack of rainfall. They result from the interaction of hazards, exposure, and vulnerability to drought (IPCC, 2012; IPCC, 2014). Following the definitions of the Intergovernmental Panel on Climate Change (IPCC), exposure refers to the "presence of people, livelihoods, species or ecosystems, environmental functions, services, and resources, infrastructure, or economic, social, or cultural assets in places that could be adversely affected" (IPCC, 2014: p. 5) by the drought or any hazard. Vulnerability is defined as the predisposition to be adversely affected, resulting from the sensitivity or susceptibility of a system and its elements to be harmed combined with a lack of short-term coping capacity and long-term adaptive capacity (IPCC, 2014). Impacts can be categorized as social, economic, and ecological impacts. While RS applications previously predominantly focused on monitoring of biophysical determinants of drought hazard characteristics, links to social, economic, and ecological consequences of droughts are increasingly made (National Research Council, 1998; Elvidge et al., 2009; Graw and Husmann, 2014; Walz et al., 2018; Graw et al., 2017).

Drought impacts on agricultural land can be estimated using vegetation condition, which provides an essential input variable to assess yield losses. Yield estimation using satellite RS imagery is conducted for decades and improved by nowadays hyperspectral sensor data, which allows the distinction of different crop types. Higher resolution imagery further allows detection of yield loss on small fields (Bhuiyan et al., 2017; Burke and Lobell, 2017; Unganai and Kogan, 1998; Yu et al., 2010). Different land management strategies can further impact the vulnerability of a crop to droughts. While irrigation practices and good water management on large-scale farms can sustain crop growth also during dry periods, small-scale farms often depending on rainfall might be much earlier affected (Graw et al., 2017). Integration of time series analysis for the monitoring of vegetation conditions and its phenology as well as monitoring of climate variables can help to draw conclusions on drought affectedness of agricultural fields due to different land management strategies or land tenure rights but also during different stages of crop growth (Cui et al., 2017; Graw et al., 2017; Jolly and Running, 2004).

Spatial data of people and settlements are a key variable in the context of risk assessment, as this information allows to measure their exposure to drought hazard conditions, e.g., with regard to detection of remote places and those difficult to access, whereas the structural characteristics, demographic information, and densities are providing information about vulnerability characteristics. Several RS-based approaches exist in order to provide georeferenced information products on population, human settlements, and infrastructure. Examples of relevant products are, e.g., the Global Urban Footprint (GUF) settlement information developed by the German Aerospace Center (DLR) (Esch et al., 2017), the population density information from different providers compiled by the POPGRID Data Collective or the nighttime light detections spearheaded by the Center for International Earth Science Information Network (CIESIN, https://www.popgrid.org/).

Geospatial modeling and a combination of different data sources will be guiding future analyses of exposure, vulnerability, and risk as well as disaster impacts to support drought risk management.

Drought early warning

In recent decades, many countries around the world are experiencing more frequent and severe droughts. Some of the worst droughts in the last decades occurred in the Sahel or the Horn of Africa, as well as in Brazil, the United States, or Australia. In Central America, the droughts of the last two decades led governments in six countries of this region to designate a particular geographic region affected by droughts: the "Central American Dry Corridor" (Maurer et al., 2017; Müller et al., 2018).

In the context of early warning systems (EWS), in many developing countries, data on daily accumulated rainfall are used as a parameter to characterize the severity of droughts. Typically, meteorological departments or offices would calculate the deviations or anomalies in rainfall in terms of both reductions in the amount of rainfall or number of days without any rainfall. The SPI (WMO, 2005) is being promoted as well to compare archived and up-to-date data on daily accumulated rainfall to characterize the severity of droughts. In the United States, for drought EW, the Palmer drought severity index (PDSI) (Alley and Alley, 1984; Vicente-Serrano et al., 2010) is used. As stated by the NOAA, the PDSI can be used to characterize the scope, severity, and frequency of prolonged dry periods.

In 2006, during the Third International Early Warning Conference held in Bonn, Germany, the notion of people-centered EWS was introduced (Heppen et al., 2008). This updated view suggested the incorporation of risk knowledge to improve the effectiveness of EWS. The use of information on risks should lead to prioritize early warning needs and guide preparations for response. Risks are assessed as the probability of impacts and therefore also combine information on hazard probability, exposure, and vulnerability, including coping and adaptive capacities (Birkmann, 2006; UNISDR, 2016).

In recent years, experts have recommended improvements to warnings leading to migrating from hazard- to impact-based warnings (WMO, 2015). To develop such impact-based warnings, information on exposure and vulnerability is required. One way to incorporate information on vulnerability is to address vegetation, which is vulnerable to droughts. It is well known that some crops such as corn are more vulnerable to droughts than others such as sorghum (Hadebe et al., 2017).

Taking advantage of the availability of historic and up-to-date data on the EVI, UN-SPIDER has been promoting the use of the SVI generated on the basis of the EVI. The SVI allows operators of drought EWS to compare the effects of a current drought on vegetation with the effects of historic droughts in the last two decades. Efforts are underway to encourage government agencies in Central American countries, in Sri Lanka, and in selected African countries to strengthen their EWS for drought through the incorporation of the routine use of the SVI[1] and the routine use of ASIS. In this fashion, government agencies in countries affected by drought can monitor both, the changes in weather and the impacts of such changes on vegetation.

In addition, UN-SPIDER is encouraging countries to incorporate additional information in drought EWS regarding social and economic vulnerability, crop types per region if available, as well as information on geographic areas benefiting from irrigation. In the near future, once there is sufficient data compiled on soil moisture derived from satellite observations, this information will be incorporated into drought EWS. Fig. 9.2 presents an overview of the UN-SPIDER decision support system for drought EWS.

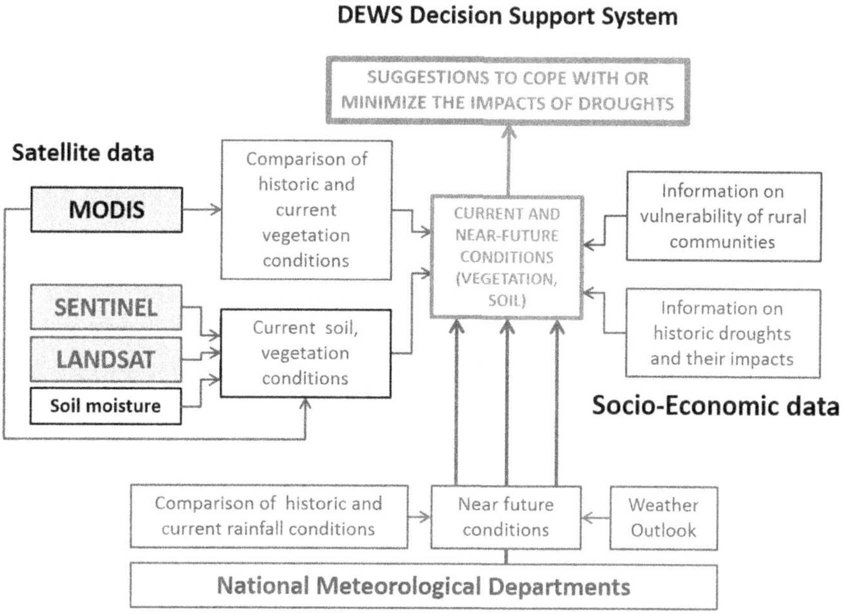

FIGURE 9.2

Decision support system for drought early warning systems proposed by UN-SPIDER.

[1]More information on this effort can be found in this link: http://www.un-spider.org/projects/SEWS-D-project-caribbean.

RS as support tool for the Global Development Agenda

RS is referred to as relevant support tool in the 2030 Agenda as well as in associated global agendas, conventions, and frameworks such as the Paris agreement, the convention to combat desertification, and the Sendai framework. RS supports systematic observation necessities in order to help nations and the international community to evaluate progress in sustainable development. RS can provide different possibilities to strengthen the assessment for global policy advice by monitoring elements and support of the 17 SDGs and its 169 associated targets (Paganini et al., 2018; Post et al., 2017).

Measuring development indicators at global, regional, and national levels is traditionally being performed by national statistical institutions. From a more integrative perspective, data collection tools and technologies, such as RS, geospatial mapping, innovative sensors, cell phone—based data, and crowd sourcing shall be incorporated in future endeavors.

Internationally, several relevant initiatives, programs, and activities are working on accelerating science and technological developments in the field of applied satellite RS and also on facilitating access to and providing data and services dedicated to support sustainable development. The space and RS community is working together with development actors and policy makers at national and international levels. Examples are the cooperation of the ESA with International Financial Institutes (IFIs) with its Earth Observation for Sustainable Development (EO4SD) initiative (Lorenzo-Alonso et al., 2018); the European Copernicus Programme with open access to its Sentinel satellite data, dedicated services, and provided products; the activities and dedicated programs of the Group on Earth Observation (GEO) and the Committee on Earth Observation satellites (CEOS, see Paganini et al., 2018); or the United Nations Office for Outer Space Affairs with its UN-SPIDER programme.

We can recognize good progress in connecting RS-based services and products with development cooperation efforts supporting the implementation of the 2030 Agenda and other global conventions as mentioned.

Drought preparedness is at the core of various global commitments including the 2013 High-level Meeting on National Drought Policy (HMNDP), the 2015 Sendai Framework for Disaster Risk Reduction (SFDRR), the 2015 Paris Agreement on Climate Change, the 2015 UN Sustainable Development Goals of Agenda 2030 (UN General Assembly, 2015), the Africa We Want, and the 2016 UN Environment Assembly resolution to combat desertification, land degradation, and drought and promote sustainable pastoralism and rangelands.

The turning point for drought management was the HMNDP which was held in Geneva in March 2013 with a focus to shift from a reactive to a proactive approach in the context of drought management (Pulwarty and Sivakumar 2014). The HMNDP resulted in a series of practical recommendations, which have been synthesized into three pillars of drought management: (1) monitoring and early warning systems; (2) vulnerability and impact assessment; and (3) mitigation and response. These pillars are now widely used as the basis of international cooperation for supporting the development of drought policies.

The first African Drought Conference was then held in Windhoek, Namibia, in 2016, adopting the implementation of a strategic framework for a Drought Resilient and Prepared Africa (DRAPA) guided by 6 principles (Tadesse, 2016): (1) drought policy and governance for drought risk management; (2) drought monitoring and early warning; (3) drought vulnerability and impact assessment; (4)

drought mitigation, preparedness, and response; (5) knowledge management and drought awareness; and (6) reducing underlying factors of drought risk.

The need for a policy framework emerging from the many initiatives since 2013 is an urgent need for national drought management policies to establish a clear set of principles and/or operating guidelines to govern the management of drought and to mitigate its impacts (Crossman, 2018). To reduce societal vulnerability to droughts, a paradigm shift of drought management approaches is required to overcome the prevailing structures of reactive, posthazard management and move toward proactive, risk-based approaches of disaster management (Tsegai et al., 2015). Drought management policy should also support the development of comprehensive drought monitoring and EWS and outline ways to better communicate and disseminate information about drought onset and risks. But above all, drought management policy should motivate action and change to reduce risk and increase resilience (Crossman, 2018).

The "Earth Observation–based Information Products for Drought Risk Reduction at the National Level" (EviDENz) project,[2] funded by the German Federal Ministry of Economic Affairs and Energy (BMWi), is one example, which demonstrated the use of RS-based data for the implementation of the SFDRR using the example of agricultural drought in the case study areas of Ukraine and South Africa. The main objectives of this project were (1) to understand risk of the social-ecological system to agricultural drought and (2) to provide a measure for indicators of the Sendai Targets B (=number of people affected) and C (=economic loss) (Fig. 9.3).

In the context of the EviDENz project, RS data provided essential input for assessing the probability of hazard severity in the context of agricultural drought using MODIS and NOAA-AVHRR time series data of the VCI (Graw et al., 2017; Graw et al., *in prep.*). These were an integral part of a comprehensive risk assessment through the spatial explicit integration of exposure and vulnerability characteristics using an indicator-based approach (Walz et al., 2018). The EviDENz project developed and demonstrated a processing chain to measure the indicator B-5 of the SFDRR on "number of people whose livelihoods were disrupted or destroyed," attributed to agricultural drought (Walz et al., n.d.), by integrating RS-based drought hazard assessments for a given year with exposed elements such as the number of agricultural dependent population or the hectares of cropland or grassland in a spatially explicit way. The majority of countries have to deal with the lack of loss and damage data and the capacity to measure the indicators of the SFDRR and to establish the baseline for monitoring progress of the Sendai Targets retrospectively. With the help of RS-based information, approaches can be developed which allow countries to monitor loss and damage information to achieve DRR.

The comprehensive risk assessment in the context of the EviDENz project has demonstrated that moderate drought hazard characteristics measured by RS-based information products can lead to extreme impacts in specific regions, whereas the very severe drought hazard of 2015/2016 has caused only low or no impacts in others (Walz et al., 2018). The reason for this nonlinear relationship between drought hazard severity and impact is the strong influence of the two additional dimensions of drought risk. However, these dimensions are of utmost importance to better understand the contribution of underlying drivers to the manifested impact of drought events and identify entry points for drought risk management and drought risk reduction.

---

[2]Project Website: https://www.zfl.uni-bonn.de/research/projects/evidenz.

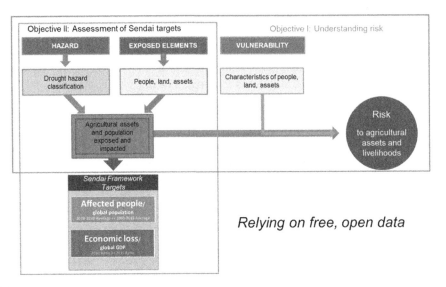

FIGURE 9.3

Approach conducted in the EvIDENz project. Integration of satellite remote sensing data and socioeconomic information to report for two of the Sendai Framework Targets.

Source: EvIDENz Project, https://www.zfl.uni-bonn.de/research/projects/evidenz.

Conclusion: strengths and limitations of satellite RS in the context of droughts

RS-based data have a very high potential for the assessment, monitoring, and early warning of droughts by measuring land surface characteristics in response to drought conditions. Particularly relevant are monitoring of (1) climate-related variables, such as precipitation, LST, or their combination as measure of ET, (2) vegetation condition measured through specific RS-based indices, and (3) land cover changes, such as dynamics of water bodies. Near real-time data further allow the monitoring of land surface dynamics related to the slow onset hazard and provide the essential information for drought EWS (see Chapter 7).

The unique values of RS-based data in this regard are manifold: First, drought assessments based on RS data provide objective measures and spatially consistent data covering large areas up to the global level. This is particularly relevant for drought monitoring, as often large areas are affected, and no other data source allows to get a full picture on drought-related land surface characteristics and its spatially explicit heterogeneity. Second, satellite RS data provide multispectral information covering the visible light, and also the near, middle, and TIR spectrum as well as microwave data which allows that land cover can be assessed unaffected by weather conditions. This technology has introduced an unprecedented advantage due to its responsiveness to dielectric properties. The vitality of vegetation is one critical measure to assess drought conditions and can be directly measured through the combination of spectral reflectance between the visible and NIR light where different indices have been developed that measure the vitality and condition of vegetation (e.g., NDVI, VHI, or

SVI). Third, RS has made the monitoring of drought over time possible through time series analysis. This is particularly relevant as drought is a slow onset hazard, and hazard severity is region-specific, which requires to set the current measure into reference of the retrospective mean.

In contrast, there are still limitations which have to be considered. Most RS-based drought information products are based on ratios, providing a relative measure that has to be brought into context. For example, a VCI can change as a result of drought, due to removal of vegetation, or as a result of crop rotation. Therefore, validation data is of utmost importance to understand RS-based products and its level of uncertainty. But this data are often not easy to acquire, especially when it comes to past drought events.

The definition of drought and classification of its severity is another barrier to overcome. Thresholds for declaring a drought or distinguishing drought severity are often arbitrary and lack validation which would make policy decisions based on these less efficient.

We have shown that RS-based products can measure key variables for assessing and monitoring droughts with a specific focus on drought hazard characteristics showing many strengths and advantages but still also limitations. Many indices exist, monitoring different aspects of drought conditions but also serving as input variables used for drought EWS. The dense time series of available RS data and increasing spatial resolutions allow a fast and detailed analysis of predominantly biophysical variables and land cover impacts over large areas. To assess and monitor impacts of drought conditions and better understand underlying drivers of drought risk, drought hazard characteristics need to be complemented with information on exposure and vulnerability of the system. Where RS-based information products can also contribute in this regard, e.g., through data of people and settlements, there remains the urgent need to add value to these remotely sensed measures by characterizing the relevant exposed elements, their susceptibility to drought, and their coping and adaptive capacity. Despite still some remaining challenges, the use of RS for drought monitoring and even early warning of droughts has been widely adopted. Nevertheless, drought events differ in their characteristics, and even if a number of drought indices serve for, e.g., early warning or assessment of yield loss, there is still a need to improve these assessments and work closely together with national authorities and institutions to adjust monitoring and modeling approaches for an effective and successful drought reporting also with regard to the Global Development Agenda 2030.

References

AghaKouchak, A., Farahmand, A., Melton, F.S., Teixeira, J., Anderson, M.C., Wardlow, B.D., Hain, C.R., 2015. Remote sensing of drought: progress, challenges and opportunities. Reviews of Geophysics 53 (2), 452–480.

Alley, W.M., Alley, W.M., 1984. The palmer drought severity index: limitations and assumptions. Journal of Climate and Applied Meteorology 23 (7), 1100–1109.

Anyamba, A., Tucker, C.J., 2012. Historical perspectives on AVHRR NDVI and vegetation drought monitoring. In: Wardlow, B.D., Anderson, M.C., Verdin, J.P. (Eds.), Remote Sensing of Drought.Innovative Monitoring Approaches, first ed., Drought and Water Crises

Aparício, S., 2017. A Golden Age for Earth Observation — Google Earth and Earth Engine — Medium. https://medium.com/google-earth/a-golden-age-for-earth-observation-f8b281cec4b7.

Bhuiyan, C., Saha, A.K., Bandyopadhyay, N., Kogan, F.N., 2017. "Analyzing the impact of thermal stress on vegetation health and agricultural drought — a case study from Gujarat, India. GIScience and Remote Sensing 54 (5), 678–699.

Birkmann, J., 2006. Measuring Vulnerability to Natural Hazards : Towards Disaster Resilient Societies. United Nations University.

Bolten, J.D., Crow, W.T., Zhan, X., Jackson, T.J., Reynolds, C.A., 2010. Evaluating the utility of remotely sensed soil moisture retrievals for operational agricultural drought monitoring. IEEE Journal of Selected Topics in Applied Earth Observations and Remote Sensing 3 (1), 57–66.

Burke, M., Lobell, D.B., 2017. Satellite-based assessment of yield variation and its determinants in smallholder African systems. Proceedings of the National Academy of Sciences of the United States of America 114 (9), 2189–2194.

Ceccato, P.N., Dinku, T., 2010. Introduction to Remote Sensing for Monitoring Rainfall, Temperature, Vegetation and Water Bodies.

Cheng, J., Ren, H., 2012. Land-surface temperature and thermal infrared emissivity. In: Liang, S., Li, X. (Eds.), Advanced Remote Sensing.Terrestrial Information Extraction and Applications. Jindi Wang, pp. 235–271.

Cristóbal, J., Jiménez-Muñoz, J., Prakash, A., Mattar, C., Skoković, D., Sobrino, J., 2018. An improved single-channel method to retrieve land surface temperature from the landsat-8 thermal band. Remote Sensing 10 (3), 431.

Crossman, N.D., 2018. Drought Resilience, Adaptation and Management Policy (DRAMP) Framework. https://www.unccd.int/sites/default/files/relevant-links/2018-08/DRAMP_Policy_Framework.pdf.

Cui, T., Martz, L., Guo, X., Cui, T., Martz, L., Guo, X., 2017. Grassland phenology response to drought in the Canadian Prairies. Remote Sensing 9 (12), 1258.

Deering, D.W., Haas, R.H., 1980. Using Landsat Digital Data for Estimating Green Biomass, NASA Technical Memorandum #80727, Greenbelt, MD, 21.

Deus, D., Gloaguen, R., 2013. Remote sensing analysis of lake dynamics in semi-arid regions: implication for water resource management. Lake Manyara, East African Rift, Northern Tanzania. Water 5 (2), 698–727.

Du, L., Tian, Q., Yu, T., Meng, Q., Jancso, T., Udvardy, P., Huang, Y., 2013. A comprehensive drought monitoring method integrating MODIS and TRMM data. International Journal of Applied Earth Observation and Geo-information 23, 245–253.

Elvidge, C.D., Sutton, P.C., Ghosh, T., Tuttle, B.T., Baugh, K.E., Bhaduri, B., Bright, E., 2009. A global poverty map derived from satellite data. Computers & Geosciences 35 (8), 1652–1660.

Esch, T., Heldens, W., Hirner, A., Keil, M., Marconcini, M., Roth, A., Zeidler, J., Dech, S., Strano, E., 2017. Breaking new ground in mapping human settlements from space – the global urban footprint. ISPRS Journal of Photogrammetry and Remote Sensing 134 (December), 30–42.

FAO, 2015. Protocol for Country-Level ASIS. Calibration and National Adaptation Process. http://www.fao.org/3/a-i5246e.pdf.

FAO, 2018. Country-Level ASIS: An Agricultural Drought Monitoring System. Rome. http://www.fao.org/publications/card/en/c/CA0986EN.

Funk, C., Peterson, P., Landsfeld, M., Pedreros, D., Verdin, J., Shukla, S., Husak, G., et al., 2015. The climate hazards infrared precipitation with stations—a New environmental record for monitoring extremes. Scientific Data 2 (December), 150066.

Giardino, C., Bresciani, M., Stroppiana, D., Oggioni, A., Morabito, G., 2013. Optical remote sensing of lakes: an overview on Lake Maggiore. Journal of Limnology 73 (s1).

Graw, V., Husmann, C., 2014. Mapping marginality hotspots. In: Gatzweiler, F., von Braun, J. (Eds.), Marginality: Addressing the Nexus of Poverty, Exclusion and Ecology, pp. 69–83 (Chapter 5).

Graw, V., Ghazaryan, G., Dall, K., Delgado Gómez, A., Abdel-Hamid, A., Jordaan, A., Piroska, R., et al., 2017. Drought dynamics and vegetation productivity in different land management systems of Eastern Cape, South Africa-a remote sensing perspective. Sustainability 9 (10).

Hadebe, S.T., Modi, A.T., Mabhaudhi, T., 2017. Drought tolerance and water use of cereal crops: a focus on sorghum as a food security crop in sub-Saharan Africa. Journal of Agronomy and Crop Science 203 (3), 177–191.

Hayes, M.J., Svoboda, M.D., Wardlow, B.D., Anderson, M.C., Kogan, F.N., 2012. Drought monitoring. In: Wardlow, B.D., Anderson, M.C., Verdin, J.P. (Eds.), Remote Sensing of Drought.Innovative Monitoring Approaches, first ed., Drought and Water Crises T4 — Historical and Current Perspectives M4 — Citavi. Taylor and Francis, Hoboken, pp. 1—22.

Heppen, J.B., Therriault, Bowles, S., 2008. Developing Early Warning Systems to Identify Potential High School Dropouts. Issue Brief. National High School Center. July. https://eric.ed.gov/?id=ED521558.

Hoolst, R. Van, Eerens, H., Haesen, D., Royer, A., Bydekerke, L., Rojas, O., Li, Y., Racionzer, P., 2016. FAO's AVHRR-based agricultural stress index system (ASIS) for global drought monitoring. International Journal of Remote Sensing 37 (2), 418—439.

Huete, A., Didan, K., Miura, T., Rodriguez, E.P., Gao, X., Ferreira, L.G., 2002. Overview of the radiometric and biophysical performance of the MODIS vegetation indices. Remote Sensing of Environment 83 (1—2), 195—213.

IAC 2018, 2018. Special Sessions | Iaf. http://www.iafastro.org/events/iac/iac-2018/technical-programme/special-sessions/.

IPCC, 2012. Managing the risks of extreme events and disasters to advance climate change adaptation. In: Field, C.B., Barros, V., Stocker, T.F., Qin, D., Dokken, D.J., Ebi, K.L., Mastrandrea, M.D., Mach, K.J., Plattner, G.-K., Allen, S.K., Tignor, M., Midgley, P.M. (Eds.), A Special Report of Working Groups I and II of the Intergovernmental Panel on Climate Change. Cambridge University Press, Cambridge, UK and New York, USA.

IPCC, 2014. Climate change 2014: impacts, adaptation, and vulnerability. Part B: regional aspects. In: Contribution of Working Group II to the Fifth Assessment Report of the Intergovernmental Panel on Climate Change.

Jolly, W.M., Running, S.W., 2004. Effects of precipitation and soil water potential on drought deciduous phenology in the Kalahari. Global Change Biology 10 (3), 303—308.

Kaplan, G., Avdan, Z.Y., Avdan, U., 2019. Mapping and monitoring wetland dynamics using thermal, optical, and SAR remote sensing data. In: Wetlands Management — Assessing Risk and Sustainable Solutions. IntechOpen.

Karnieli, A., Agam, N., Pinker, R.T., Anderson, M., Imhoff, M.L., Gutman, G.G., Panov, N., et al., 2010. Use of NDVI and land surface temperature for drought assessment: merits and limitations. Journal of Climate 23 (3), 618—633.

Kogan, F.N., 1995. Application of vegetation index and brightness temperature for drought detection. Advances in Space Research 15 (11), 91—100.

Kogan, F.N., 2002. World droughts in the new millennium from AVHRR-based vegetation health indices. EOS, Transactions American Geophysical Union 83 (48), 557.

Kogan, F.N., 2019. Remote Sensing for Food Security. Springer.

Liu, W.T., Kogan, F.N., 1996. Monitoring regional drought using the vegetation condition index. International Journal of Remote Sensing 17 (14), 2761—2782.

Liu, W.T., Negrón Juárez, R.I., 2001. ENSO drought onset prediction in northeast Brazil using NDVI. International Journal of Remote Sensing 22 (17), 3483—3501.

Liu, Y., Fu, Q., Zhao, X., Dou, C., 2012. Precipitation. In: Liang, S., Li, X., Wang, J. (Eds.), Advanced Remote Sensing.Terrestrial Information Extraction and Applications. Elsevier, Amsterdam, pp. 533—556. M4—Citavi.

Loon, A. F. Van, Gleeson, T., Clark, J., Van Dijk, A.I.J.M., Stahl, K., Hannaford, J., Di Baldassarre, G., et al., 2016. Drought in the anthropocene. Nature Geoscience 9 (2).

Lorenzo-Alonso, A., Utanda, Á., Aulló-Maestro, M., Palacios, M., 2018. Earth observation actionable information supporting disaster risk reduction efforts in a sustainable development framework. Remote Sensing 11 (1), 49.

Maurer, E.P., Roby, N., Stewart-Frey, I.T., Bacon, C.M., 2017. Projected twenty-first-century changes in the Central American mid-summer drought using statistically downscaled climate projections. Regional Environmental Change 17 (8), 2421—2432.

Mishra, A.K., Singh, V.P., 2010. A review of drought concepts. Journal of Hydrology 391 (1—2), 202—216.

Müller, A., Mora, V., Rojas, E., Díaz, J., Fuentes, O., Giron, E., Gaytan, A., van Etten, J., 2019. Emergency drills for agricultural drought response: a case study in Guatemala. Disasters 43 (2), 410—430.

National Research Council, 1998. People and Pixels. National Academies Press, Washington, D.C.

Paganini, M., Dyke, G., Steventon, M., Harry, J., 2018. Satellite Earth Observations in Support of the Sustainable Development Goals. Special 2018 Edition. CEOS ESA. http://eohandbook.com/sdg/files/CEOS_EOHB_2018_SDG.pdf.

Peters, A.J., Walter-Shea, E.A., Ji, L., Vina, A., Hayes, M., Svoboda, M.D., 2002. Drought monitoring with NDVI-based standardized vegetation index. Photogrammetric Engineering & Remote Sensing 68 (1), 71−75.

Petropoulos, G.P., Islam, T., 2017. Remote Sensing of Hydrometeorological Hazards.

Poméon, T., Jackisch, D., Diekkrüger, B., 2017. Evaluating the performance of remotely sensed and reanalysed precipitation data over West Africa using HBV light. Journal of Hydrology 547, 222−235.

Post, J., de Léon Villagrán, J.C., St-Pierre, L., 2017. Some remarks on making remote sensing-based mapping of elements at risk useable in international development cooperation. Natural Hazards 86 (S1), 189−191.

Pulwarty, R.S., Sivakumar, M.V.K., 2014. Information systems in a changing climate: early warnings and drought risk management. Weather and Climate Extremes 3 (June), 14−21.

Rhee, J., Im, J., J Carbone, G., 2010. Monitoring agricultural drought for arid and humid regions using multi-sensor remote sensing data. Remote Sensing of Environment 114 (12), 2875−2887.

Riordan, B., Verbyla, D., McGuire, A.D., 2006. Shrinking ponds in subarctic Alaska based on 1950−2002 remotely sensed images. Journal of Geophysical Research: Biogeosciences 111 (G4).

Schwalm, C.R., Anderegg, W.R.L., Michalak, A.M., Fisher, J.B., Biondi, F., Koch, G., Litvak, M., et al., 2017. Global patterns of drought recovery. Nature 548 (7666), 202−205.

Senay, G., Bohms, S., Verdin, J.P., 2012. Remote sensing of evapotranspiration for operational drought monitoring using principles of water and energy balance. In: Wardlow, B.D., Anderson, M.C., Verdin, J.P. (Eds.), Remote Sensing of Drought. Innovative Monitoring Approaches, first ed., Drought and Water Crises. Taylor and Francis, Hoboken. 123−44.

Senay, G.B., Velpuri, N.M., Bohms, S., Budde, M., Young, C., Rowland, J., Verdin, J.P., 2015. Drought monitoring and assessment. In: Hydro-Meteorological Hazards, Risks and Disasters. Elsevier, 233−62.

Skidmore, A.K., Bijker, W., Schmidt, K., Kumar, L., 1997. Use of remote sensing and GIS for sustainable land management. ITC Journal (3-4). https://pdfs.semanticscholar.org/77c7/b048fd8af700cbd9f5b8bd6ed5ccb7cba11b.pdf.

Stephens, G.-L., Kummerow, C.D., 2007. The remote sensing of clouds and precipitation from space: a review. Journal of the Atmospheric Sciences 64 (11), 3742−3765.

Tadesse, T., 2016. Strategic Framework for Drought Risk Management and Enhancing Resilience in Africa. https://www.unccd.int/sites/default/files/relevant-links/2018-04/DRAPA_strategic_framework_final_ttadesse2.pdf.

Tian, Y., D Peters-Lidard, C., Eylander, J.B., J Joyce, R., J Huffman, G., Adler, R.F., Hsu, K., Turk, F.J., Garcia, M., Zeng, J., 2009. Component analysis of errors in satellite-based precipitation estimates. Journal of Geophysical Research 114 (D24), 1377.

Tsegai, D., Liebe, J., Ardakanian, R., 2015. Capacity Development to Support National Drought Management Policies: Synthesis (Bonn, Germany).

UN, 2015. The 2030 Agenda for Sustainable Development. http://www.un.org/ga/search/view_doc.asp?symbol=A/RES/70/1&Lang=E.

UN General Assembly, 2015. Transforming our world: the 2030 Agenda for Sustainable Development. A/RES/70/1. https://www.refworld.org/docid/57b6e3e44.html.

Unganai, L.S., Kogan, F.N., 1998. Drought monitoring and corn yield estimation in southern Africa from AVHRR data. Remote Sensing of Environment 63 (3), 219−232.

UNISDR, 2015. Sendai Framework for Disaster Risk Reduction 2015−2030.

UNISDR, 2016. Report of the Open-Ended Intergovernmental Expert Working Group on Indicators and Terminology Relating to Disaster Risk Reduction. https://www.preventionweb.net/files/50683_oiewgreportenglish.pdf.

Urban, M., Berger, C., Mudau, T., Heckel, K., Truckenbrodt, J., Onyango Odipo, V., Smit, I., et al., 2018. Surface moisture and vegetation cover analysis for drought monitoring in the southern Kruger national park using sentinel-1, sentinel-2, and landsat-8. Remote Sensing 10 (9), 1482.

Vicente-Serrano, S.M., Beguería, S., López-Moreno, J.I., Angulo, M., El Kenawy, A., 2010. A New global 0.5° gridded dataset (1901−2006) of a multiscalar drought index: comparison with current drought index datasets based on the palmer drought severity index. Journal of Hydrometeorology 11 (4), 1033−1043.

Walz, Y., Dall, K., Graw, V., Villagran de Leon, J.C., Kussul, N., Jordaan, A., 2018. Understanding Agricultural Drought Risk: Examples from South Africa and Ukraine, Policy Report No. 3. Bonn, Germany.

Walz, Y., Min, A., Dall, K., Duguru, M., Villagran de Leon, J.C., Graw, V., Dubovyk, O., Sebesvari, Z., Jordaan, A., Post, J., (n.d.). Monitoring progress of the Sendai Framework for Disaster Risk Reduction: Status quo, key challenges and a way forward using a geospatial model to estimate the number of people affected by agricultural droughts in Eastern Cape, South Africa. Progress in Disaster Science.

Wang, K., Dickinson, R., Ma, Q., 2012. Terrestrial evapotranspiration. In: Liang, S., Li, X., Wang, J. (Eds.), Advanced Remote Sensing.Terrestrial Information Extraction and Applications. Elsevier, Amsterdam.

Wardlow, B.D., Tadesse, T., Brown, J., Callahan, K., Swain, S., Hunt, E., 2012. Vegetation drought response index. In: Wardlow, B.D., Anderson, M.C., Verdin, J.P. (Eds.), Remote Sensing of Drought.Innovative Monitoring Approaches, first ed., Drought and Water Crises. Taylor and Francis, Hoboken.

Wilhite, D.A., Glantz, M.H., 1985. Understanding: the drought phenomenon: the role of definitions. Water International 10 (3), 111−120.

Wilhite, D.A., Svoboda, M.D., Hayes, M.J., 2007. Understanding the complex impacts of drought: a key to enhancing drought mitigation and preparedness. Water Resources Management 21 (5), 763−774.

Wilhite, D., 2011. National Drought Policies: Addressing Impacts and Societal Vulnerability. Drought Mitigation Center Faculty Publications, Washington D.C., USA. http://digitalcommons.unl.edu/droughtfacpub/80.

WMO, 2005. Drought Assessment and Forecasting. Draft Report. WGH/RA VI/Doc. 8 (26.IV.2005). http://www.wmo.int/pages/prog/hwrp/documents/regions/DOC8.pdf.

WMO, 2012. In: Hayes, M., Wood, D., Svoboda, M. (Eds.), Standardized Precipitation Index User Guide (Geneva).

WMO, 2015. WMO Guidelines on Multi-Hazard Impact-Based Forecast and Warning Services. https://www.wmo.int/pages/prog/www/DPFS/Meetings/ET-OWFPS_Montreal2016/documents/WMOGuidelinesonMulti-hazardImpact-basedForecastandWarningServices.pdf.

Xing, L., Tang, X., Wang, H., Fan, W., Wang, G., 2018. Monitoring monthly surface water dynamics of Dongting lake using sentinal-1 data at 10 M. PeerJ 6 (June), e4992.

Yu, L., Zhou, L., Liu, W., Zhou, H., 2010. Using remote sensing and GIS technologies to estimate grass yield and livestock carrying capacity of alpine grasslands in golog prefecture, China. Pedosphere 20 (3), 342−351.

Zhang, L., Jiao, W., Zhang, H., Huang, C., Tong, Q., 2017. Studying drought phenomena in the continental United States in 2011 and 2012 using various drought indices. Remote Sensing of Environment 190, 96−106.

Zhangyan, J., Huete, A.R., Didan, K., Miura, T., 2008. Development of a two-band enhanced vegetation index without a blue band. Remote Sensing of Environment 112, 3833−3845. https://doi.org/10.1016/j.rse.2008.06.006, 2008.

Development of a system for drought monitoring and assessment in South Asia

10

Giriraj Amarnath[a], Peejush Pani[a,b], Niranga Alahacoon[a], Jeganathan Chockalingam[c],
Saptarshi Mondal[c], Karthikeyan Matheswaran[d], Alok Sikka[e], K.V. Rao[f], Vladimir Smakhtin[a,g]

[a]*International Water Management Institute (IWMI), Colombo, Sri Lanka,*
[b]*Institute of Remote Sensing and Digital Earth, Beijing, China,*
[c]*Birla Institute of Technology (BIT), Mesra, India,*
[d]*Stockholm Environment Institute (SEI), Bangkok, Thailand,*
[e]*International Water Management Institute (IWMI), New Delhi, India,*
[f]*Central Research Institute for Dryland Agriculture (CRIDA), Hyderabad, India,*
[g]*United Nations University — Institute for Water, Environment and Health (UNU-INWEH), Hamilton, ON, Canada*

Introduction

Drought is a complex hazard that can have significant agricultural, ecological, and socioeconomic impacts. Although drought indices derived from remote sensing data have been used to monitor meteorological or agricultural droughts, it is difficult to generate indices that can comprehensively capture drought's meteorological and agricultural aspects. This chapter describes an innovative approach for monitoring and assessment of drought risks in South Asia that integrates meteorological data, vegetation conditions from satellite imagery, and targeted collection of ground truthed moisture and crop yield data. We have created an integrated drought severity index (IDSI) that reflects the effects of drought as observed through both 1) satellite-derived vegetation data and 2) the level of dryness expressed by traditional climate-based drought indices. Additional biophysical/environmental characteristics such as ecoregion, elevation, land use/land cover (LULC) type, and soil type are also incorporated because they can influence climate—vegetation interactions.

Our IDSI combines a vegetation condition index (VCI), temperature condition index (TCI), precipitation condition index (PCI), and soil condition index (SCI). This IDSI combines multisource remote sensing data from Moderate Resolution Imaging Spectroradiometers (MODIS) with Tropical Rainfall Measuring Mission (TRMM) and ESA Soil Moisture (ASCAT) products to identify precipitation deficits, soil thermal stress, and vegetation growth status during the drought process. The IDSI has been tested extensively in India and Sri Lanka using in situ observation and government records of drought agricultural impacts and drought relief. In validating the IDSI, we found that it is not only strongly correlated with a 3-month scale standardized precipitation index (SPI3) for the study region but also with the crop yield differences in drought-affected areas. The IDSI therefore appears to be an effective comprehensive drought monitoring indicator, capturing both meteorological drought information and drought's influence on agriculture. The remainder of this chapter provides technical details on various work packages in development of this drought management system and IDSI; the process of integrating satellite data sources; major drought case studies in Sri Lanka, Maharashtra, and other Indian states used in this project; and summarized results from a field assessment carried out for 2014 droughts in India and Sri Lanka.

Background

The study region for our project, South Asia, consists of six countries: India, Bangladesh, Nepal, Pakistan, Sri Lanka, and Bhutan. According to the FAO (2011), there are 2.6 million square kilometers of agricultural land in South Asia, accounting for 54.66% of total land use. More than 60% of the region's population is employed in agriculture, contributing 22% of the regional GDP. Being able to accurately identify and monitor drought is therefore of considerable importance.

In developing our drought monitoring system, the project team members evaluated research papers that compared various drought indices in terms of usability and applicability in the study region, the availability of datasets, and the procedure to calculate several indices. Drought monitoring indices derived from remote sensing data appeared to be more suitable than in situ indices. Among many drought indices based on remote sensing techniques, the VCI based on the normalized difference vegetation index (NDVI) (Kogan and Sullivan, 1993; Kogan, 1990) and the TCI based on land surface temperature (LST) (Kogan, 1995b) have shown to be useful tools for monitoring the intensity, duration, and impact of drought on regional or global levels (Singh et al., 2003).

Earlier researchers have found that the VCI is suitable for monitoring large-scale drought impacts on vegetation, including agricultural drought, and that VCI has a strong correlation with the crop yield (Liu and Kogan, 1996; Unganai and Kogan, 1998; Kogan et al., 2005; Salazar et al., 2007, 2008). A study in a semi-arid region of the Iberian Peninsula found that the interpretation of the VCI was more complicated than other drought indices because it provides an indirect measure of moisture (drought) conditions. Anything that stresses vegetation, including things other than soil moisture, such as insects, diseases, and lack of nutrients, will be captured by VCI (Vicente-Serrano, 2007). Some studies find that VCI and TCI when taken together are better than separately, and combining them developed a vegetation health index (VHI) (Kogan, 1997; Kogan et al., 2004). However, in humid high-latitude regions, where vegetation growth is primarily limited by lower temperatures, using VHI to monitor drought condition has to be undertaken with caution (Karnieli et al., 2006).

VHI and TCI do not directly include measures of precipitation and soil moisture variation, which are key influencing factors of drought in the semi-arid areas. In South Asia, drought is a slow process which begins with precipitation deficit, creating a soil moisture deficit and a higher LST, which builds until vegetation growth is affected. To comprehensively monitor drought, we must account for parameters that include precipitation, soil, and vegetation. In this study, we used TRMM, the Global Precipitation Measurement (GPM) data as a component from precipitation to calculate PCI; MODIS LST as a component to calculate TCI; and, lastly the NDVI as a component from vegetation to calculate VCI. A problem is that drought does not have a linear relation with NDVI, LST, and TRMM anomalies in different season and regions, and there is always a correlation between them in some case. IDSI is an integrated drought severity index using a data fusion technique by aggregating and combining data from multiple sources such as the individual outputs of different drought indices obtained from precipitation, temperature, and vegetation parameters. The overall goal of this procedure is to provide more accurate and reliable solutions according to some measure of evaluation, in comparison with usage of single-source data alone (Dasarathy, 1997).

Materials and methods

Our project and resulting drought monitoring index bring together multiple dataset from a variety of sources (Table 10.1) which were combined in a systematic fashion according to the workflow shown in (Fig. 10.1). The treatment of these various datasets is now summarized briefly.

Surface reflectance

MODIS surface reflectance data product MOD09A1 8-day composite of 500m spatial resolution was downloaded for a period of 2001–15 via ftp link using a python script. Twelve MODIS tiles of 70 mb on average provide complete coverage of South Asia (Fig. 10.2). The basic input data in hdf format provided by the archive were then converted to ERDAS Imagine (.img) files format that consists of a layer stack of all 13 bands present in the HDF. The projection was kept same as the input data format (World Sinusoidal). Band 1 (red) and Band 2 (near infrared) were then used to calculate the NDVI in the Agriculture Drought Assessment and Monitoring System (ADAMS) tool (described in Integrated drought severity index (IDSI) development Section) using the following equation:

$$NDVI = \frac{NIR - Red}{NIR + Red} \tag{10.1}$$

Since MODIS NDVI products with the highest resolution of 500 m are available only at 16-day intervals, we instead used surface reflectance data to monitor changes at 8-day temporal resolution. The 8-day NDVI data were converted to 46 annual layer stacks using ERDAS Imagine software. A three-step error correction method was adopted to overcome the gaps and noises in the raw data, such as cloud cover.

Land surface temperature

MODIS LST product (MOD11A2) was also downloaded and converted by the same procedure as the surface reflectance (Fig. 10.3). The average size of the downloaded file is around 10 MB and the study area is covered by the same number of tiles. It is available at a comparatively coarser spatial resolution of 1 km, but the temporal resolution is the same. The LST 8-day composite was also clubbed together to form annual layer stack of 46 layers. Since the number of dropouts is more in LST compared to the reflectance data, a different error correction procedure was adopted to fill those gaps which are described in detail in the following sections.

Rainfall data

Satellite-based rainfall estimates

Tropical rainfall measuring mission

The TRMM is a well-known source for providing daily rainfall estimates from 1998 to March 2015. It provides a global dataset of daily rainfall estimates in gridded format at a spatial resolution of 0.25° (Fig. 10.4). This is accessed from the TRMM data archive of NASA in NetCDF file format, which is then converted to Geotiff file format using R program. The daily dataset is then converted to an 8-day composite by computing an 8-day accumulation to correspond temporally with the MODIS 8-day composites on a Julian-day basis.

Table 10.1 List of data sources used in this research.

S. No	Data	Detail	Resolution	Duration	Source	Link
1	MODIS reflectance	MOD09A1 surface reflectance 8-day composite	500 m	2001–15	NASA	http://reverb.echo.nasa.gov/reverb/
2	MODIS surface temperature	MOD11A2 land surface temperature 8-day composite	1 km	2001–15	NASA	http://reverb.echo.nasa.gov/reverb/
3	TRMM rainfall	3B42 daily precipitation estimates	0.25 deg	1998-Mar 2015	NASA-JAXA	http://disc.sci.gsfc.nasa.gov/SSW/#keywords=TRMM_3B42_daily%207
4	GPM rainfall	IMERG late run daily precipitation estimates	0.1 deg	Mar-Dec 2015	NASA-JAXA	ftp://jsimpson.pps.eosdis.nasa.gov/NRTPUB/
5	PERSIANN rainfall	Daily global rainfall estimates	0.25 deg	1983–2015	CHRS, University of California	http://chrs.web.uci.edu/persiann/data.html
6	CRISP rainfall	Daily global rainfall estimates	0.05 deg	1981–2015	Climate Hazard Group, USA	ftp://ftp.chg.ucsb.edu/pub/org/chg/products/CHIRPS-2.0/
7	IMD rainfall	Daily gridded data	0.25 deg	1901–2015	India Meteorological Department	CD
8	Bangladesh rainfall	Station data	36 stations	1958–2015	Department of Meteorology, Bangladesh	
9	NRSC land use	Thematic map for India	56m	2013–14	NRSC, India	http://bhuvan.nrsc.gov.in/gis/thematic/index.php
10	GLC 2010 land use	Global tthematic map	30m	2010	National Geomatics Centre of China	http://www.globallandcover.com/
11	Hensen water body	Hansen GFC 2013 Datamask	30m	2013	University of Maryland	http://commondatastorage.googleapis.com/earthenginepartners-hansen/GFC2013/
12	District level crop yield	Crop yield for all districts of India	–	1998–2012	Directorate of Economics and Statistics, India	http://aps.dac.gov.in/APY/Public_Report1.aspx
13	District level crop yield	Crop yield for all districts of Pakistan	–	1998–2008	Directorate of Agriculture Crop Report Service, Punjab Lahore	Punjab Agricultural Statistics 1998–2002 and 2003–08

GPM, *global precipitation measurement*; MODIS, *moderate resolution imaging spectroradiometers*; TRMM, *tropical rainfall measuring mission*.

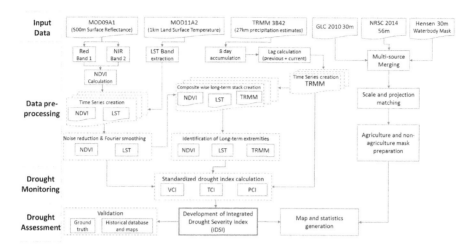

FIGURE 10.1

Flow diagram of the methodology.

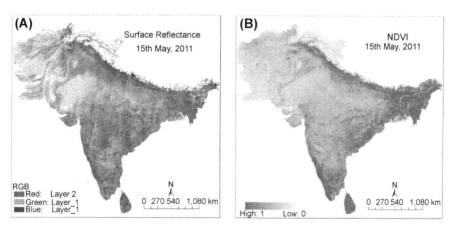

FIGURE 10.2

(A) MODIS surface reflectance as FCC (MOD09A1), (B) NDVI computed from the surface reflectance. *MODIS,* moderate resolution imaging spectroradiometers; *NDVI,* normalized difference vegetation index.

Global Precipitation *Measurement*

The Global Precipitation Measurement (GPM) Core Observatory, launched in 2014 to succeed the TRMM, is the first Space-Bourne Ku/Ka band Dual-frequency Precipitation Radar and provides high and moderate rainfall estimates at a resolution of 0.1°. Accumulated rainfall estimates are recorded on 30-minute, 3-hourly, daily, weekly, and monthly bases. We added daily rainfall estimates from the GPM's IMERG late run dataset of the GPM to the TRMM-derived data to extend our rainfall

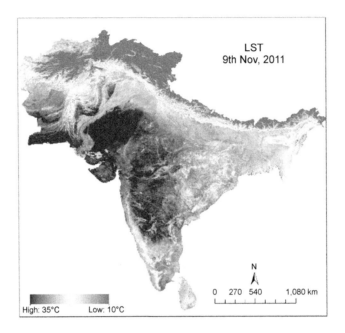

FIGURE 10.3

MODIS land surface temperature (LST) at a spatial resolution of 1 km (MOD11A2).

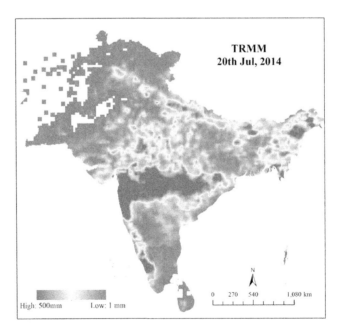

FIGURE 10.4

Tropical Rainfall Measuring Mission (TRMM) precipitation estimates at a spatial resolution of 0.25° (3B42).

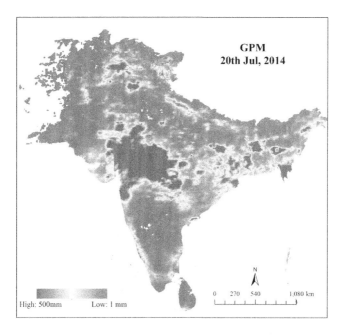

FIGURE 10.5

GPM rainfall estimates at a spatial resolution of 0.1° (IMERG).

sequence beyond 2014. Geotiff global images were downloaded from the GPM archive, projection was defined as WGS-1984 and the scale factor of 10 was applied (Fig. 10.5).

PERSIANN-CDR
The Climate Data Record of Precipitation Estimation from Remotely Sensed Information used Artificial Neural Network (PERSIANN) provides global precipitation estimates at 0.25-degree spatial resolution using a combination of infrared and passive/active microwave information from multiple geostationary and low Earth orbiting satellites, respectively. This dataset was used for computing the standardized precipitation index (SPI), which requires a long-term (i.e., 30 years) dataset to identify climatic variability.

CHIRPS (rainfall estimates)
A 0.05-degree high-resolution quasi-global (50 degrees N−50 degrees S) rainfall estimate from Climate Hazard group InfraRed Precipitation with Station data (CHIRPS) by the United States Agency for International Development Famine Early Warning Systems Network (FEWS NET) was used in the bias correction procedure of satellite estimates.

In situ rainfall data
Daily rainfall records for the last 115 years in gridded (.grd) file format were obtained from the India Meteorological Department. These gridded data come with a spatial resolution of 0.25° and are developed from in situ rainfall recorded by more than 6000 meteorological stations throughout the country (Fig. 10.6). The gridded data was converted to a compatible raster data format by decoding it through

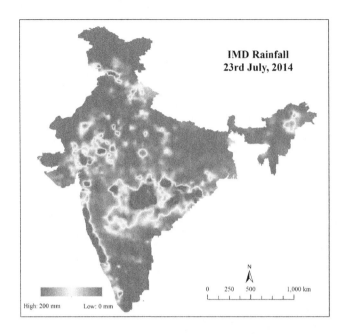

FIGURE 10.6

India Meteorological Department (IMD) daily rainfall at a spatial resolution of 0.25°.

an integrated method using C++, Python, MS Excel, and ArcMap. This database of in situ rainfall was used in different steps of the project, including bias correction of remotely sensed satellite estimates of rainfall and computing the long-term standard precipitation index.

In situ rainfall data from 37 stations for a period of 1958−2015 was obtained from the Bangladesh Meteorological Department and converted from discrete point data into a 1 km continuous raster database using an inverse distance weighting interpolation technique. This database was also used for bias correction of remotely sensed satellite rainfall estimates.

Land use/land cover

NRSC land use/land cover maps

To identify agricultural land in India, we used a national land use/land cover map at a scale of 1:50,000, using 2013−14 Resources at-1 Linear Imaging Self-Scanning Sensor (LISS-III) satellite data at 56m resolution, developed by the National Remote Sensing Centre, in collaboration with the Indian Space Research Organization of the Department of Space under National Resources Census (Fig. 10.7).

Global land cover (2010) 30 m

GLC 2010 is a 30-m-resolution, Landsat-based global land use/land cover mapping project at the National Geomatics Centre of China (Fig. 10.8). It provides a global land use map for the years 2000 and 2010 using the pixel-object-knowledge-based method for classifying the globe into 10 basic land use classes, with an accuracy level of up to 83.5%. We used it to identify agricultural areas in study region countries other than India.

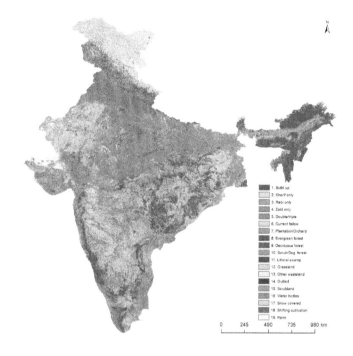

FIGURE 10.7

National Remote Sensing Centre (NRSC) land use/land cover product at 56 m resolution (2013–14).

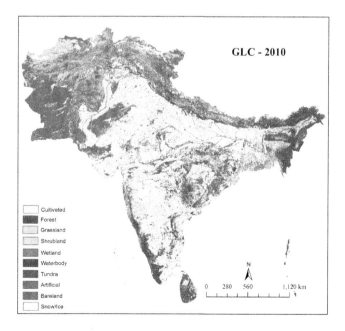

FIGURE 10.8

GLC 2010 land use/land cover map 30 m.

Hansen 30 m water body mask

Hansen Data (Water) mask GFC-2015 is one of the products by the Global Forest Change 2000−14 project initiated at the Department of Geographical Science, University of Maryland (Hansen et al., 2013). The global EOS Landsat data are used to map temporal changes in forest cover and surface water at a very high spatial resolution of 30 m. The whole globe is divided into $10 \times 10°$ granules (or tiles), and each granule is indexed based on the geographic coordinate system (e.g., 40N_80W).

Data preprocessing

Our project is based on time series using satellite imageries from 2001 to 2015. We processed our datasets as follows. The composite NDVI was calculated using red and near-infrared surface reflectance at 8-day intervals at 500m resolution from MODIS (MOD09A1). LST data were extracted at 8-day intervals at 1 km resolution from MODIS (MOD11A2). Daily accumulated rainfall estimates from TRMM (3B42) were converted into 8-day intervals using a temporal aggregation method to make them temporally compatible the NDVI and LST data. Annual time series of NDVI, LST, and TRMM data from 2001 to 2015 were created for further processing purpose by using 46 composites per year.

The time series MODIS NDVI and LST product contain data gaps and noise caused by cloud contamination and atmospheric instability during satellite overpass. Those noises affect the accurate measurements of vegetation and temperature dynamics. A linear interpolation technique was used to reduce major noise, but high-frequency noise still remains in the data series. From various smoothing techniques available to reduce high-frequency noise (e.g., Atberger et al., 2011Chen et al., 2011; Jonsson and Eklundh, 2004, 2002; Velleman, 1980), we adopted a Discrete Fourier Transformation (DFT)−based Fourier smoothing technique due to its advantage in less number of assumption, ease of implementation, and smoothening out high-frequency noise by reflecting genuine variation within time series (Atkinson et al., 2012; Jeganathan et al., 2010; Dash et al., 2010). This approach is applied to the time series MODIS NDVI and LST over South Asia during 2001−15.

For rainfall data, noise correction was not performed as a data gap may be normal if rainfall does not continuously occur across multiple weeks, which is common given the region's monsoonal climate. However, vegetation growth does not depend on only the current week's rainfall, but is a cumulative result of a longer period. Therefore, cumulative rainfall accumulation was aggregated into 2 week intervals.

Vegetation time series gap filling

The raw time series NDVI product comes with inherent noise due to cloud cover, atmospheric conditions, and variable illumination and viewing geometry. This noise could reduce the original VI value to "0" (or even a negative value), which is generally considered to be a "data dropout." We used a moving average function to correct the dropouts in MODIS raw VI. For each pixel, the current MODIS VI value at time "t_c" ($VI(t_c)$) was checked to identify the real value. If the value is less than or equal to zero (or a likely dropout), then the average of immediately preceding neighbor VI value ($VI(t_{c-1})$) and immediate succeeding neighbor VI value ($VI(t_{c+1})$) is calculated (($VI(t_{c-1})$ + $VI(t_{c+1})$)/2) to replace the dropout value. If any value of preceding neighbor or succeeding neighbor is again a dropout, then the previous immediate preceding ($VI(t_{c-2})$) or next immediate succeeding ($VI(t_{c-2})$) value is considered, respectively, for average calculation, as follows:

$$
\begin{aligned}
\mathrm{VI}(t_c) &= (((\mathrm{VI}(t_{c-1}) + \mathrm{VI}(t_{c+1}))/2) &&: \text{if } \mathrm{VI}(t_c) = 0 \\
&= ((\mathrm{VI}(t_{c-1}) + \mathrm{VI}(t_{c+2}))/2) &&: \text{if } \mathrm{VI}(t_c) = 0 \quad \mathrm{VI}(t_{c+1}) = 0 \\
&= ((\mathrm{VI}(t_{c-2}) + \mathrm{VI}(t_{c+1}))/2) &&: \text{if } \mathrm{VI}(t_c) = 0 \quad \mathrm{VI}(t_{c-1}) = 0 \\
&= ((\mathrm{VI}(t_{c-2}) + \mathrm{VI}(t_{c+2}))/2) &&: \text{if } \mathrm{VI}(t_c) = 0 \quad \mathrm{VI}(t_{c-1}) = 0 \quad \mathrm{VI}(t_{c+1}) = 0
\end{aligned}
$$

$$(10.2)$$

where, $\mathrm{VI}(t_c) = $ NDVI value at the current temporal band, $\mathrm{VI}(t_{c-1}) = $ NDVI value at the previous immediate temporal band, and $\mathrm{VI}t_{c+1} = $ NDVI value at the next immediate temporal band.

If 0 values in any pixel occur continuously for 20 composites, it is assuming that the respective pixel is not valid for further analysis, and the preceding and succeeding neighbor searching is restricted to up to 20 composites; the output of all bands of the respecting pixel will consequently be 0. However, in most cases, the anomaly was resolved within two to four follow-up composites.

Thermal time series gap filling

MODIS provides LST having 1000 m spatial and 8-day temporal resolution (**MOD11A2**). However, gaps in the composite quality of pixel (where the LST value is 0) may occur depending on atmospheric condition, the variable illumination and viewing geometry, and amount of cloud cover at the time of data acquisition. A continuous data gap in time series LST is a common scenario, as the noise can persist through a number of composites. A linear interpolation-based technique is adopted whenever the noise persists up to more than one preceding or succeeding neighbor, as follows:

$$
\mathrm{LST}(t_c) = ((\mathrm{LST}(t_{c-n})) + (r * ((t_c) - (t_{c-n})))) \quad \text{If } \mathrm{LST}(t_c) \leq 0 \tag{10.3}
$$

The "r" is representing the rate of change in VI value between one time ranges,

$$
r = \frac{\Delta h}{\Delta t}
$$

where
$$
\Delta h = (\mathrm{LST}(t_{c+n}) - (\mathrm{LST}(t_{c-n}))
$$
$$
\Delta t = (t_{c+n}) - (t_{c-n})
$$

where, $\mathbf{LST}(t_c)$ is the LST value of current time; $\mathbf{LST}(t_{c-n})$ is the LST value of preceding nonnoisy (nonzero or nonnegative) neighbor; $\mathbf{LST}(t_{c+n})$ is the LST value of succeeding nonnoisy neighbor; t_c is the current composite time; (t_{c-n}) is the location of the preceding nonnoisy neighbor value; (t_{c-n}) is the location of succeeding nonnoisy neighbor value; Δh is the LST value difference between succeeding and preceding nonnoisy neighbor; and Δt is the positional (time) difference between succeeding and preceding nonnoisy neighbor.

Annual cyclicity is assumed to maintain the continuity if the preceding neighbor search crosses the first composite or the succeeding neighbor search crosses the last composite accordingly (Fig. 10.9). In this regard, locational difference and value difference are calculated considering the total number of bands in the loop. However, the estimation of new data is fully dependent on the number of invalid pixels and gap length in a temporal profile (Justice et al., 2002). If the amount of noise is at maximum, proper reconstruction will be underestimated.

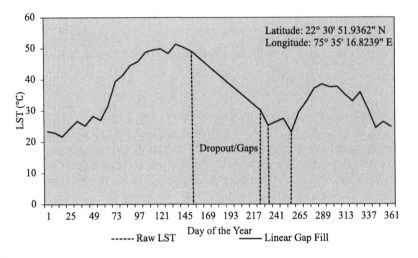

FIGURE 10.9

Graph showing thermal time series gap fill using linear method.

Noise elimination in the temporal series

A temporal moving average window (considering two neighbors on both the sides) was used to identify a temporal outlier value in comparison to its two preceding neighbor value and two succeeding neighbor value (Fig. 10.10). For each pixel, mean (μ) and standard deviation (σ) have been calculated by using two immediate preceding neighbor value ($VI(t_{c-1})$, $VI(t_{c-2})$) and two immediate succeeding neighbor value ($VI(t_{c+1})$, $VI(t_{c+2})$) of current NDVI or LST value ($VI(t_c)$). By using the mean and

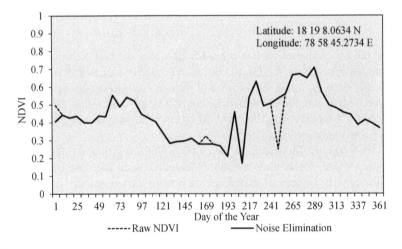

FIGURE 10.10

Graph showing noise elimination of vegetation time series data.

standard deviation, the upper fence ($\mu + 1.5\ \sigma$) and lower fence ($\mu - 1.5\ \sigma$) are created to check the current NDVI or LST value ($VI(t_c)$). The current VI value would be considered as an extreme outlier if it falls beyond either upper fence ($\mu + 1.5\ \sigma$) or lower fence ($\mu - 1.5\ \sigma$) and the current NDVI or LST value has been replaced by the average of immediate preceding ($VI(t_{c-1})$) and succeeding ($VI(t_{c+1})$) neighbor value in order to maintain the annual trend.

$$VI(t_c) = ((VI(t_{c-1}) + VI(t_{c+1}))/2) \qquad (10.4)$$

If

$$VI(t_c) > (\mu + 1.5\ \sigma) \text{ or } VI(t_c) < (\mu - 1.5\ \sigma)$$

where

μ = the mean of two preceding and succeeding neighbor value,
σ = the standard deviation of two preceding and succeeding neighbor value,
n = the total number of a neighbor,
x_i = NDVI or LST value of ith temporal band,
VI_{t_c} = NDVI or LST value of current time,
$VI(t_{c-1})$ = NDVI or LST value at the previous immediate time, and
$VI(t_{c+1})$ = NDVI or LST value at next immediate time.

Temporal fourier smoothing

Gap filling and noise correction are only able to reduce major noise perturbations in time series data. To suppress high-frequency subannual noise resulting from atmospheric turbidity and sensor fluctuation, and to obtain the meaningful smoothed temporal profile, a time series smoothing technique is generally used, such as Harmonic series (e.g., fast and discrete Fourier transformation (Atkinson et al., 2012), Savitzky—Golay (Chen et al., 2004)), double logistic (Beck et al., 2006), asymmetric Gaussian (Jonsson and Eklundh, 2004), best index slope extraction (Lovell and Graetz, 2001), weighted least squares windowed regression (Swets et al., 1999), and Whittaker smoother (Atzberger and Eilers, 2010). We used a Harmonic series—based DFT (Figs. 10.11 and 10.12). The minimum user input requirement (in terms of a number of harmonics) to reconstruct the time series is the main advantage of this approach (Dash et al., 2010). The DFT decomposes any complex time series phenological data into a series of sinusoids of different frequency (Jeganathan et al., 2010 and Mondal et al., 2017). Inverse Fourier transformation is used to reconstruct the phonological data, as follows:

$$F_{(u)} = a_0 + \sum_{n=1}^{N} \left(a_n \cos\frac{\pi n u}{N} + b_n \sin\frac{\pi n u}{N} \right) \qquad (10.5)$$

The equation consists of two parts: real or cosine part, and imaginary or sine part.
The cosine part is:

$$F_{c(u)} = \frac{2}{N} \sum_{n=0}^{N-1} \left(f(x) * \cos\frac{2\pi u n}{N} \right)$$

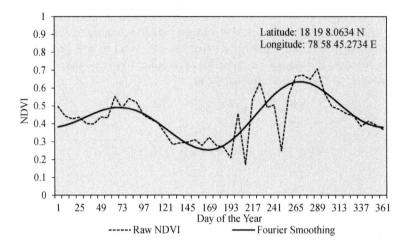

FIGURE 10.11

Graph showing Fourier smoothing of vegetation time series data.

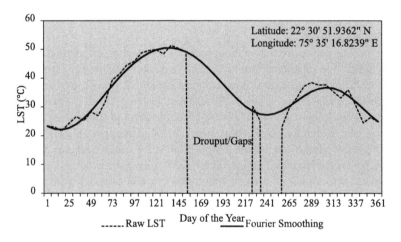

FIGURE 10.12

Graph showing Fourier smoothing of thermal time series data.

And the sine part is:

$$F_{s(u)} = \frac{2}{N} \sum_{n=0}^{N-1} \left(f(x) * \sin \frac{2\pi un}{N} \right)$$

where, $f(x)$ is the xth value in the time series, u is the harmonic component of Fourier transformation, n is the time of $f(x)$, and N is the length of the total time period.

For a complete reconstruction of phenological time series, consideration of appropriate harmonics number is very important. The first two components of inverse Fourier transformation can account for

50%−90% of the variability of a time series (Jakubauskas et al., 2001) but cannot represent the semi-annual or bimodal time series. Therefore, a four to six part harmonic is usually recommended for correct identification of various phenological metrics (e.g., number of growing season, sowing, harvesting time, etc.) (Jeganathan et al., 2010).

Long-term extreme event computation

After the three-step error correction of MODIS NDVI and LST products, long-term minimums and maximums for each composite were calculated. For this, a long-term layer stack (2001−14) was made for each composite using ERDAS Imagine. A program in the ADAMS tool (discussed in Integrated drought severity index (IDSI) development section) was used to compute the long-term min-max for all the 7728 total composites of noise-corrected NDVI and LST for all the 12 tiles of South Asia. The TRMM 8-day accumulated rainfall estimate for the whole South Asia tile was also long-term stacked in a similar manner, and extremes were computed in the ADAMS tool. These extremes were used for the indices calculation described in the following section.

Standardized indices computation

Remotely sensed NDVI, LST, and TRMM rainfall accumulation was converted into three standardized indices (e.g., VCI, TCI, and PCI) to remove scaling uncertainty in value range and to get meaningful and interpretable information.

Vegetation condition index

To identify the drought condition in an ecosystem, minute changes in the NDVI are not sufficient. Kogan (1995a) developed VCI that monitors the vegetation vigor by normalizing the NDVI with reference to long-term extremities. It is a pixel-based calculation that efficiently separates the short-term changes in NDVI from the long-term ecological changes (Fig. 10.13). In a given climatic region and season, extreme drought conditions will hamper the growth of vegetation and ultimately lead to low NDVI when compared with long-term observations of the same time of the year. Similarly, high NDVI will be reflected when there is optimum vegetation growth. VCI has proved to be more efficient in identifying drought condition compared to NDVI. Following is the equation for calculating VCI:

$$\text{VCI}_{ijk} = \frac{\text{NDVI}_{ijk} - \text{NDVI}_{ijn}}{\text{NDVI}_{ijx} - \text{NDVI}_{ijn}} * 100 \tag{10.6}$$

where, VCI_{ijk} is VCI for pixel i in composite j of year k; NDVI_{ijk} is NDVI for pixel i in composite j of year k; NDVI_{ijn} is the long-term minimum of NDVI for pixel i in composite j; and NDVI_{ijx} is the long-term maximum of NDVI for pixel i in composite j. The value range varies between "0" to "100." The value nearby "0" reveals extreme drought situation and nearby 100 expresses healthy situation (Kogan, 1997).

Temperature condition index

Surface temperature plays an equal role in the progression of drought conditions. Thus Kogan et al. (1995b) introduced the TCI, which is a remote sensing−based thermal stress indicator that is efficient in identifying drought-like situations by relating it to temperatures. This index assumes that when the

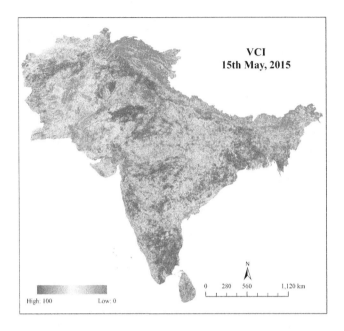

FIGURE 10.13

Vegetation condition with reference to long-term (2001—15) data.

soil has less moisture content the surface temperature tends to increase. Drought years tend to have higher LST compared to other years. In contrast to NDVI, where values are higher during above normal vegetation conditions, the LST shows higher values during extremely dry conditions (Singh et al., 2003). The TCI, similar to VCI, compares the current thermal condition with historical (long-term) extreme events and assigns it a normalized value ranging from 0 to 100, where values close to 0 depict stress condition for vegetation growth and those close to 100 reflect extremely favorable condition (Fig. 10.14). The equation of TCI is as follows:

$$\text{TCI}_{ijk} = \frac{\text{LST}_{ijx} - \text{LST}_{ijk}}{\text{LST}_{ijx} - \text{LST}_{ijn}} * 100 \tag{10.7}$$

where, TCI_{ijk} is the TCI for pixel i in composite j of year k; LST_{ijk} is LST for pixel i in composite j of year k; LST_{ijn} is a long-term minimum of LST for pixel i in composite j; and LST_{ijx} is a long-term maximum of LST for pixel i in composite j.

Precipitation condition index

Meteorological drought plays a direct role in the current soil moisture availability for vegetation growth, irrespective of all the secondary sources like irrigation through canals and underground water. Rainfall estimates from sources like TRMM 3B42 and GPM IMERG were used to compute the PCI. The data were calibrated using the inherent scale factor mentioned in the metadata of the products and validated using meteorological station data. The PCI, like VCI and TCI, compares the present rainfall

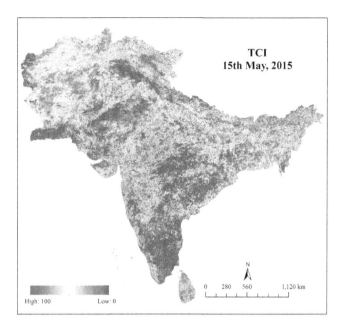

FIGURE 10.14

Temperature condition with reference to long-term (2001−15) data.

scenario with long-term extreme events and assigns it a percentage value where, values close to 0 infer scarcity of rainfall compared to the historical scenario of the same time frame, and a value close to 100 depicts favorable condition for vegetation growth (Fig. 10.14). Following is the equation for PCI:

$$\text{PCI}_{ijk} = \frac{\text{TRMM}_{ijk} - \text{TRMM}_{ijn}}{\text{TRMM}_{ijx} - \text{TRMM}_{ijn}} * 100 \tag{10.8}$$

where, PCI_{ijk} is the PCI for pixel i in composite j of year k; TRMM_{ijk} is the 2 weeks accumulated (previous and current 8-day composites) TRMM and GPM rainfall estimates for pixel i in composite j of year k; TRMM_{ijn} is the long-term minimum of 2 weeks accumulated TRMM and GPM rainfall estimates for pixel i in composite j; and TRMM_{ijx} is the long-term maximum of 2 weeks accumulated TRMM and GPM rainfall estimates for pixel i in composite j.

Integrated drought severity index development

Each of the three drought indices we developed (e.g., VCI, TCI, and PCI) has its own attributes in terms of spatiotemporal coverage, data acquisition, and development techniques. Because no single index is able to provide complete synoptic and temporal drought severity information, we combined them to create an IDSI using a data fusion technique (Fig. 10.15). The IDSI is calculated based on the following equation:

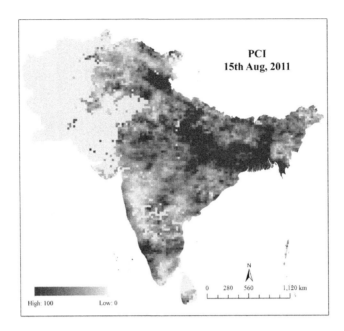

FIGURE 10.15

Precipitation condition with reference to long-term (2001–15) data.

$$\text{IDSI}_{ijk} = \left[L * \text{VCI}_{ijk} * \left\{ c + \frac{1}{(L * (\text{VCI}_{ijk} + \text{TCI}_{ijk} + \text{PCI}_{ijk} + c)} * (\text{TCI}_{ijk} + \text{PCI}_{ijk}) \right\} \right] \quad (10.9)$$

where, IDSI_{ijk}, VCI_{ijk}, TCI_{ijk}, and PCI_{ijk} are IDSI, VCI, TCI, and PCI values, respectively, for pixel i in composite j of year k. The IDSI value ranges from 0 to 100; L is the normalization factor to keep the output value in the expected range; and c is a constant to avoid null in the denominator. The values close to 0 reveal extreme drought situations where vegetation is under stress, precipitation is very low, and the temperature is very high. Likewise, the values closer to 100 reveal a normal situation where vegetation growth is good, precipitation is high, and the temperature is favorable. The interrelationship of these four indices can be better reflected in the graph (Fig. 10.16):

Agriculture drought assessment and monitoring system toolbox

Time series drought monitoring over a large area is a challenging task due to many aspects such as the need to process a huge volume of data, addressing the inherent noise in time series data, integration of multiscale biophysical variables (VCI, TCI, and PCI) and availability of efficient data processing platforms. Most of the available and widely used GIS and image processing packages such as ArcMap, ERDAS, ENVI, IDRISI, GRASS work well for discreet data processing, but do not provide efficient and dedicated time series data processing functions that can provide continuous time series monitoring options.

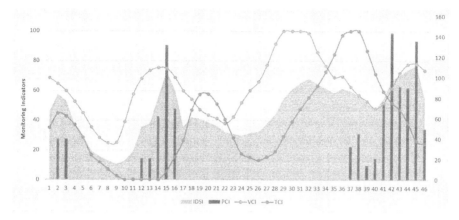

FIGURE 10.16

Graph showing the comparison of VCI, TCI, PCI, and IDSI for a sample year. *IDSI*, integrated drought severity index; *PCI*, precipitation condition index; *TCI*, temperature condition index; and *VCI*, vegetation condition index.

We therefore developed ADAMS, an Agriculture Drought Assessment and Monitoring System toolset, which is an integrated time series drought analysis and monitoring platform developed on ArcObject VBA of ArcMap for South Asia drought monitoring (Fig. 10.17). During the developmental phase, we concentrated on three main components: (1) dedicated time series algorithm and toolset development, (2) integration of multisource and multiresolution drought-related variables, and (3) efficient processing and data handling by considering memory limitation. In the version 1.0, a total of 50 different programs were developed with batch processing facilities to process MODIS NDVI, MODIS LST, and TRMM data, and those programs were further regrouped into three sections: (1) data preprocessing, (2) drought monitoring, and (3) drought assessment. The batch processing technique is fully automated with respect to file input and output perspective and is quite different from conventional batch processing technique available in ArcMap. The main architecture of ADAMS tool in ArcMap environment is represented in Fig. 10.18.

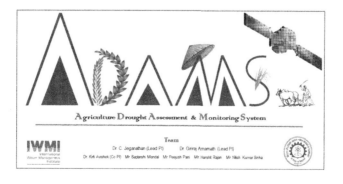

FIGURE 10.17

Logo of ADAMS tool.

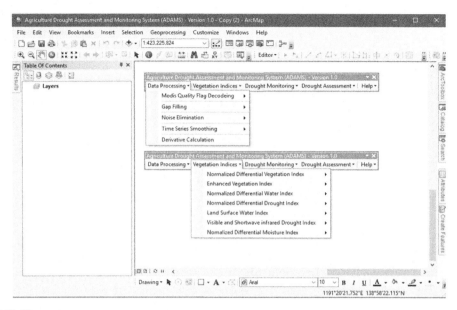

FIGURE 10.18

The structure of ADAMS tool in ArcGIS platform.

IDSI versus crop yield anomaly

We sought to link the newly developed IDSI with actual agricultural productivity. Scarcity of water from different sources like rain, irrigation schemes, wells, and so forth ultimately affects vegetation cover. The IDSI considers all such parameters, and it is therefore assumed that there should be some relationship between the IDSI and crop yield. For this, the crop yield statistics for different crops were obtained at district levels from the Directorate of Economics and Statistics, India; Directorate of Agriculture Crop Reporting Service, Punjab Lahore, Pakistan; Department of Agriculture, Sri Lanka. The relationship between IDSI and crop yield anomalies due to droughts was also computed using the long-term statistical information as follows:

$$CYA_i = \frac{CY_i - CY_m}{CY_{std}} \tag{10.10}$$

where, CYA_i is the crop yield anomaly of a particular year, CY_i is the crop yield of a district for a particular year, CY_m is the long-term mean of crop yield of that district, and CY_{std} is the long-term standard deviation of crop yield for that district. Here, the number of years for long-term statistics varies depending upon the availability of data. In the majority of cases, it is a minimum of 15 years (1998−2012).

A linear correlation was obtained between the CYA and IDSI of different districts in a state/province by considering both good as well as bad years. Here, the good and bad year are referred to as a year of water scarcity and a year of surplus rainfall, respectively. The results are broadly discussed below.

Results and discussion
Drought monitoring using NDVI

Long-term NDVI data for the entire crop seasons (January—December) were used in the present study to monitor agriculture stress. Though NDVI has been used by many authors for studying drought condition, it has been recommended by a number of studies to use VCI rather than using NDVI alone. The stacked NDVI layers were visualized to identify differences in vegetation health during drought years and wet years (Fig. 10.19). In order to assess the performance of MODIS Terra—derived NDVI, two consecutive years, 2002 and 2003, were chosen for their distinct NDVI characteristics. While visually comparing the NDVI of 2002 and 2007, a difference was observed in vegetation stress in a drought year (2002) and normal year (2007) for the states of Gujarat and Rajasthan, as compared with the wider study area. The diversity in NDVI indicates spatial variations in vegetation health within the area, due mainly to uneven distribution of monsoonal rainfall, which is significantly less in the western part than the eastern part of the region.

Also, the NDVI images of both the years show considerable temporal variation within the study area. The NDVI of the western part of Gujarat and Rajasthan was relatively lower, which can be explained by the presence of great Indian Thar Desert and the semi-arid climatic region surrounding it. However, the NDVI during the year 2002 was found to be far lower than 2007 in most of the areas, indicating vegetation stress conditions during the peak monsoon season.

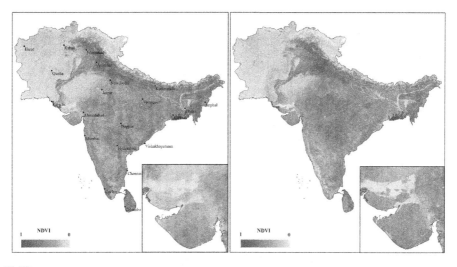

FIGURE 10.19

Normalized difference vegetation index (NDVI) map of drought year (2002) and normal year (2007) for South Asia with a closer view of Gujarat state in India.

Monitoring vegetation health using VCI

In order to quantify drought from long-term observations from space, the NDVI-derived drought index, VCI, was used. Fig. 10.20 illustrates vegetation conditions for August fortnights of summer crops for the years 2012 and 2015. It was found that severe drought conditions prevailed during the summer crop season of 2012 over a large area of South Asia. The onset and extent of drought can be clearly observed from the VCI maps of consecutive fortnights of 2012 for Sri Lanka (Fig. 10.21). Acute water stress is evident all over the region during the first fortnight of August 2012. From the first fortnight of September 2002, the conditions had improved in the Southern and Northeastern provinces. Several states in India also experienced severe droughts during this year, and it was the fifth consecutive drought in Rajasthan, Gujarat, and Maharashtra states.

Drought monitoring through SPI

The SPI is a very popular meteorological drought index which has been frequently used by decision-makers for measuring and monitoring the intensity of meteorological drought events. SPI is also useful for identifying the spatiotemporal extent of long-term historical droughts. In this study, SPI was used to identify the incidence of meteorological drought, its intensity, and spatiotemporal extent, allowing us to compare the results with the VCI agricultural drought index. The spatiotemporal pattern of SPI shows that prolonged dry conditions prevailed during the monsoon season of 2002 (Fig. 10.22). SPI values ranged from −1.00 to −2.00 during the months of July and August, indicating the presence of a moderately to extremely dry condition over the selected area.

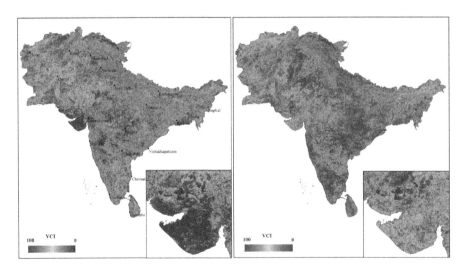

FIGURE 10.20

Spatial pattern of vegetation condition index (VCI) for South Asia and closer view of Gujarat state in India in a drought year (2012) and normal year (2015).

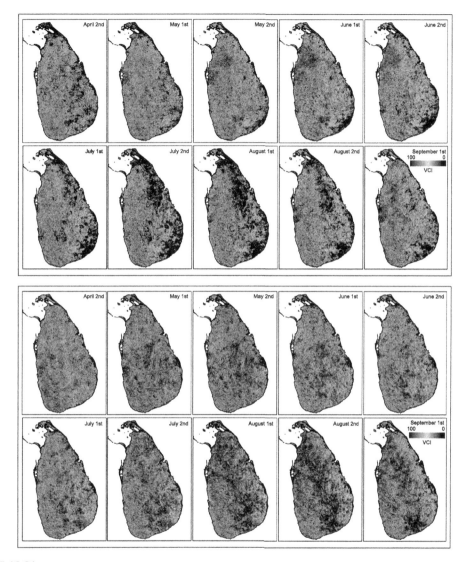

FIGURE 10.21

Spatiotemporal (fortnightly) pattern of vegetation condition index (VCI) for Sri Lanka. Top = drought year (2014) and bottom = normal year (2015).

Integrated drought severity index

VCI, PCI, and TCI variables were integrated to derive our IDSI. Fig. 10.23 shows the spatial distribution of monthly drought progression for the drought year 2012. IDSI classes were evaluated for two consecutive years 2014 and 2015 in India and Sri Lanka. Compared to stand-alone drought indices (i.e., VCI, PCI, SPI), the IDSI performed well in capturing historical drought events.

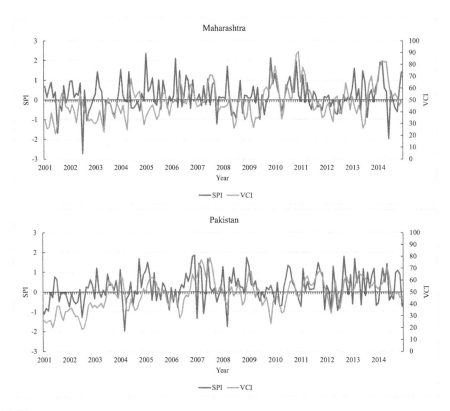

FIGURE 10.22

Comparison of vegetation condition index (VCI) and standardized precipitation index (SPI) between 2001 to 2015 for Maharashtra (India) and Punjab Province (Pakistan).

Comparison of IDSI and SPI drought index

In order to evaluate the performance of IDSI, spatial and temporal assessments were carried out for VCI, SPI, and anomalies of rainfall and crop yield for countries in South Asia in drought years and normal years. Fig. 10.24 compares the performance of IDSI and VCI with crop yield anomalies. To compare the VCI with yield-based drought index, a yield anomaly index (YAI) for major crops was calculated using long-term yield data from 1990 to 2014 as a baseline. The YAI for maize was compared for years 2009 and 2010 (Fig. 10.25); we found that most districts in India had a positive YAI in 2010, but a negative YAI in 2009.

VCI, which is obtained through remote sensing, indicates an agricultural drought occurred in 2009, which is consistent with the YAI outcome. Failure of the monsoon in the state of Rajasthan, India, that year caused shortages of food, drinking water, ground water, and fodder and adversely affected employment opportunities. India as a whole experienced a 33% deficit of rainfall for the year, with July being especially dry.

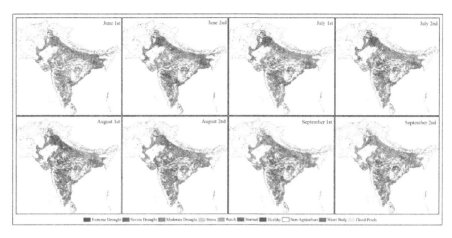

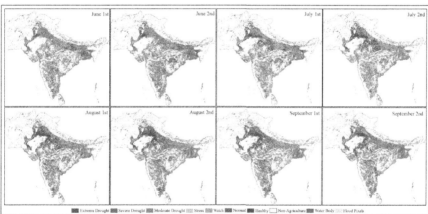

FIGURE 10.23

Spatiotemporal pattern of integrated drought severity index for South Asia during the normal year (2007) and drought year (2002).

SADMS system development
SADMS portal capabilities

The end result of our project, the South Asia Drought Monitoring System (SADMS), integrates open source satellite imagery at finer scales (500 m) to create vegetation and precipitation indices to assess the severity and spatial extent of drought across South Asia on 8-day and monthly bases for the period 2001 to 2015. As shown in Fig. 10.26, in particular, the map product displays three drought indicators for monitoring SPI, 8-day drought extent, and monthly drought extent based on IDSI. This regional scale map product provided in web-based interactive format has been designed to meet the needs of

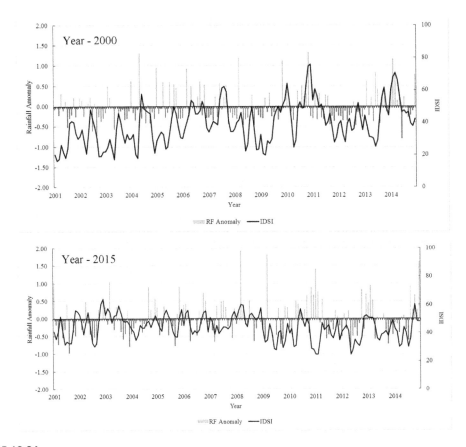

FIGURE 10.24

Comparison of integrated drought severity index (IDSI) and rainfall anomaly for India and Sri Lanka between 2000 and 2015.

a wide range of users in South Asia for data sharing, analysis, communication, and content delivery on drought conditions.

The DMS offers essential information to track drought conditions, derive historical trends for long-term planning, design preemptive measures and mitigation strategies. From the dense spatiotemporal datasets available in SADMS, user-specified time series graphs of drought characteristics and deviation from the long-term mean can be obtained for any administrative boundary up to district level and be analyzed at the pixel level. Once the unit is selected in DMS, a default time series graph appears showing the current and long-term average VCI, TCI, PCI. Drought indices "IDSI" or "SPI" for South Asia can be downloaded in image format for inclusion in various reports or presentations.

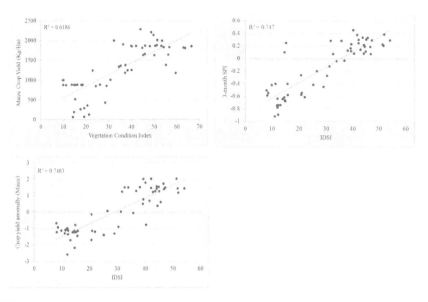

FIGURE 10.25

Comparison of integrated drought severity index (IDSI) with vegetation condition index (VCI), 3-month SPI for the state of Rajasthan, India, during the drought year (2009) and normal year (2010).

The advantage of SADMS is in its ability to represent drought conditions at the administrative scale most relevant for policy- and decision-makers, enabling timely and appropriate mitigation and emergency response measures. It is envisaged that the primary users of the DMS product will be national disaster management authorities in Bangladesh, India, Nepal, Pakistan, and Sri Lanka; the SAARC Disaster Management Centre; provincial and state governments; multilateral agencies; private insurance companies; and NGOs. The project team is sharing the toolbox used to create SADMS with national, state, and local organizations involved in drought monitoring, which will enable them to facilitate constant updating of SADMS.

Conclusion

This chapter described the development and successful implementation of an IDSI to monitor the onset, duration, extent, and severity of drought, which resulted in the 2014 launch of the SADMS online platform. Reducing current vulnerability to climate risks such as drought is necessary for adapting to climate change in the future. SADMS is an important tool in this regard. Vulnerable farmers experience climate change primarily through increases in the frequency and severity of extreme events. These climate shocks such as drought, flooding, or heat waves erode smallholder farmers' livelihoods through loss of assets, impaired health, and destroyed infrastructure. The uncertainty imposed by climate variability is an impediment to the adoption of improved agricultural technology and market opportunities and to the community-wide transformations required to adapt to climate change.

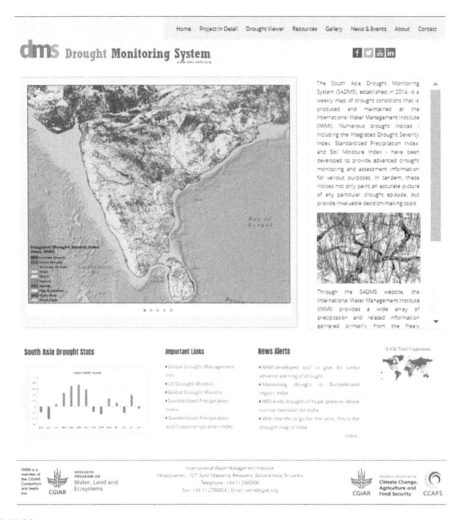

FIGURE 10.26

The GUI of drought monitoring system (SADMS).

Apart from effective intervention, projected increases in the frequency and severity of extreme climate events (e.g., precipitation extremes, drought, heat waves, coastal inundation) are expected to undermine development gains and intensify the cycle of rural poverty and vulnerability.

In risk-prone environments, efforts to foster climate-smart and resilient agricultural systems must be underpinned by effective climate information services and climate-informed safety nets. As agriculture becomes increasingly information-dependent, traditional knowledge struggles to keep pace, and greater climatic extremes challenge the capacity and experience of small-scale farmers. SADMS addresses critical gaps in knowledge, methodology, evidence, and capacity by improving the use of climate-related information to manage climate-related risk. In doing so, we foster the development of an enabling environment for smallholder farmers to adopt climate-smart production practices and develop climate-resilient livelihoods, all while protecting them from damaging climatic extremes.

Only two of eight South Asian countries (India and Pakistan) possess drought monitoring systems along with necessary experience and technical capability for constant updating and maintenance. Even in these two countries there exists a vast difference in capabilities across regions and governments. SADMS improves drought monitoring capabilities across South Asia. Ongoing maintenance of the SADMS is required and involves routine downloads of MODIS land surface reflectance data from USGS data gateway every 8 or 16 days, and subsequent processing in ArcGIS environment with the aid of ADAMS toolbox for deriving multitude of indicators representing drought. The resultant drought maps have to be constantly uploaded to the database from which latest drought information can be accessed via SADMS web interface. Compatibility of SADMS with other regional systems under development, such as Sri Lankan Drought Monitoring initiative by UNESCAP, will be explored for long-term SADMS sustainability. Further avenues will be explored to integrate information from global systems such as FAO's Global Information and Early Warning System, NASA's near real-time drought map, University of California's Global Integrated Drought Monitoring and Prediction System (GIDMaPS), and SPEI Global Drought Monitor.

Acknowledgments

The authors would like to thank the CGIAR Research Program on Water, Land and Ecosystems (WLE) which is carried out with support from the CGIAR Trust Fund and through bilateral funding agreements. For details, please visit https://wle.cgiar.org/donors. We also thank the funding support from the World Meteorological Organization (WMO), Global Water Partnership (GWP) through the Integrated Drought Management Programme. IWMI would like to thank Robert Stefanski, World Meteorological Organization; Frederik Pischke, Global Water Partnership; Avinash. C. Tyagi, International Commission on Irrigation and Drainage (ICID); and Ravinder Kaur, Indian Agriculture Research Institute (IARI) and Country Water Partnership of India and Sri Lanka for their valuable inputs and cooperation. We would like to thank various agencies including Indian Meteorological Agency (IMD), Departments of Meteorology, Irrigation Department, Agrarian in Sri Lanka, Disaster Management Center (DMC), of Sri Lanka, and Department of Disaster Management, Maharashtra (India), for data sharing and their inspiring guidance and encouragement.

References

Atkinson, P.M., Jeganathan, C., Dash, J., Atzberger, C., 2012. Inter- comparison of four models for smoothing satellite sensor time-series data to estimate vegetation phenology. Remote Sensing of Environment 123, 400−417.

Atzberger, C., Eilers, P.H.C., 2010. A smoothed 1-km resolution NDVI time series (1998−2008) for vegetation studies in South America. International Journal of Digital Earth. https://doi.org/10.1080/17538947.2010.505664.

Beck, P.S.A., Atzberger, C., Høgda, K.A., Johansen, B., Skidmore, A.K., 2006. Improved monitoring of vegetation dynamics at very high latitudes: a new method using MODIS NDVI. Remote Sensing of Environment 100 (3), 321−334. https://doi.org/10.1016/j.rse.2005.10.021.

Chen, J., Jönsson, P., Tamura, M., Gu, Z., Matsushita, B., Eklundh, L., 2004. A simple method for reconstructing a high-quality NDVI time-series data set based on the Savitzky–Golay filter. Remote Sensing of Environment 91 (3–4), 332–344. https://doi.org/10.1016/j.rse.2004.03.014.

Chen, I.C., Hill, J.K., Ohlemüller, R., Roy, D.B., Thomas, C.D., 2011. Rapid range shifts of species associated with high levels of climate warming. Science 333, 1024–1026.

Dasarathy, B.V., 1997. Sensor fusion potential exploitation-innovative architectures and illustrative applications. Proceedings of the IEEE 85 (1), 24–38.

Dash, J., Jeganathan, C., Atkinson, P.M., 2010. The use of MERIS Terrestrial Chlorophyll Index to study spatio-temporal variation in vegetation phenology over India. Remote Sensing of Environment 114 (7), 1388–1402. https://doi.org/10.1016/j.rse.2010.01.021.

FAO, 2011. The State of the World's Land and Water Resources for Food and Agriculture (SOLAW) – Managing Systems at Risk. Food and Agriculture Organization of the United Nations, Rome and Earthscan, London.

Hansen, M.C., Potapov, P.V., Moore, R., Turubanova, M.S.A., Tyukavina, A., Thau, D., Stehman, S.V., Goetz, S.J., Loveland, T.R., Kommareddy, A., Egorov, A., Chini, L., Justice, C.O., Townshend, J.R.G., 2013. High-resolution global maps of 21st-century forest cover change. Science 342, 850.

Jakubauskas, M.E., Legates, D.R., Kastens, J.H., 2001. Harmonic analysis of time-series AVHRR NDVI data. Photogrammetric Engineering and Remote Sensing 67 (4), 461–470.

Jeganathan, C., Dash, J., Atkinson, P.M., 2010. Mapping the phenology of natural vegetation in India using a remote sensing derived chlorophyll index. International Journal of Remote Sensing 31 (22), 5777–5796. https://doi.org/10.1080/01431161.2010.512303.

Jonsson, P., Eklundh, L., 2002. Seasonality extraction by function fitting to time-series of satellite sensor data. IEEE Transactions on Geoscience and Remote Sensing 40, 1824–1832. https://doi.org/10.1109/TGRS.2002.802519.

Jonsson, P., Eklundh, L., 2004. TIMESAT – a program for analysing time series of satellite sensor data. Computational Geosciences 30 (8), 833–845.

Justice, C.O., Townshend, J.R.G., Vermote, E.F., Masuoka, E., Wolfe, R.E., Saleous, N., Roy, D.P., Morisette, J.T., 2002. An overview of MODIS land data processing and product status. Remote Sensing of Environment 83 (1–2), 3–15. https://doi.org/10.1016/S0034-4257(02)00084-6.

Karnieli, A., Bayasgalan, M., Bayarjargal, Y., Agam, N., Khudulmur, S., Tucker, C.J., 2006. Comments on the use of the vegetation health index over Mongolia. International Journal of Remote Sensing 27 (10).

Kogan, F.N., 1990. Remote sensing of weather impacts on vegetation in non-homogeneous areas. International Journal of Remote Sensing 11 (8), 1405–1419.

Kogan, F.N., 1995a. Droughts of the late 1980s in the United States as derived from NOAA polar-orbiting satellite data. Bulletin of the American Meteorological Society 76 (5), 655–668.

Kogan, F.N., 1995b. Application of vegetation index and brightness temperature for drought detection. Advances in Space Research 15 (11), 91–100. https://doi.org/10.1016/0273-1177(95)00079-T.

Kogan, F.N., 1997. Global drought watch from space. Bulletin of the American Meteorological Society 78 (4), 621–636.

Kogan, F.N., Sullivan, J., 1993. Development of global drought-watch system using NOAA/AVHRR data. Advances in Space Research 13, 219–222. https://doi.org/10.1016/0273-1177(93)90548-P.

Kogan, F., Stark, R., Gitelson, A., Jargalsaikhan, L., Dugrajav, C., Tsooj, S., 2004. Derivation of pasture biomass in Mongolia from AVHRR-based vegetation health indices. International Journal of Remote Sensing 25 (14), 2889–2896. https://doi.org/10.1080/01431160410001697619.

Kogan, F.N., Yang, B., Wei, G., Pei, Z., Jiao, X., 2005. Modelling corn production in China using AVHRR-based vegetation health indices. International Journal of Remote Sensing 26, 2325–2336. https://doi.org/10.1080/01431160500034235.

Liu, W.T., Kogan, F.N., 1996. Monitoring regional drought using vegetation condition index. International Journal of Remote Sensing 17, 2761—2782.

Lovell, J.L., Graetz, R.D., 2001. Filtering pathfinder AVHRR land NDVI data for Australia. International Journal of Remote Sensing 22 (13), 2649—2654. https://doi.org/10.1080/01431160116874.

Mondal, S., Jeganathan, C., Amarnath, G., Pani, P., 2017. Time-series cloud noise mapping and reduction algorithm for improved vegetation and drought monitoring. GIScience and Remote Sensing 54 (2), 202—229.

Salazar, L., Kogan, F., Roytman, L., 2007. Use of remote sensing data for estimation of winter wheat yield in the United States. International Journal of Remote Sensing 28 (17), 3795—3811.

Salazar, L., Kogan, F., Roytman, L., 2008. Using vegetation health indices and partial least squares method for estimation of corn yield. International Journal of Remote Sensing 29 (1), 175—189.

Singh, R.P., Roy, S., Kogan, F.N., 2003. Vegetation and temperature condition indices from NOAA-AVHRRA data for drought monitoring over India. International Journal of Remote Sensing 24 (22), 4393—4402.

Swets, D.L., Reed, B.C., Rowland, J.D., Marko, S.E., 1999. A weighted least-squares approach to temporal NDVI smoothing. In: Paper presented at the Proceedings of the 1999 ASPRS Annual Conference: From Image to Information, Portland, Oregon.

Unganai, L.S., Kogan, F.N., 1998. Drought monitoring and corn yield estimation in Southern Africa from AVHRR data. Remote Sensing of Environment 63, 219—232.

Velleman, P.F., 1980. Definition and comparison of robust nonlinear data smoothing algorithms. Journal of the American Statistical Association 75 (371), 609—615. https://doi.org/10.2307/2287657.

Vicente-Serrano, S.M., 2007. Evaluating the impact of drought using remote sensing in a Mediterranean, semi-arid region. Natural Hazards 40 (1), 173—208.

Early warning systems for drought and violent conflict—toward potential cross-pollination

11

Lars Wirkus[a], Jinelle Piereder[b]

[a]*Data and Geomatics, Bonn International Center for Conversion (BICC), Bonn, Germany,*
[b]*Global Governance at the Balsillie School of International Affairs (BSIA), University of Waterloo, Canada*

Introduction

Among all natural hazards, drought is one of the most severe due to its long-lasting negative impacts such as loss of life and livelihoods, economic losses, and adverse effects on social and ecological systems. Droughts kill proportionately more people than other disasters and are particularly deadly in Africa, with over 800,000 deaths directly attributable to drought between 1970 and 2010 (FAO et al., 2018; Sanghi et al., 2010). In addition to its severity, drought is a "creeping" or slow-onset disaster and usually affects larger land areas than other types of disasters, making mitigation and adaptation strategies difficult to implement. Many of the negative effects of drought often accumulate slowly and may persist for years after the event has ended (Wilhite and Svoboda 2000; Wilhite et al. 2000).

Drought is partly responsible for the fact that "more than 124 million people in 51 countries are threatened by food insecurity" (Food Security Information Network 2018). Furthermore, over half of these people are in situations affected by conflict, adding more dimensions to their present and ongoing insecurity. Droughts are particularly devastating for those whose livelihoods depend on rain-fed agriculture and livestock farming, especially people living in arid areas of developing countries. In these arid and mostly rural areas, vulnerable people are also exposed to an increased risk of food insecurity and famine.

As part of disaster prevention, risk management, and mitigation strategies, drought monitoring and early warning systems (EWSs) have become an important tool for policy makers, humanitarian organizations, and the entire international community. Drawing on new and existing research on climate and meteorological processes, as well as knowledge from various scales of decision networks (international, national, subnational, tribal, and local), a drought early warning system (DEWS) serves to improve the capacity of relevant stakeholders to assess, monitor, forecast, plan for, cope with, and respond to the risk and impacts of drought.

In parallel, conflict EWS has helped to inform local, regional, and international conflict response since the 1970s and 1980s. Following the failures of the international community to respond to the Rwandan genocide and the conflicts in the Balkans, the interest in early warning as a prevention tool gained new energy in both governmental and nongovernmental sectors (Nyheim, 2009). By closely monitoring various human security indicators and potential conflict triggers, then rapidly

Drought Challenges. https://doi.org/10.1016/B978-0-12-814820-4.00011-0

165

communicating these data to decision-makers, the hope among early warning advocates was that the international community could prevent another violent catastrophe.

Despite decades of advancement on both drought and conflict EWS, these systems remain largely siloed. Most DEWS information stays within the scientific community and disaster reduction spheres; most conflict EWS information remains in the foreign policy, peace, and security, as well as the humanitarian crisis spheres. This seeming isolation has often led to responses that fail to take account of significant inhibiting conditions including conflict-related recommendations that miss important cross-border and socioecological dimensions of drought and its impacts, on the one hand, and drought-related mitigation and adaptation recommendations that ignore the historical and potential conflict fault lines in the drought-affected region, on the other. Without acknowledging the coupled nature of these socioecological factors, both kinds of EWS are missing key insights that are important for providing the most relevant advice to governments, civil society, and the international community.

As the majority of drought-prone and food insecure regions are within or span across postconflict countries and so-called Fragile and Conflict-Affected States (FCAS) (Fig. 11.1), the challenges and impacts of fragility and violent conflicts must be more seriously taken into account in monitoring and addressing drought risk and vulnerability. This is more so because such states are often characterized by weak or missing governance structures. As such, they are often unable to control their entire territory and thus are confronted with the challenge of effectively implementing suitable and sustainable drought adaptation and emergency measures (Grävingholt et al., 2018; United Nations and World Bank, 2018; Crawford, 2015). At the same time, droughts as potentially conflict-triggering events must be better integrated into existing regional and national conflict EWS.

Against the foregoing background, we argue for further inclusion and integration of conflict EWS in drought and famine EWS. In other words, conflict assessments should inform disaster response to drought, but more importantly, entire drought risk management strategies. At the minimum, recommendations made to policy makers regarding drought preparedness and response should include and reflect the conflict history for the region in question.

This chapter proceeds with a foundational discussion of the relationship between drought vulnerability and climate change, as well as the climate—conflict link. We then turn to a brief history of both drought and conflict EWS while providing key examples of each that are specific to or partly focused on Africa. Afterward, this chapter presents a series of new and emerging conflict databases and EWS—including ones run by nongovernmental organizations (NGOs)—that could help inform various drought risk management and reduction efforts. These initiatives were chosen especially because of their part or full open-source nature, as well as their use of leading-edge data approaches and early warning technologies. Finally, we point to insights in the extensive early warning literature to highlight trends and opportunities for drought/conflict EWS cross-pollination, before concluding.

Climate change, drought vulnerability, and the climate—conflict link

Although the causes of droughts are essentially natural, anthropogenic climate change and environmentally degrading activities are increasingly triggering processes that exacerbate the severity, frequency, duration, and spatial extent of droughts. Many regions are significantly more vulnerable to drought now than before due to a host of factors including deforestation, overgrazing, mining, degradation of land and vegetation. Other activities related to pollution, overexploitation of water resources,

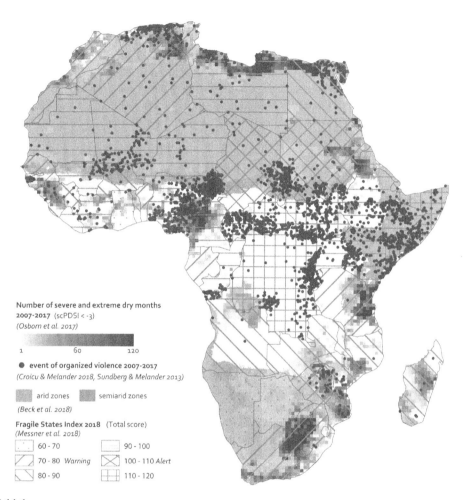

Number of severe and extreme dry months
2007-2017 (scPDSI < -3)
(Osborn et al. 2017)

1   60   120

● event of organized violence 2007-2017
(Croicu & Melander 2018, Sundberg & Melander 2013)

arid zones   semiarid zones

(Beck et al. 2018)

Fragile States Index 2018 (Total score)
(Messner et al. 2018)

60 - 70   90 - 100
70 - 80 *Warning*   100 - 110 *Alert*
80 - 90   110 - 120

FIGURE 11.1

Droughts, violent conflicts, and fragility in Africa 2007–17.

Source: Own illustration by Wirkus & Blitza 2019.

and the increasing drying of soils accentuate the risks and effects of drought on populations. These and other factors are contributing to a decline of resilience in both landscapes and people. Many scholars argue that the impacts of adaptation and mitigation failures (Dabelko et al., 2013), including those related to land use (Magnan et al., 2016), are likely to result in serious social and environmental insta-bility (Scheffran et al., 2012; Barnett, 2007; Barnett and Adger, 2007; Smit and Wandel, 2006), includ-ing mass migration and resource-related violent conflicts (Dalby, 2013; Bernauer et al., 2010; German Advisory Council on Climate Change, 2007; Wisner et al., 2003; Bohle et al., 1994).

Today, the likely consequences of climate change for society are largely well known. According to the IPCC, areas that alternate between rainy and dry seasons will experience more extreme seasonal contrasts with more floods and droughts (IPCC, 2013). More short-term droughts are expected to be seen in dry regions (IPCC, 2013, 2014). The combined effects of changes in temperature and precipitation patterns will severely affect dry subtropical regions. The Sahel and East Africa regions, where livelihoods are mainly dependent on rain-fed agriculture and pastoralism, would be particularly highly vulnerable to drought-induced food shortages. Global warming has already significantly reduced the Sahel's crop production period, thereby constraining its agricultural potential (IPCC, 2007). Production reduction especially affects households that rely heavily on subsistence farming (Smidt and Theisen, 2018).

Importantly, drought impact on agriculture depends not only on low precipitation levels and/or soil moisture deficit but also on the "exposure and the vulnerability of the agricultural system and the people depending on it" (Walz et al., 2018, p. 5). Food production is a particularly central factor in understanding how climate change might relate to violent conflict (Smidt and Theisen, 2018), in addition to the robustness and adaptive capacities of social orders and their (in)formal institutions. But in many FCAS, agriculture-related exposure, and vulnerability are higher than in other regions, which lead to more severe impacts for people in these states than in less conflict-prone areas. For too long, the socioeconomic and political contexts in which climate extremes—such as droughts—undermine social stability and increase the risk of conflict have not been properly understood (Uexkull et al., 2016; Buhaug 2015; Linke et al., 2015; Bernauer et al., 2012).

There continues to be much academic debate—ongoing since the 1990s—regarding the causal relationships between climate change, drought, and conflict (Gleditsch, 2012; Schnurr and Swatuk 2012; Matthew et al., 2009). Some authors argue that there is a direct causal link between warming, low precipitation levels, and the occurrence of conflict. Especially, in societies that heavily depend on agriculture, drought often leads to economic uncertainties around yield declines that tend to exacerbate existing tensions which may trigger conflict (Burke et al., 2009). Other scholars, however, insist that while political opportunists and warlords may exploit drought and other disasters in their strategies, the primary causes of conflict remain political and economic (Nordas and Gleditsch, 2007; Popovski, 2017). Nonetheless, Uexkull et al. (2016) show that for people who live in developing countries, who are often marginalized or dependent on agriculture, "…the occurrence and duration of droughts, besides central drivers of violent conflicts such as ethnopolitical exclusion, temporal and spatial proximity to violence, and various country-specific risk factors, increase the likelihood of sustained conflict involvement" (Uexkull et al., 2016, p. 12394). Therefore, much depends on the local context, culture, and initial conditions.

Schleussner et al. (2016, p. 9201) conclude that the disagreement in this field could be partly attributed to the fact that "most studies investigating the relation between climate change and armed conflicts focus solely on meteorological indices such as temperature or precipitation time series, thereby neglecting the crucial importance of vulnerability and exposure for the impacts of climate hazards." Yet, we know that local vulnerabilities and coping capacities condition the effects of drought. Hence, long-term exposure to drought undermines people's alternative coping mechanisms (Uexkull, 2014).

But regardless of the existence of a direct causal link between drought and conflict, levels of conflict affect the capacity to prepare for and effectively respond to drought (Sneyers, 2017; Uexkull, 2014). What can be said for certain is that the human-felt consequences of climate change, drought, and conflict are similar: dislocation and migration, periods of acute and chronic humanitarian need, loss of arable land to support livelihoods, breakdown of local institutions, and coping practices among others. All of these tend to increase people's vulnerability and exposure.

Considering fragile and conflict-affected states

FCAS, especially, are caught in a vicious cycle when it comes to drought. First, drought acts as a threat multiplier in FCAS that are already threatened by, for example, political tensions, group rivalries, and economic turbulence. The loss of agricultural wages, subsidies, or other knock-on economic impacts of drought often contribute to widespread dissatisfaction with the government, as well as deepen existing cleavages between social groups. In the face of these struggles, the risk of violent conflict increases. In times of conflict, the presence of cropland increases the frequency of violence against civilians (Koren and Bagozzi, 2017, p. 1). Combatants, be they government or rebel actors, often resort to local agricultural resources and products and thus use violence in favor of immediate food procurement (Koren and Bagozzi, 2017).

Second, fragile settings make it difficult or even impossible for governmental or aid workers to (re) act appropriately and swiftly to food insecurity resulting from drought vulnerability. Some regions may be under the control of nonstate armed groups, or the state may simply not have a far enough reach to access them. In some cases, access to food aid may even be weaponized by a state government or by an antigovernment faction, thereby drawing hungry citizens into a political chess game.

Third, as people try to adapt, in the face of drought, existing social orders are disrupted and provide increasing opportunities for conflict. For example, migration is often an adaptation strategy against both drought-induced food insecurity and violent conflict. Yet, new resource pressures from migrating populations can also aggravate food insecurity in the receiving countries/regions. Tensions due to ethnic and demographic changes in these regions may also lead to violent conflict.

A brief history of early warning
Drought monitoring and early warning systems

EWSs have a dual heritage. The first can be traced back to the 1950s and was centrally about predicting natural disasters and their effects. EWSs were developed to alert local governments, agencies, and citizens of impending hazards like floods, earthquakes (though these are notoriously hard to predict), tsunamis, tornadoes, etc. Today, drought EWSs are "the foundation of effective proactive drought policies" (WMO and GWP, 2019). The purpose of these systems is to continually monitor meteorological and geological variables, and when a certain threshold of hazard is predicted or detected, they rapidly diffuse this information to emergency services and the public so that they can evacuate or engage in mitigation measures (WMO and GWP, 2019). Frequently, DEWSs are connected to intergovernmental water management and agricultural agencies, which increasingly use satellite imagery-derived vegetation indexes to strengthen national drought EWSs through the incorporation of space-based information.

Disaster-related EWSs have matured significantly since their early iterations, benefiting from rapid technological advancements in computational power and increased provision of data obtained by remote sensing, as well as from international information sharing and global cooperation. As a result of the UN Sendai Framework for Disaster Risk Reduction (SFDRR), adopted by the United Nations member states in 2015, countries have started to change their policy approach from simply drought response toward drought risk assessments. The SFDRR, which stressed the necessity to shift both thinking and application from reactive hazard-centered disaster management to proactive disaster risk management and risk reduction approaches, led to the inclusion of vulnerability assessments (Walz at al. 2018). In many cases the vulnerability to drought is assessed by sets of social-ecological vulnerability indicators on subnational level such as unemployment, social dependency, income, soil quality and surface water supply, in and/or out migration. However, conflict-related information is not typically included in DEWS.

Generally, there are many types of drought early warning platforms and products, which are often collaborations between government agencies and research institutes or private companies. While not specifically a drought EWS, the *Famine Early Warning Systems Network (FEWS NET)* is one of the leading providers of research and analysis related to food insecurity and includes significant climate-related data in its assessments. It provides objective, evidence-based analysis to help government decision-makers and relief agencies plan for and respond to humanitarian crises. FEWS was established in 1985 by USAID, but today consists of five US government agencies (National Aeronautics and Space Administration (NASA), National Oceanic and Atmospheric Administration (NOAA), US Department of Agriculture (USDA), and US Geological Survey (USGS)) and two private contractors, who manage FEWS NET's field offices and provide tools for knowledge and data management.

Another key example is the FAO's *Early Warning Early Action (EWEA)*[1] System, which translates natural disaster warnings into anticipatory actions to reduce the impact of specific disaster events, such as droughts. Through consolidating forecasts, the FAO initiates responses before a disaster has actually happened or reached its peak. The EWEA approach consists of three components: (1) early warning, (2) early action, and (3) a financing mechanism (FAO, 2019). On global level, the EWEA team uses both FAO and external early warning sources to monitor main threats to agriculture and food security. At country level, the team works closely with country offices to develop EWEA systems tailored to the local context. In addition to a dedicated Early Action Fund, these systems enable the FAO to monitor and mitigate major risks to the agriculture sector and the livelihoods it supports.

Conflict early warning systems

The second heritage of early warning is rooted in Cold War Era nuclear attack warning-and-response systems. At the height of the War, the threat of a surprise nuclear attack was felt almost constantly by the core states. While these types of systems still exist, the end of the Cold War meant that the security-related interest in early warning shifted in the 1990s from mutually assured destruction to humanitarian crisis. In the brutal wake of civil war in Somalia, genocide in Rwanda, and ethnic cleansing in the Balkans, the international community sought to develop new strategies for conflict prevention, leading to the first generation of conflict EWSs.

At a general level, conflict early warning includes collection of data—which includes monitoring human security indicators—risk analysis, and the sharing of information and recommendations with decision-makers. EWS can operate from (inter-)governmental, and/or nongovernmental sectors, and can focus on correlational, sequential, conjunctural, and/or response models or approaches (Wondema-gegnehu, 2009). As Nyheim (2009) explains, early EWS focused on prediction and analysis and were most often based outside the potential conflict region. Second-generation EWS shifted closer to the regions being observed and relied much more on local field monitors. But in recognizing the need

---

[1]See http://www.fao.org/emergencies/fao-in-action/ewea/en/.

Table 11.1 Generations of early warning systems by David Nyheim.

	Time period	Structure	Focus
First generation	1995–99 (and continued today)	Centralized	Prediction, providing analysis to inform decision-making
Second generation	1999–2003 (and continued today)	Closer to regions they monitor/observe	Prediction and analysis, also response proposals
Third generation	2003–Present	Localized	Using information as a response; preventing violence in specific localities

for better connections between warning and response, third-generation EWS became even more local-ized and integrated into national and regional peace and security strategies. The warning/response gap is, however, still a key challenge that is often linked less to data issues and more to (inter)organiza-tional politics (Rohwerder, 2015) (Table 11.1).

Some scholars have pointed to the emergence of a fourth generation of EWS, following new efforts to crowdsource conflict-related information via mobile devices, social media, and other Internet-based platforms (Bock, 2015; Mancini, 2013). Some systems include trend analyses or other pattern recog-nition. The *Early Warning Project*, discussed below, is an example of a system with a crowdsourced component. This "Internet-democratized" version of EWS has many potential benefits, especially for civil society actors and others that are outside a state's peace and security apparatus. But it also has many implications yet to be explored, not least of which is the ethical use of people's data and the potential for misuse toward inciting violence rather than preventing it (Bock, 2015). There are also concerns around data reliability, disparities in access to new technology potentially leading to data biases, and privacy rights, especially under repressive regimes (Pham and Vinck, 2012). All the same, many scholars and practitioners remain hopeful about the emergence and use of these new tech-nologies in early warning-and-response systems.

Turning to specific examples at the level of regional organizations in Africa, key EWSs (all of which would be considered third generation) are the Conflict Early Warning and Response Mechanism (CEWARN)[2] under Intergovernmental Authority on Development (IGAD), the African Union's (AU) Continental Early Warning System (CEWS),[3] the Economic Community of West African States' (ECOWAS) Early Warning and Response Network (ECOWARN), and the Economic Community of Central African States's (ECCAS) Mecanisme d'Alerte Rapide pour l'Afrique Centrale (MARAC).[4] As one of the five pillars of the AU's Peace and Security Architecture, CEWS has received significant attention from scholars of EWSs. It has undergone dramatic improvements and broad harmonization in the last decade. The system's broad mandate encompasses collecting and analyzing data that are

[2]See http://www.igadregion.org/cewarn/.
[3]See http://www.peaceau.org/en/page/28-continental-early-warning.
[4]See http://www.operationspaix.net/DATA/DOCUMENT/3654 ~ v ~ Mecanisme_d_alerte_rapide_de_l_Afrique_Centrale_MARAC_.pdf.

relevant to conflict prevention, as well as providing advisory services to the Peace and Security Council on recommended responses (Wondemagegnehu, 2009).

Most of these regional systems also serve as hubs or umbrellas over the individual national EWS of each country in the region, with IGAD's CEWARN and country-level CEWARN Units—which focus mainly on pastoralist conflict and cattle raiding in the border zones of the Horn of Africa—being the most developed.[5] ECOWARN, on the other hand, has a much larger mandate and monitors "all aspects that affect peace and security" in the region, including issues related to human rights, food shortages, droughts and floods, unemployment, arms flows, and others (Wondemagegnehu, 2009). ECOWARN, especially, collaborates extensively with nongovernmental organizations' early warning efforts. The West Africa Network for Peacebuilding (WANEP) runs the Early Warning and Early Response Program (WARN),[6] which provides options for early warning response to policy makers and to WANEP's peacebuilding activities.

Increasingly, national governments and NGOs have filled the gaps with their systems (some of which are described above), but adequate information sharing, unclear institutional responsibilities, insufficient technical and scientific personnel, and inadequate financial resources for adequate monitoring infrastructure and forecasting technology remain a challenge (Zenko and Friedman, 2011).

New developments in conflict data and early warning systems

So far, we have argued that violent conflicts and fragility are decisive parameters that impact the coping and adaptation capacities of local populations, especially in drought-affected FCAS. Thus, conflict databases and EWSs are sources of important information for assessing the feasibility of drought risk reduction measures. The following section looks at several exciting new developments in this area.

Conflict databases as inventories of past and/or ongoing conflicts

Current geo-referenced conflict databases function as inventories of past and/or ongoing violent incidents. Historically, only state-based data on violent conflicts and wars were available. But the development of geocoded events data from the early 2000s onwards, by, for example, ACLED[7] and UCDP[8] on violent conflicts and by SCAD[9] on social conflicts in Africa enabled a geographically disaggregated conflict analysis (Croicu and Sundberg, 2017; Petterson and Eck, 2018; Sundberg and Melander, 2013; Salehyan et al., 2012; Raleigh et al., 2010; Gleditsch et al., 2002).

The *Armed Conflict Location & Event Data Project (ACLED)* (established in 2005) is a disaggregated conflict collection, analysis, and crisis mapping project. ACLED collects and makes freely available real-time and historical data on political violence and protest events in over 75 developing countries across Africa, South Asia, South East Asia, the Middle East, Europe, and Latin America. It

---

[5]For more comprehensive information related to these and other early warning systems, see the OECD report, "Preventing Violence, War and State Collapse" available at https://www.oecd-ilibrary.org/development/preventing-violence-war-and-state-collapse_9789264059818-en.

[6]See http://www.wanep.org/wanep/index.php?option=com_content&view=section&layout=blog&id=9&Itemid=93.

[7]ACLED—Armed Conflict Location and Event Data; https://www.acleddata.com.

[8]UCDP—Uppsala Conflict Data Program; https://www.ucdp.uu.se.

[9]SCAD—Social Conflict in Africa Database; https://www.strausscenter.org/scad.html.

codes the dates, actors, types of violence, locations, and fatalities of all reported political violence and protest events. According to their definition, political violence and protest includes events that occur within civil wars and periods of instability, public protest, and regime breakdown.

The ***Global Event Dataset***[10] (***UCDP-GED,*** version 18.1) is the most aggregated conflict dataset of the Uppsala Conflict Data Program (UCDP) (Croicu and Sundberg 2018; Sundberg and Melander, 2013). It contains individual events of organized violence from 1989 to 2017 as a base unit, covering villages or days at maximum spatial or temporal resolution. As the locations are fully geocoded, they enable further geographical analysis. In addition, it is possible to create links to other compatible UCDP datasets, such as the UCDP/PRIO Armed Conflict Dataset[11] (Petterson and Eck, 2018). UCDP-GED defines an *event* as "an incident where armed force was used by an organized actor against another organized actor, or against civilians, resulting in at least one direct death at a specific location and a specific date" (Croicu and Sundberg, 2018, p. 9). In terms of data sources, the (slight) majority of the events are based on global newswire reporting. In a second step, translations of local news by the BBC and secondary sources (NGO, IGO, field reports, or books) are included.

The ***Social Conflict Analysis Database (SCAD)*** is another conflict database that is worthy to note. It is a collection of data on protests, riots, strikes, intercommunal conflicts, government violence against civilians, and other forms of social conflicts that have been collected by researchers at the University of North Texas and the University of Denver (Salehyan et al., 2012). The area of coverage spans Africa, Latin America, and the Caribbean.

Since 1990, the dataset compiles events reported by the *Associated Press* (AP) and *Agence France Presse* (AFP). In comparison with other conflict databases, SCAD also includes nonfatal conflict events such as organized and spontaneous demonstrations and riots, limited and general strikes, and various forms of government violence. However, SCAD excludes armed conflicts and events that could be considered part of an armed conflict, as defined and monitored by the UCDP database. Besides the type of conflict, SCAD also provides information on start and end dates, actors, number of participants and deaths as well as location in coordinates.

Two of the aforementioned three databases (UCDP-GED and SCAD) only provide data about conflicts from the previous year, rather than "up-to-date overviews." This is mainly due to their rigorous data collection and validation methods, and the provision of accompanying variables to the conflict event (e.g., number of battle-related deaths, involved conflict parties, and their allies). ACLED, on the other hand, provides quasi real-time information as it is mainly based upon the analysis of the most recent news reports. According to Smidt and Theissen (2018, p. 42), "these two [ACLED and UCDP-GED] datasets have the advantage of giving more exact timing and location of violence, making them better suited for analysing the effects of shocks like floods and droughts."

Conflict early warning systems as tools for conflict prediction

Turning to EWSs, we must first differentiate between those developed, used, and maintained by (inter-) governmental organizations (cf. aforementioned examples of African conflict EWSs) and those developed and run by university and nonuniversity research institutions. Whereas the latter ones are generally publicly accessible, the former are typically closed (i.e., restricted to a small circle of selected state

---

[10]Available at: https://ucdp.uu.se/downloads/#d1.
[11]Available at: https://ucdp.uu.se/downloads/#d3.

actors). What these systems have in common is that they build upon conflict incident monitoring and data collection. They are also engaged in the assessment of thematic foci such as crime, environment, gender, health, governance, and security with each often operationalized by a set of indicators. These sources are used for situational analyses either for a region, country, or a subnational unit.

The conflict databases are particularly important for the development and operation of the nongovernmental conflict EWSs. This is because their historic conflict events data are normally the most important source for developing underlying scenario and/or forecasting models. Many of the state-run EWSs also use the aforementioned conflict databases, but all of them have their own detailed conflict information obtained by their intelligence services. As most recently stated by Hegre at al. (2019, p. 16), "[l]ocal governments have much better information about what is going on than any system based on open-source data can deliver."

The following section zeroes in on the methodological details of two public conflict EWS: the *Early Warning Project*[12] of the Simon-Skjodt Center for the Prevention of Genocide (based at the United States Holocaust Memorial Museum and the Dickey Center for International Understanding at Dartmouth College) and Uppsala University's "political Violence Early Warning System (ViEWS)"[13] project.

First, the publicly accessible *Early Warning Project* provides conflict early warnings on an annual basis to prevent mass atrocities by using a variety of publicly available data and forecasting methods. Based on a statistical risk assessment, this EWS provides an annually updated list of more than 160 countries ranked by their likelihood to experience a new episode of mass killing.

The statistical risk assessment is based on the identification of historical instances of state- and nonstate-led mass killings (1945−present for state-led, 1989−present for nonstate-led), countries where mass killing did and did not take place, and the application of a logistic regression model with "elastic-net" regularization to the latest publicly available country data. Using these data and model, a 2-year forecast is calculated which results in risk estimation of a new mass killing in each of more than 160 countries and a corresponding ranking. Based on the model, factors associated with greater risk of mass killing include history of mass killing, large population size, high infant mortality, ethnic fractionalization, high battle-related deaths, ban on opposition parties, existence of politically motivated killings, lack of freedom of movement, repression of civil society, coup attempts within the last 5 years, and anocratic regime type (i.e., neither full democracy nor full autocracy). Their assessment is designed to help governments, civil society groups, and other influential actors prioritize an effective response.

In addition, the *Early Warning Project* uses two types of crowd forecasting—an ongoing public opinion pool and an annual comparison survey, both of which are publicly available—to identify countries at risk for an onset of mass killing. The opinion pool is a structured process where, throughout the year, opinions (here: forecasts) about the likelihood of mass atrocities are collected and combined from an open group of people. In contrast, each December the comparison survey asks the respondents to choose one country out of presented country pairs that they believe is more likely to experience a new mass killing in the following calendar year.

The *Violence Early Warning System (ViEWS)* by the Department of Peace and Conflict Research of Uppsala University provides the capability of forecasting political violence 36 months into the

---

[12]https://earlywarningproject.ushmm.org.
[13]http://www.pcr.uu.se/research/views/.

future—spatially limited to the African continent. ViEWS accomplishes these predictions by combining different datasets, including UCDP-GED and ACLED for conflict-related data, and many others. These are complimented by different world development indicators of World Bank (World Bank, 2015), as well as data on excluded ethnic groups provided by the Ethnic Power Relations dataset of ETH Zurich or data of the Varieties of Democracy Project (V-Dem) by the Kellogg Institute For International Studies (Hegre et al. 2019; Coppedge et al., 2011; Cederman et al., 2010). The more than 30 datasets used to forecast political violence are grouped in four themes (conflict history, natural geography, social geography, and protests) and aggregated by different models (Hegre et al. 2018, p. 5f). For example, the social geography theme consists of *distance to neighboring country, travel time to nearest city, distance to capital city, population size, gross cell product, infant mortality rate*, and *number of excluded groups*. The conflict predictions are generated for two spatiotemporal variants and three conflict types. For the spatiotemporal variants, there is one on country level, and one on the subnational geographical locations, which is based on the PRIO-GRID (Tollefsen at al. 2012, version 2.0), a standardized spatial grid structure with a resolution of 0.5×0.5 decimal degrees. In terms of conflict types, the distinction is made between armed conflicts involving states and rebel groups *(state-based conflicts)*, armed conflicts between nonstate actors (*nonstate conflicts*) and violence against civilians (*one-sided violence against civilians*). To accomplish a high temporal resolution, the forecast is updated on a monthly basis, based on the provision of monthly preliminary UCDP-GED datasets (Hegre et al. 2018, p. 6). The forecasting process is automatized as a set of SQL and Python 3.x scripts. Methodologically, the project follows an innovative approach by combining a logistic regression model (generalized linear model (GLM)) with the machine learning—based random forest model.

Still in its infancy, the current "strength of ViEWS is not the ability to forecast entirely new conflicts" (Hegre et al., 2019, p. 16), but to predict future developments of already known and ongoing conflicts. ViEWS should be seen as "a step towards a future where high-resolution forecasts of conflict at practically useful spatial and temporal scales are publicly available" (Hegre et al., 2019, p. 17).

Toward cross-pollination and integration

So far, this chapter has (1) discussed the important connections between climate, conflict, and drought vulnerability, (2) considered the history and key Africa-related examples of both drought and conflict EWSs, as well as (3) presented a series of new open-source and leading-edge conflict databases and EWS. To bring these strands together, this final section points to several emerging trends in the early warning literature regarding EWS integration. Though there are several dimensions to this subliterature, we focus on new thinking around differential vulnerability and fragility, especially as it relates to the drought—conflict link. Because drought risks are somewhat socially constructed rather than completely naturally triggered, social structuring processes heavily impact drought adaptation and mitigation capacities (Brüntrup and Tsegai, 2017). New research on differential drought vulnerability—especially those that consider temporally and spatially informed approaches—attempts to unpack and create indicators related to some of these structuring processes.

Drought forecasting and policy is difficult due to the complexity, scalability, and strong connectivity of the environmental and societal parameters involved. It requires a high level of inter- and transdisciplinary expertise, sound knowledge of the local conditions, and, ultimately, a functioning state that can take action to adequately address drought risk and impact. However, many drought EWS and risk management strategies fail to sufficiently take account of the difficult cycle that FCAS deal with. We therefore agree with a policy paper commissioned for the G7, which argues that climate change adaptation/mitigation efforts—including those dealing with drought—require a "conflict-sensitive approach." Such an approach would integrate both climate and socioeconomic—political vulnerabilities (Ruttinger et al., 2015). This recommendation stems from research findings that climate change and its consequences will overburden weak states and potentially destabilize strong states with negative implications for regional stability (cf. Moran, 2011; Conca, 2001).

However, adaptation/mitigation initiatives may also have unanticipated and unintended negative social, political, economic, and ecological consequences for local communities. This is especially the case when the vulnerabilities of these communities or regions are not well understood. Concerning EWS, systems at the global and regional level tend to follow more top-down approaches and focus on high-level decision-makers. But this can mean less attention is given to the strength and coping capacities at the local level, including those related to current and historic dynamics of violence and fragility. In a boomerang effect, some of these local unintended consequences may then constrain or even threaten the state or regional organization implementing the interventions, leading to even less capacity to address climate change—related impacts (Swatuk et al., 2018).

Because drought vulnerability is complex, it is important to incorporate social, economic, and environmental determinants of both vulnerability and drought response (Naumann et al., 2014, p. 1592). Naumann et al. (2014) propose a drought vulnerability index that takes account of not just exposure but also susceptibility, coping capacity, and adaptive capacity. But despite having a composite measure of "human and civic resources" (which includes variables for institutional capacity, government effectiveness and a measure of the displaced population and refugees), the authors neglect any sort of actual conflict or conflict history indicator. Using their index, Naumann et al. (2014) classify Somalia, Mali, Ethiopia, Niger, Burundi, and Chad—nearly all of which are located in the arid or semi-arid region of the Sahel—as having higher drought vulnerability relative to other African countries. Most interestingly though, nearly all these identified countries have a score of over 90 (out of 120) on the Fragile States Index (see Fig. 11.1), further supporting the link between drought vulnerability and conflict or fragility.

In other approach, Comes et al. (2014) point out that because vulnerability is such a dynamic concept, temporal and geographical vulnerability assessments must be understood as key threads running through an overall risk management and governance cycle. Static or aggregate indicators are not enough. They further explain that because it is difficult to directly measure, efforts at indirect measurements should focus on integrating the processes and trends that shape and influence vulnerability as part of all EWS research and data collection (Comes et al., 2014, p. 7). In practice, this means tracking diverse vulnerability-related variables over time, correlating this evolution with important events databases, and creating space for adaptive decision-making to ensure that vulnerability assessments can be integrated into each step of risk management (Comes et al., 2014).

In addition to multidimensional/structural and process/temporally oriented approaches to vulnerability, another area of research emphasizes the need for spatially informed approaches.

Toomey and Kennedy (2011), for example, advocate for the incorporation of risk-terrain modeling (RTM) with GIS software or sociological data in EWS. RTM would allow an integrated or multidisciplinary EWS to go beyond just "uniform warnings" and provide accessible, clearly understood, spatially informed risk analysis (i.e., "risk hot spots") that feeds into mitigation and response strategies (Toomey and Kennedy, 2011). Given the emphasis on spatial knowledge, there is great potential for synergy between RTM and several of the conflict databases and EWS discussed in the previous section.

Overall, we suggest that while these recent efforts are promising, they still do not sufficiently consider conflict and fragility as part of a holistic understanding of vulnerability. This leaves some important gaps in EWS and drought risk management. Space constraints here prohibit a comprehensive discussion or proposal of a new framework for integrated EWS, but this is a promising area for future research. Future systems should consider important structural dynamics; incorporate holistic and differential vulnerability assessments; and include measures, indicators, and trends regarding conflict. These EWSs would be able to provide more timely and relevant insights and recommendations for all stakeholders in dealing with drought response and impact—the international community, national and local governments, organizations, households, and individuals. Much remains to be done, but the shift toward differential vulnerability assessments and more integrated EWS methodologies is essential.

Conclusion

If EWSs are to become proactive and inclusive, as urged by the SFDRR, systems should focus not only on crisis response but also the integration of a more comprehensive assessment of local and regional vulnerabilities.

In this chapter, we have called for increased communication and integration of conflict assessments using conflict events datasets and conflict EWS, drought early warning and response systems as part of a more holistic approach to risk management, reduction, and mitigation. By taking better account of the societal dynamics and trends, as well as important structural factors, drought and disaster risk management strategies will be better suited to the unique needs of the most drought-affected regions, which happen to often overlap with FCAS. Furthermore, practitioners who work in these fields will be better equipped to provide contextually relevant policy advice to local, national, and regional governments.

Research has shown that both fragility and violence lower the possibilities and capacities to implement measures for risk reduction and management. Thus, information on current violence incidents, as well as degrees of fragility should be factored into DEWS. For example, a conflict prediction system, such as ViEWS could be coupled with existing DEWS. More comprehensively, disaster prevention, preparedness, and response strategies should incorporate conflict analysis when considering state capacity, aid accessibility, availability of coping mechanisms, unique migration, and mobility risks and more.

As Hegre et al. (2019, p. 17) stated with regard to their ViEWS conflict EWS, "any such system for violence will necessarily be less precise than those for physical systems, the goal is to improve outcomes relative to a world where these forecasts do not exist. Even an imperfect future system has the potential to inform the placement of peacekeepers, the deployment of NGO resources and even the decisions of private citizens." Taking this thought further, one could imagine a system of information integration on both the history and future risks of violent conflict combined with drought monitoring systems. A platform of this kind could provide (1) a better basis not only for a more comprehensive assessment of the drought risk but (2) also for a more realistic assessment of the opportunities to implement drought risk reduction and drought management measures, particularly in fragile and conflict-affected settings. While a system like this is certainly years away, the current progress being made with regard to conflict analysis and forecasting, especially among public research institutions, is enlightening. Still, much can be done now to better fulfill the spirit of the SFDRR.

References

Barnett, J., 2007. Environmental security and peace. Journal of Human Security 3 (1), 4—16.

Barnett, J., Adger, W.N., 2007. Climate change, human security and violent conflict. Political Geography 26 (6), 639—655.

Bernauer, T., Böhmelt, T., Koubi, V., 2012. Environmental changes and violent conflict. Environmental Research Letters 7 (1), 1—8.

Bernauer, T., Kalbhenn, A., Koubi, V., Ruoff, G., 2010. Climate change, economic growth, and conflict. Paper Presented at the Conference Climate Change and Security, Trondheim, Norway 21—24 June.

Bock, J.G., 2015. Firmer footing for a policy of early intervention: conflict early warning and early response comes of age. Journal of Information Technology & Politics 12, 103—111. https://doi.org/10.1080/19331681.2014.982265.

Bohle, H.G., Downing, T., Watts, M.J., 1994. Climate change and social vulnerability: toward a sociology and geography of food insecurity. Global Environmental Change 4 (1), 37—48.

Brüntrup, M., Tsegai, D., 2017. Drought Adaptation and Resilience in Developing Countries. DIE/GDI Briefing Paper 23.

Buhaug, H., 2015. Climate-conflict research. Wiley Interdisciplinary Reviews: Climate Change 6 (3), 269—275.

Burke, M.B., Miguel, E., Satyanath, S., Dykema, A., Lobell, D.B., 2009. Warming increases the risk of civil war in Africa. Proceedings of the National Academy of Sciences 106 (46), 10670—20674. Retrieved from: http://www.pnas.org/content/106/49/20670.full.pdf.

Crawford, A., 2015. Climate Change and State Fragility in the Sahel. Policy brief no. 205. FRIDE, Madrid. Available at: https://www.iisd.org/sites/default/files/publications/climate-change-and-state-fragility-in-the-Sahel-fride.pdf.

Cederman, L.-E., Wimmer, A., Min, B., 2010. Why do ethnic groups rebel? New data and analysis. World Politics 62 (1), 87—119.

Comes, T., Mayag, B., Negre, E., 2014. Decision support for disaster risk management: integrating vulnerabilities into early-warning systems. In: ISCRAM-med 2014: Information Systems for Crisis Response and Management in Mediterranean Countries, pp. 178—191.

Conca, K., 2001. Environmental cooperation and international peace. In: Diehl, P., Gleditsch, N.P. (Eds.), Environmental Conflict. Westview Press, Boulder and Oxford, pp. 225—247.

Coppedge, M., Gerring, J., Altman, D., Bernhard, M., Fish, S., Hicken, A., Kroenig, M., Lindberg, S.I., McMann, K., Paxton, P., Semetko, H.A., Skaaning, S.-E., Staton, J., Teorell, J., 2011. Conceptualizing and measuring democracy: a new approach. Perspectives on Politics 9 (2), 247—267.

Croicu, M., Sundberg, R., 2017. UCDP GED Codebook Version 17.1. Department of Peace and Conflict Research, Uppsala University.

Dabelko, G., Herzer, L., Null, S., Parker, M., Sticklor, R., 2013. Backdraft: The Conflict Potential of Climate Change Adaptation and Mitigation. Environmental Change and Security Program, Washington, DC.

Dalby, S., 2013. Biopolitics and climate security in the Anthropocene. Geoforum 49, 184—192.

Drought early warning systems in the context of drought preparedness and mitigation, Early Warning Systems for Drought Preparedness and Drought Management. In: Wilhite, D.A., Svoboda, M.D. (Eds.), 2000. Proceedings of an Expert Group Meeting, Lisbon, Portugal. World Meteorological Organization, Geneva, Switzerland, pp. 1—21.

FAO (Food and Agriculture Organization of the United Nations), 2019. Early Warning Early Action. Available at: http://www.fao.org/3/CA3127EN/ca3127en.pdf.

FAO, IFAD, UNICEF, WFP, WHO, 2018. The State of Food Security and Nutrition in the World 2018. In: Building Climate Resilience for Food Security and Nutrition. FAO, Rome.

Food Security Information Network (FSIN), 2018. Global Report on Food Crisis. https://www1.wfp.org/publications/global-report-food-crises-2018.

German Advisory Council on Global Change, 2007. Climate Change as a Security Risk. Earthscan, London and Sterling.

Gleditsch, N.P., 2012. Whither the weather? Climate change and conflict. Journal of Peace Research 49 (1), 3—9.

Gleditsch, N.P., Wallensteen, P., Eriksson, M., Sollenberg, M., Strand, H., 2002. Armed conflict 1946—2001: a new dataset. Journal of Peace Research 39 (5).

Grävingholt, J., Ziaja, S., Ruhe, C., Fink, P., Kreibaum, M., Wingens, C., 2018. Constellations of State Fragility v1.0. German Development Institute/Deutsches Institut für Entwicklungspolitik (DIE). https://doi.org/10.23661/CSF1.0.0.

Hegre, H., Allansson, M., Basedau, M., Colaresi, M., Croicu, M., Fjelde, H., Hoyles, F., Hultman, L., Högbladh, S., Mouhleb, N., Muhammad, S.A., Nilsson, D., Nygård, H.M., Olafsdottir, G., Petrova, K., Randahl, D., Rød, E.G., von Uexkull, N., Vestby, J., 2019. ViEWS: a political violence early-warning system. Journal of Peace Research. https://doi.org/10.1177/0022343319823860. Online First Published February 15, 2019.

IPCC, 2007. Climate Change 2007 — The Physical Science Basis. Contribution of Working Group I to the Fourth Assessment Report of the Intergovernmental Panel on Climate Change. Cambridge University Press, Cambridge, UK.

IPCC, 2013. Climate Change 2013 — The Physical Science Basis. Contribution of Working Group I to the Fifth Assessment Report of the Intergovernmental Panel on Climate Change. Cambridge University Press, Cambridge, UK.

IPCC, 2014. Climate Change 2014: Impacts, Adaptation, and Vulnerability. Contribution of Working Group II to the Fifth Assessment Report of the Intergovernmental Panel on Climate Change. Cambridge University Press, Cambridge, UK.

Koren, O., Bagozzi, B.E., 2017. Living off the land: the connection between cropland, food security, and violence against civilians. Journal of Peace Research 54 (3), 351—364. https://doi.org/10.1177/0022343316684543.

Linke, A.M., O'Loughlin, J., McCAbe, T.J., Tir, J., Witmer, F.D.W., 2015. Rainfall variability and violence in rural Kenya. Global Environmental Change 34, 35—47.

Magnan, A.K., Schipper, A.L.F., Burkett, M., Bharwani, S., Burton, I., Eriksen, S., Gemenne, F., Schaar, J., Ziervogel, G., 2016. Addressing the risk of maladaptation to climate change. Wiley Interdisciplinary Reviews: Climate Change 7 (5), 646—665.

Mancini, F. (Ed.), 2013. New Technology and the Prevention of Violence and Conflict. UNDP & UNSAID. Retrieved from: http://reliefweb.int/sites/reliefweb.int/files/resources/ipi-e-pub-nw-technology-conflict-prevention-advance.pdf.

Matthew, R.A., Barnett, J., McDonald, B., O'Brien, K.L., 2009. Global Environmental Change and Human Security. MIT Press, Boston, MA.

Moran, D. (Ed.), 2011. Climate Change and National Security. Georgetown University Press, Washington, DC.

Naumann, G., Barbosa, P., Garrote, L., Iglesias, A., Vogt, J., 2014. Exploring drought vulnerability in Africa: an indicator-based analysis to be used in early warning systems. Hydrology and Earth System Sciences 18, 1591—1604.

Nordas, R., Gleditsch, N.P., 2007. Climate change and conflict. Political Geography 26, 627—638. Retrieved from: http://n.ereserve.fiu.edu/010034599-1.pdf.

Nyheim, D., 2009. Preventing Violence, War and State Collapse: The Future of Conflict Early Warning and Response. OECD.

Pettersson, T., Eck, K., 2018. Organized violence, 1989—2017. Journal of Peace Research 55 (4), 535—547.

Pham, P.N., Vinck, P., 2012. Technology, conflict early warning systems, public health, and human rights. Health and Human Rights 14 (2), 106−117. Available at: http://www.hhrjournal.org/2013/08/14/technology-conflict-early-warning-systems-public-health-and-human-rights/.

Popovski, V., 2017. Viewpoint: Does Climate Change Cause Conflict? Foresight Africa 2017. Africa Growth Initiative at Brookings, pp. 84−85.

Raleigh, C., Linke, A., Hegre, H., Karlsen, J., 2010. Introducing ACLED − armed conflict location and event data. Journal of Peace Research 47 (5), 651−660.

Rohwerder, B., 2015. Conflict Early Warning and Early Response. GSDRC Helpdesk Research Report 1195. GSDRC, University of Birmingham, Birmingham, UK.

Ruttinger, L., Smith, D., Stang, G., Tänzler, D., Vivekananda, J., 2015. A New Climate for Peace. Adelphi, International Alert, Woodrow Wilson International Center for Scholars, European Union Institute for Security Studies, Berlin.

Salehyan, I., Hendrix, C.S., Hamner, J., Case, C., Linebarger, C., Stull, E., Williams, J., 2012. Social conflict in Africa: a new database. International Interactions 38 (4), 503−511.

Sanghi, A., Ramachandran, S., de la Fuente, A., Tonizzo, M., Sahin, S., Adam, B., 2010. Natural Hazards, Unnatural Disasters: The Economics of Effective Prevention. The World Bank. Available at: http://documents.worldbank.org/curated/en/620631468181478543/pdf/578600PUB0epi2101public10BOX353782B.pdf.

Scheffran, J., Brzoska, M., Brauch, H.G., Link, P.M., Schilling, J. (Eds.), 2012. Climate Change, Human Security and Violent Conflict. Springer Science and Business Media, Berlin, Heidelberg.

Schleussner, C.-F., Donges, J.F., Donner, R.V., Schellnhuber, H.J., 2016. Armed-conflict risks enhanced by climate-related disasters in ethnically fractionalized countries. Proceedings of the National Academy of Sciences 113 (33). Available at: www.pnas.org/cgi/doi/10.1073/pnas.1601611113.

Schnurr, M., Swatuk, L.A., 2012. Towards critcal environmental security. In: Schnurr, M., Swatuk, L.A. (Eds.), Natural Resources and Social Conflict. Towards Critical Environmental Security. Palgrave, London, pp. 1−14.

Smidt, M., Theisen, O.T., 2018. Climate change and conflict: agriculture, migration and institutions. In: Zurayk, R., Woertz, E., Bahn, R. (Eds.), Crisis and Conflict in Agriculture. CAB International, pp. 40−52.

Smit, B., Wandel, J., 2006. Adaptation, adaptive capacity and vulnerability. Global Environmental Change 16 (3), 282−292.

Sundberg, R., Melander, R., 2013. Introducing the UCDP georeferenced event dataset. Journal of Peace Research 50 (4), 523−532.

Sneyers, A., 2017. Food, Drought and Conflict Evidence From a Case Study on Somalia. HiCN Working Paper 252. Available at: http://www.hicn.org/wordpress/wp-content/uploads/2012/06/HiCN-WP252.pdf.

Swatuk, L., Wirkus, L., Krampe, F., Bejoy, K.T., Batista da Silva, L.P., 2018. The Boomerang effect: overview and implications for climate governance. In: Swatuk, L., Wirkus, L. (Eds.), Water Climate Change and the Boomerang Effect. Unintentional Consequences for Resource Insecurity. Routledge, New York, pp. 1−19.

Tollefsen, A.F., Strand, H., Buhaug, H., 2012. PRIO-GRID: a unified spatial data structure. Journal of Peace Research 49 (2), 363−374.

Toomey, M., Kennedy, L.W., 2011. An analysis of Modern Early Warning Systems: How might Risk-Terrain Modeling Contribute to the Development of an Optimal System? Rutgers Center on Public Security, New York, NJ.

Uexkull, N. von, 2014. Sustained drought, vulnerability and civil conflict in Sub-Saharan Africa. Political Geography 43 (11), 16−26. https://doi.org/10.1016/j.polgeo.2014.10.003.

Uexkull, N. von, Croicu, M., Fjelde, H., Buhaug, H., 2016. Civil conflict sensitivity to growing-season drought. Proceedings of the National Academy of Sciences 113 (44), 12391−13296.

United Nations, World Bank, 2018. Pathways for Peace: Inclusive Approaches to Preventing Violent Conflict. World Bank, Washington, DC. Available at: https://openknowledge.worldbank.org/handle/10986/28337. License: CC BY 3.0 IGO.

Walz, Y., et al., 2018. Understanding and Reducing Agricultural Drought Risk: Examples from South Africa and Ukraine. Policy Report No. 3. United Nations University - Institute for Environment and Human Security (UNU-EHS), Bonn.

Wilhite, D.A., Hayes, M.J., Knudson, C., Smith, K.H., 2000. Planning for Drought: Moving from Crisis to Risk Management. Drought Mitigation Center Faculty Publications, p. 33. http://digitalcommons.unl.edu/droughtfacpub/33.

Wirkus, L., Blitza, H., 2019. Droughts, violent conflicts, and fragility in Africa 2007−2017, Thematic Map unpublished.

Wisner, B., Blaikie, P., Cannon, T., Davis, I., 2003. At Risk: Natural Hazards, People's Vulnerability and Disasters, second ed. Routledge, London and New York.

WMO, GWP, 2019. Integrated Drought Monitoring Program. Monitoring and Early Warning. (IDMP). Available at: http://www.droughtmanagement.info/pillars/monitoring-early-warning/.

Wondemagegnehu, D.Y., 2009. An Exploratory Study of Harmonization of Conflict Early Warning in Africa. Master Thesis. University of Vienna and University of Leipzig.

World Bank, 2015. World Development Indicators. Available at: http://data.worldbank.org/products/wdi.

Zenko, M., Friedman, R.R., 2011. UN early warning for preventing conflict. International Peacekeeping 18 (1), 21−37. https://doi.org/10.1080/13533312.2011.527504.

Further reading

Bavinck, M., Pellegrini, L., Mostert, E. (Eds.), 2014. Conflict on Natural Resources in the Global South: Conceptual Approaches. CRC Press, Boca Raton, FL.

Beck, H.E., Zimmermann, N.E., McVicar, T.R., Vergopolan, N., Berg, A., Wood, E.F., 2018. Present and future Köppen-Geiger climate classification maps at 1-km resolution. Scientific Data 5. https://doi.org/10.1038/sdata.2018.214, 180214 EP -.

Messner, J.J., Haken, N., Taft, P., Onyekwere, I., Blyth, H., Fiertz, C., Murphy, C., Quinn, A., Horwitz, Mc, 2018. Fragile State Index 2018 − Annual Report. Retrieved from: http://fundforpeace.org/fsi/2018/04/24/fragile-states-index-2018-annual-report/.

Osborn, T.J., Barichivich, J., Harris, I., van der Schrier, G., Jones, P.D., 2018. Monitoring global drought using the self-calibrating palmer drought severity index. In "state of the climate in 2017". Bulletin of the American Meteorological Society 99, S36−S37. https://doi.org/10.1175/2018BAMSStateoftheClimate.1.

Salehyan, I., Hendrix, C., 2016. Social Conflict Analysis Database (SCAD) Version 3.2. Robert S. Strauss Center for International Security and Law, Austin, USA.

Schnurr, M.A., Swatuk, L.A. (Eds.), 2012. Environmental Change, Natural Resources and Social Conflict: Towards Critical Environmental Security. Palgrave, London.

United Nations Office for Disaster Risk Reduction (UNISDR), 2015. The Sendai Framework for Disaster Risk Reduction 2015−2030. Available at: https://www.unisdr.org/files/43291_sendaiframeworkfordrren.pdf.

Making weather index insurance effective for agriculture and livestock forage: lessons from Andhra Pradesh, India

Krishna Reddy Kakumanu[a,*], Palanisami Kuppanan[b], Udaya Sekhar Nagothu[c], Gurava Reddy Kotapati[d], Srikanth Maram[a], Srinivasa Rao Gattineni[e]

[a]*National Institute of Rural Development and Panchayati Raj, Hyderabad, India,*
[b]*International Water Management Institute, New Delhi, India,*
[c]*Norwegian Institute of Bioeconomy Research, Ås, Norway,*
[d]*Acharya N G Ranga Agricultural University, Guntur, India,*
[e]*eeMAUSAM, Weather Risk Management Services Private Limited, Hyderabad, India*
*Corresponding author

Introduction

Although agriculture contributes less than 15% of India's gross domestic product, its importance to the country's economic, social, and political fabric goes well beyond this indicator. Rural areas are home to 69% of India's population of 1.3 billion people (Census of India, 2011), the majority of which are poor and depend on agriculture and related activities for their livelihoods. India's agricultural sector is vulnerable to extreme events, especially droughts and floods, and to future changes in climate. To address these risks, the Government of India has launched various initiatives over the last five decades, including the General Insurance Corporation (1972–79), the pilot crop insurance scheme (1979–85), the comprehensive crop insurance scheme (1985–99), the national agricultural insurance scheme (1999–2010), the modified national agricultural insurance scheme (2010–16), the weather-based crop insurance scheme (WBCI) (2003–16), and Pradhan Mantri Fasal Bima Yojana (Prime Minister's Crop Insurance Scheme (PMFBY)) (2016) (Singh, 2010; BartonKakumanu et al., 2012; Kakumanu et al., 2013; Greatrex et al., 2015), just to name a few. Of these, the WBCI became popular with Indian farmers for the risk management solutions it offers (Giné et al., 2008). Beginning in 2003, the WBCI covered two crops (groundnut and castor) and was later extended to 40 crops, covering various climatic risks such as deficit in rainfall, extreme temperatures, excessive humidity, and harmful winds.

In addition to government schemes, private sector companies and nongovernmental organizations have also offered crop insurance in India. However, unlike life and general insurance sectors, agricultural insurance has not been popular in India, for a number of reasons, including a lack of location-specific insurance contracts; uncertainty among potential buyers about premium structures; slowness

Drought Challenges. https://doi.org/10.1016/B978-0-12-814820-4.00012-2

183

in the settlement and payment of claims; and a general perceived lack of transparency. A particular complaint about the government's WBCI scheme has been the limited access to and quality of weather data in the Reference Unit Area (RAU) used in the scheme (Parthasarathy, 2014).

In this chapter, we look at steps needed to develop location-specific weather index insurance products in India and how to facilitate their implementation in close coordination with the farmers and insurance companies in order to increase the level of awareness and understanding of weather-based insurance products. Specifically, we look at weather-based insurance products for insuring rice, cotton, and chili crops against floods and low temperatures, and for livestock forage against drought conditions in arid and semi-arid districts.

Methodology
Study area

Two districts in Andhra Pradesh (AP) state were selected to pilot the study of weather index insurance products. The first of these is Guntur district, where three villages—V. Bonthiralla, Yerraguntla, and Rangareddypalem from Narasaraopet, Nadendla, and Chilakaluripet blocks (mandal) were selected to study the impacts of wet condition (see Table 12.1). The villages are located within a 10 km radius and are found in a hot subhumid to semi-arid ecological region that receives most of its annual rainfall of approximately 400 mm during monsoon periods. The modified national agricultural insurance scheme and WBCI were in practice until 2016 in the region, using the block/mandal as reference unit.

Pilot villages

The second district is Kurnool, where index insurance for forage crops was implemented to protect small and marginal farmers against a deficit in rainfall. Two villages—V. Bonthiralla and Yerraguntla—in Dhone block or mandal of Kurnool district were selected for study. Historically, Dhone is a drought-prone mandal with 360 mm of mean annual rainfall. The forage crop insurance plan protects farmers if there is a deficit in rainfall between June and September that necessitates the purchase of supplementary fodder to avoid reduced milk yield and maintain the average weight of their cattle.

Table 12.1 Pilot villages.

Sr. No.	State	District	Block/Mandal	Village	Geo-coordinates	
					Latitude (N)	Longitude (E)
1	AP	Kurnool	Dhone	V. Bonthiralla	15 degrees 25'27"	77 degrees 42'19"
2	AP	Kurnool	Dhone	Yerraguntla	15 degrees 23'7"	77 degrees 55'54"
3	AP	Guntur	Narasaraopet	Rangareddypalem	16 degrees 15'18"	80 degrees 4'57"
4	AP	Guntur	Nadendla	Kanaparru	16 degrees 12'30"	80 degrees 6'39"
5	AP	Guntur	Chilakaluripet	Kavur	16 degrees 8'25"	80 degrees 7'23"

Implementation of weather index insurance

To implement the weather index insurance for the specified crops, the following steps were followed, which are described here in some detail so as to provide a template for how a similar process might be implemented elsewhere.

Step 1: opportunity assessment

The first step is to determine if weather index insurance is actually feasible in the study villages. Unless all of the essential elements are in place, it is unlikely that crop insurance will be successful. Opportunity assessment usually requires a field visit with all the stakeholders involved, to determine that sufficient weather stations are in place and that they are working properly. Historical data for a period of 30 years is the standard requirement for designing an insurance contract/product (Mapfumo et al., 2017).

An opportunity assessment survey was conducted in all of the five villages from both districts in the months of May and June in 2014. Automatic weather stations had previously been installed in the two villages of Kurnool district by International Crop Research Institute for Semi-Arid Tropics (ICRISAT). The weather stations were in good working condition and thus provided the data used for the experiment. In the villages located in Guntur district, there was no automatic weather station already installed. Therefore, National Collateral Management Services Private Limited (NCML), a third-party data provider, was appointed to install a station in Kanaparru village. This is an ideal location to serve the same purpose for the other two villages as they are both situated within a 10 km radius of Kanaparru village. Historical weather data were obtained from revenue offices in Kurnool and Guntur for the respective mandals. Farmers in all five study villages showed keen interest to participate in the program, and Bajaj Allianz General Insurance Company agreed to underwrite the risks for the selected sites and the crops.

Step 2: contract design

The actual product design starts with collecting detailed climatic and agronomic data as well as developing prototype contracts. After conducting the feasibility study, local farmers and scientists were contacted to understand the risks involved in raising livestock fodder and other field crops. A detailed product note was prepared and submitted to the insurer, Bajaj Allianz, in respect to fodder crop in Kurnool district and rice, chili, and cotton crops in Guntur district. Based on the inputs, Bajaj Allianz prepared the term sheet and finalized the pricing/premiums after discussing with the farmers.

Step 3: contract pricing

The next step is to establish a process for developing the proper term sheets and spreadsheets that allow the insurer to price the contracts. Here, we need to consider commissions, taxes, etc. in addition to activity-based costing with the contract distributor to identify normal administrative expenses. Bajaj Allianz prepared the contract pricing and term sheet for all the crops of interest.

Insurance is a mechanism where farmers make fixed payments known as "premiums" in exchange for the promise to be reimbursed for contingent future losses (Wang, 1995). The actuarial rationale for determining premiums is that these need to be sufficient to cover future losses on average. To adjust for inflation, the present value of premiums should be set to equal the present value of expected future losses. Estimating the total losses "X" from an insured risk in a specified time period is done through a stochastic process. The insurer has little or no influence on the random variable

"X," with mean "μ" and standard deviation "σ." Since it is the objective of an insurance mechanism to define insurance coverage and the corresponding premium "π," it is necessary to provide an estimate of "π" in advance (Bühlmann, 1985). Thus, the estimate of mean losses "μ," is included in the actuarial pure insurance premium (Lu et al., 2008). Since future losses are random and not known with certainty, premium "π" is set prospectively. The pure insurance premium may not be sufficient to cover all losses and costs in the future with a certain probability. Hence, the insurer can control the risk of insolvency "α" by adding a risk-loading "θ" to the premium that is dependent on the distribution of losses "X." The required actuarial premium "π" for insurance risks controls the risk of insolvency and can be defined by

$$\pi = (1 + \theta)\mu \tag{12.1}$$

The risk-loading "θ" can be derived by various principles, all of which aim at limiting the risk of insolvency to a sufficiently small value. If a large enough number of insured "n" (farmers) is assumed, the central limit theorem yields $\theta = \frac{(z1 - \alpha\sigma)}{(\mu\sqrt{n})}$. Assuming the number of farmers insured is independent of the premium, the insurer may control the risk of insolvency by increasing the risk-loading "θ" to a sufficient level (Kliger and Levikson, 1998). Besides the total cost of future losses, the insurer has to incur significant additional costs while running the organization (i.e., distribution, management, settlement, and cost of capital). These costs need to be recovered through premium income and are reflected in a cost-loading "c" that equals the present value of expected costs. Thus, the required actuarial premium "π" for insurance risks controlling for risk of insolvency and including cost is (Biener, 2011)

$$\pi = (1 + \theta)\mu + c \tag{12.2}$$

The premiums for the study areas of wetland and dryland were established using Eq. (12.2). The term sheet, which includes the premium cost, payout information, rainfall and temperature limits during the season, was developed after having discussions with farmers. The term sheets were then tested for the Kharif season (i.e., July to January months in the study area) 2014 and 2015 in the study areas. In case of paddy, three growth phases were considered for the payout during October, November, and December with an assured sum of Rs. 25,000 per hectare (Table 12.2). Trigger points, which correspond to the initiation of crop losses, were decided based on the average weather conditions in Guntur and Kurnool districts during the last 20–25 years. The payouts in each phase were calculated using the trigger points. Since excess rainfall on any given day has direct and negative impact on the production of rice, cotton, and chili crops, higher weightage was given to it on the allocation of sum insured when compared to the temperature variation. The sum insured was increased in each subsequent phase of the crop to compensate the increasing cost of cultivation. Premium rates were arrived at by adding the risk cost with the risk-loading and management expenses. Detailed term sheets for the rice, chili, and cotton crops for Kharif 2014 and 2015 are presented in Tables 12.2–12.4.

Similarly, the term sheet for forage crops was developed for Dhone block/mandal in Kurnool district. During Kharif 2014, the average rainfall of 360 mm was considered on a cumulative basis during the period of July 16th to October 15th, 2014, with an assured sum of Rs. 12,500 per ha at a premium of Rs. 1875 per ha. However, in Kharif (2015), the premium was revised for monthly cumulative rainfall, as the total rainfall is captured in the last weeks of coverage time. The details of the term sheet for the two seasons are presented in Table 12.5.

Table 12.2 Term sheet for rice.

	Kharif 2014			Kharif 2015		
	October	**November**	**December–January 2015**	**October**	**November**	**December–January 2016**
Maximum payout (Rs/acre)	800	2400	4800	1200	2800	4000
Cover: Single event of two consecutive days maximum rainfall above the trigger						
Trigger (mm)	100	60	40	110	70	50
Exit (mm)	180	120	80	190	140	100
Notational payout (Rs/mm)	10	40	120	15	40	80
Maximum payout under excess rainfall (Rs/ha)	20,000			20,000		
Daily minimum temperature	<20°C for eight or more consecutive days			<18°C for eight or more consecutive days		
Cover period minimum temperature	November			November		
Maximum payout for minimum temperature (Rs/ha)		5000			5000	
Payment slabs (Rs/ha)	8 consecutive days: 1000 9 consecutive days: 2000 10 consecutive days: 3000 11 consecutive days: 4000 12 consecutive days: 5000			8 consecutive days: 1000 9 consecutive days: 2000 10 consecutive days: 3000 11 consecutive days: 4000 12 consecutive days: 5000		
Total sum insured (Rs/ha)		25,000			25,000	
Premium (Rs/acre)		1910			1937	

Step 4: development of contract administration tool kit

An administration tool kit was developed for contract workflows, manuals, and monitoring sheets. A password-protected spreadsheet was created for users to enter their inputs including weather data received from the weather stations and to record whether a payout has been generated or not.

Step 5: client education

It is a normal practice to develop printed materials in the form of brochures to be used by field officers to train farmers and other stakeholders about the principles of insurance, premiums, and the process of a claim settlement. In many cases, farmers possess limited knowledge about the insurance, making them suspicious of the products, but this can be overcome through proper awareness campaigns.

Table 12.3 Term sheet for chili.

	Kharif 2014			Kharif 2015		
	October	November	December	October	November	December
Maximum payout (Rs/acre)	1200	2800	4000	1500	3000	4500
Cover: Single event of two consecutive days maximum rainfall above the trigger						
Trigger (mm)	100	60	40	120	100	70
Exit (mm)	180	120	80	220	200	170
Notational payout (Rs/mm)	15	46.67	100	15	30	45
Maximum payout under excess rainfall (Rs/ha)	20,000			22,500		
Maximum temperature	>36°C for six or more consecutive days			>36°C for six or more consecutive days		
Cover period minimum temperature	October			August to October		
Maximum payout for temperature (Rs/ha)	5000			2500		
Payment slabs (Rs/ha)	6 consecutive days: 625 7–10 consecutive days: 1875 11–15 consecutive days: 3125 16 consecutive days: 5000			More than 6 consecutive days: 2500		
Total sum insured (Rs/ha)	25,000			25,000		
Premium (Rs/ha)	2527			2565		

Table 12.4 Term sheet for cotton.

	Kharif 2014		Kharif 2015	
	October	November–December	October	November–December
Maximum payout (Rs/ha)	12,500	12,500	12,500	12,500
Cover: Single event of three consecutive days maximum rainfall above the trigger				
Trigger (mm)	180	125	180	125
Exit (mm)	280	225	280	225
Notational payout (Rs/mm)	250	250	250	250
Maximum payout under excess rainfall (Rs/ha)	20,000		20,000	
Total sum insured (Rs/ha)	25,000		25,000	
Premium (Rs/ha)	2807		2565	

Table 12.5 Term sheet for livestock forage.

| | Kharif 2014 | | | Kharif 2015 | | |
	Cumulative			Phase I	Phase II	Phase III
Cover period	16/7/2014 to 15/10/2014			15/7–31/8/ 2015	01/9–30/9/ 2015	01/10–31/10/ 2015
Sum insured (Rs/ha)	12,500			5000	3750	3750
Benchmark rainfall	350			110	50	32
Deficit range	Payout %	Rainfall (mm)	Amount (Rs/ha)	Phase-wise payout (%) for deficit rainfall		
20%–30%	10	252–288	1250	10	10	10
30%–50%	30	180–252	3750	30	30	30
50%–75%	70	90–180	8750	70	70	70
>75%	100	<90	12,500	100	100	100
Premium (Rs/ha)	1875			1875		

The present study experience suggests one of the most effective ways of developing such tools is to organize focus group meetings where the insurance products are explained to farmers and other stakeholders, who are then requested to identify features they think should be included in the insurance product. Training programs were then organized in all the participating villages to explain the prospective insurance contract and index procedures to be implemented in each village.

Step 6: farmer recruitment
Once the contract is ready, key staff and field officers undertake training of trainers sessions to create a cascading effect. Accordingly, they hold client-training programs to identify and register the prospective clients. Training was also provided to local NGOs who are already working in the villages, to help educate the farmers and explain the program.

Step 7: risk transfer
Memoranda of Understanding were signed with key stakeholders and the data providers to ensure appropriate and timely data delivery to all concerned parties. The contract distributor then uses the farmer registry to develop premium schedules, which are then forwarded to the insurer. Premiums collected from 109 participating farmers (50 livestock forage insurance and 59 for the remaining three crops) in the five villages were paid to the insurer, and policies issued by the insurer were distributed among the farmers.

Step 8: contract monitoring
Contract monitoring sheets and other product tools were distributed to relevant stakeholders to enter appropriate data during the contract period. Weather data from Automatic Weather Stations (AWS) were provided by ICRISAT in both villages of the Kurnool district, while NCML provided the same for the Guntur district. The weather data recorded during the contract period was analyzed for the claim process against the term sheet.

Step 9: claim process

If a claim is generated, the contract monitoring sheet will reveal the amount, and the insurer will make the payments directly to the insured farmers within an agreed period mentioned in the Product Note. Such payments could go directly to the bank accounts of the insured farmer to reduce the transaction cost and make delivery of the payment faster.

Results and discussion
Weather index insurance for agricultural crops

The study areas for rice, chili, and cotton are irrigated locations within the Krishna river basin. The main risks we observed there are excess rainfall and crop disease associated with extreme temperatures. Hence, the key weather parameters for developing the term sheet were excess rainfall and temperature. Different triggers were designed for different growth phases of rice, chili, and cotton crops using weather data recorded by NCML for September 1, 2014, to January 31, 2015, and September 1, 2015, to January 31, 2016 (see Figs. 12.1 and 12.2).

The contract period for the wetland area (Guntur) was matured (i.e., completed) in different time periods for each crop. For rice crops, excess rainfall insurance coverage was matured by January 31 in both 2015 and 2016, respectively, while minimum temperature was matured by November 30 in both 2014 and 2015, respectively. For chili crops, excess rainfall cover was matured by December 31st, while maximum temperature risk was completed by October 31. For cotton crops, the risk period was matured by December 31 in 2014 and 2015, respectively.

In 2014, there were two recorded periods of rainfall, one in October (maximum rainfall for 2 days was 69 mm) and the other in November (maximum rainfall for 2 days was 53 mm), and there were no insurance payouts during either of these two periods. No crop damage was reported by any farmers in

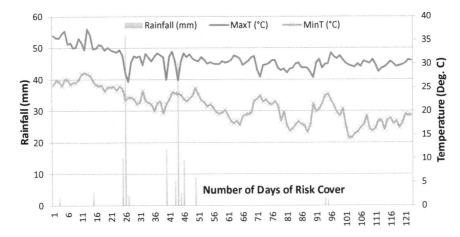

FIGURE 12.1

Rainfall (mm) recorded in Kanaparru village, Guntur district, during Kharif 2014.

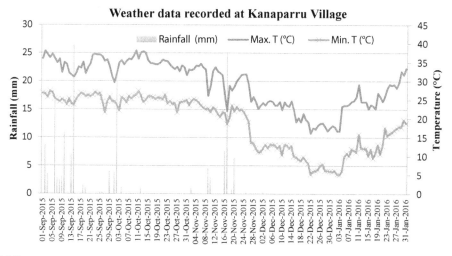

FIGURE 12.2

Rainfall (mm) recorded in Kanaparru village, Guntur district, during Kharif 2015.

2014, and the general crop condition was reported to be "good." In 2015, the total amount of rainfall received during the crop growing season was 272 mm (in comparison with 400 mm average rainfall). There were only three events where daily rainfall exceeded 20 mm, and only 25 rainy days were observed during the entire growing season. Analysis of weather data revealed that there is no payout. Information from farmers also revealed that there was a lack of water in local irrigation canals, leading farmers to not sow crops in areas served by canals. However, a few farmers were able to successfully raise crop through use of alternative irrigation sources, although this resulted in greater production costs for purchasing water and labor.

In the case of rice, a minimum temperature trigger was used as a proxy to assume a risk of disease outbreaks and was set as being a period of eight or more consecutive days in the month of November with temperatures below 18°C. Lower temperatures at booting or heading stage can also cause a high percentage of sterility in the rice (Ishiguro et al., 2014). The weather data show that temperatures fell below 20°C for only 4 days during Kharif 2014. The minimum temperatures during Kharif 2015 remained below 18°C continuously for 65 days starting from November 26. However, as per the term sheet, it was only below for five consecutive days in November (not the required eight consecutive days in November), and therefore insurance claims were not paid out.

For chili crops, temperatures above 36°C for six or more consecutive days in the month of October were used as a proxy for chili diseases. It is observed from the weather data that maximum temperature above 36°C was recorded only for a couple of days during the phase in Kharif 2014 and only 5 days during Kharif 2015. Coincidentally, there was no widespread disease reported during the month of October on for chili crops, and no claims were paid out.

In the case of cotton, excess rainfall over a period of three consecutive days during October to December was considered to be risky. For the month of October, 180 mm was set as the trigger for crop losses and 280 mm was the threshold for crop failure. Because only 80 mm rainfall fell in October

2014 and 27.42 mm in October 2015, there were no claims paid out. For the months of November and December, a cumulative rainfall trigger of 125 mm was set. However, only 98.5 mm were received in 2014 and 48.7 mm in 2015, and so there were no claims.

Weather index insurance for livestock forage

Rainfall data show that V. Bonthiralla AWS received a total rainfall of 263.6 mm and Yerraguntla AWS received a total rainfall of 316 mm during the 2014 contract period (i.e., July 16 to October 15, 2014). Both villages received less rainfall when compared with the Dhone mandal where the normal rainfall was 360 mm. There was a deficiency of total rainfall to the amount of 27% in V. Bonthiralla and 12% in Yerraguntla. As per the term sheet, V. Bonthiralla farmers received Rs. 1250 per ha as the deficit rainfall fell in the 20%–30% range, while there was no payout in Yerraguntla as the village received the required rainfall as per the term sheet.

There is a difference of 52.4 mm of rainfall between both locations, even though the aerial distance between the two locations is about 24 km. Much of the difference is due to a single event on August 22 when 49.4 mm fell in Yerraguntla, while only 13.6 mm fell at V. Bonthiralla. It is also observed from Fig. 12.3 that rainfall distribution was well distributed throughout the season except during the month of September, where there was a break of 19 days. Because the contract for forage crops was based on the total amount of rainfall, claims were not paid despite this gap. Because of this, farmers would have preferred that the contract be structured on a phase-by-phase basis.

In 2015, Bonthiralla AWS received 29.6 mm rainfall during phase 1 (July 15 to August 31) and Yerraguntla AWS received a total rainfall of 108.5 mm during the same period. In Phases 2 and 3,

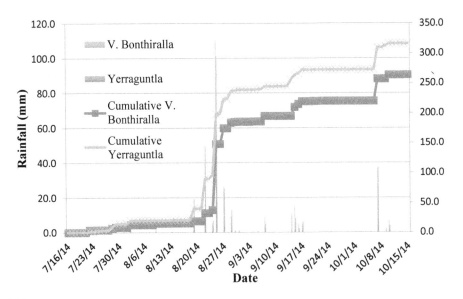

FIGURE 12.3

Rainfall (mm) recorded in V. Bonthiralla and Yerraguntla villages, Kurnool district, during 2014.

Table 12.6 Phase-wise rainfall recorded and payout for the livestock forage in Kurnool district.

Village	Phase 1 (July 15–August 31)			Phase 2 (September 1–30)			Phase 3 (October 1–31)		
	Cumulative rainfall, mm	% Deficit	Payout (Rs/ha)	Cumulative rainfall, mm	% Deficit	Payout (Rs/ha)	Cumulative rainfall, mm	% Deficit	Payout (Rs/ha)
V. Bonthiralla	29.6	73	3500	26.4	47	1125	1.0	97	3750
Yerraguntla	108.5	1	Nil	151.4	Nil	Nil	2.6	93	3750

both V. Bonthiralla and Yerraguntla received 24.6 and 151.4 mm; and 1.0 and 2.6 mm, respectively. This meant that V. Bonthiralla received a deficit rainfall during all three phases. In Phase 3, there was a total deficit of 97% in V. Bonthiralla and 93% in Yerraguntla (Table 12.6). As per the term sheet, each farmer in V. Bonthiralla received a total assured sum of Rs. 8375 per ha, and in Yerraguntla, farmers received Rs. 3750 per ha.

Although we have data for only a relatively brief period, our results can help stakeholders in developing relevant index insurances, and the resulting insurance products would help households to reduce their dependence on unfavorable coping strategies during severe weather shocks (Jensen and Barrett, 2017). A timely payout is always important to get farmers to adopt crop insurance, which is precisely what was done in our project, with the aim of motivating farmers to continue using the product in the subsequent years.

Conclusions and policy implications

The weather index insurance provides a sense of security against the risks that the weather often presents in the agricultural sector. It is currently being tested and implemented in most states in India, but its uptake by farmers is still poor due to a lack of awareness and an absence of village level weather stations. In the present study, AWSs were installed and tested, enabling us to develop customized weather index products in coordination with the stakeholders. It was observed that additional changes are needed for agricultural crops in wetland areas, such as extending dry period coverage accounting to address consecutive days in a month without rain and adjusting minimum temperature thresholds in the second half of November until the end of December.

We also found that extended dry periods that affect forage yields in the dryland system would not necessarily trigger a claim payment because the monthly average rainfall could be close to the historical average. Hence, contracts had to be revised from a cumulative rainfall total to shorter intervals of 1 month periods.

The present PMFBY program can add relevant products through the revised weather-based crop insurance scheme and develop awareness among farmers looking to overcome weather risks. Region-specific training programs would help farmers in understanding the crop and weather index insurance products while also enhancing the adoption of the products. The agricultural insurance company of India should also have continuous monitoring, as well as timely payouts, to improving the further adaptation and adoption by farmers of weather insurance products.

References

Barton, D.N., Kakumanu, K.R., Palanisami, K., Tirupataiah, K., 2012. Analysis of economic incentives to manage risk at farm level in the context of climate change. In: Water and Climate Change: An Integrated Approach to Address Adaptation Challenges. Macmillan publishers India Ltd., New Delhi, ISBN 935-059-059-X, pp. 143−168.

Biener, C., 2011. Pricing in Micro Insurance Markets. Institute of Insurance Sci., Univ. of Ulm, Germany, pp. 1−25.

Bühlmann, H., 1985. Premium calculation from top down. ASTIN Bulletin 15 (2), 89−102.

Census of India, 2011. Census info India 2011. http://www.censusindia.gov.in/2011census/population_enumeration.html.

Giné, X., Townsend, R., Vickery, J., 2008. Patterns of Rainfall Insurance Participation in Rural India. The World Bank Economic Review, pp. 1−28.

Greatrex, H., Hansen, J., Garvin, S., Diro, R., Guen, M.L., Blakeley, S., Rao, K., Osgood, D., 2015. Scaling up Index Insurance for Smallholder Farmers: Recent Evidences and Insights, CCAFS Report No. 14 Copenhagen, CGIAR Research Program on Climate Change, Agriculture and Food Security (CCAFS). Available online at: www.ccafs.cgiar.org.

Ishiguro, S., Ogasawara, K., Fujino, K., Sato, Y., Kishima, Y., 2014. Low temperature-responsive changes in the anther transcriptomes repeat sequences are indicative of stress sensitivity and pollen sterility in rice strains. Plant Physiology 164 (2), 671−682.

Jensen, N., Barrett, C., 2017. Agricultural index insurance for development. Applied Economic Perspectives and Policy 39 (2), 199−219.

Kakumanu, K.R., Palanisami, K., Gurava Reddy, K., Nagothu, U.S., Tirupataiah, K., Ashok, B., Xenarios, S., 2013. An insight on farmers' willingness to pay for insurance premium in south India: hindrances and challenges. In: Gommes, R., Kayitakire, F. (Eds.), The Challenges of Index-Based Insurance for Food Security in Developing Countries. European Union, Luxembourg, pp. 137−145.

Kliger, D., Levikson, B., 1998. Pricing insurance contracts − an economic viewpoint. Insurance: Mathematics and Economics 22 (3), 243−249.

Lu, L., Macdonald, A., Wekwete, C., 2008. Premium rates based on genetic studies: how reliable are they. Insurance: Mathematics and Economics 42 (1), 319−331.

Mapfumo, S., Groenendaal, H., Dugger, C., 2017. Risk Modelling for Appraising Named Peril Index Insurance Products: A Guide for Practitioners. World Bank, Washington, DC, 20433.

Parthasarathy, D., 2014. Strategizing a New Approach to Crop Insurance in India. CCAFS News Blog published on April 18 2014. http://ccafs.cgiar.org/blog/strategising-new-approach-crop-insurance-india.

Singh, G., 2010. Crop Insurance in India. Working paper 2010-06-01, Indian Institute of Management, Ahmedabad, India. http://www.iimahd.ernet.in/publications/data/2010-06-01Singh.pdf.

Wang, S., 1995. Insurance pricing and increased limits ratemaking by proportional hazards transforms. Insurance: Mathematics and Economics 17 (1), 43−54.

Drought risk insurance and sustainable land management: what are the options for integration?

Daniel Tsegai[a],*, Ishita Kaushik[b]

[a]*United Nations Convention to Combat Desertification (UNCCD), Bonn, Germany,*
[b]*University of California, Berkeley, CA, United States*
*Corresponding author

Background

Droughts are projected to increase in severity, frequency, duration, and spatial extent (Dai, 2011). Drought has detrimental impact on agriculture, where in sub-Saharan Africa the sector on an average contributes to a 32% of GDP and 65% of the labor force (Chijioke et al., 2011; Pye-Smith, 2012; Ali et al., 2017). Drought is a major risk to rain-fed agriculture, amounting to 83% of all the risks in sub-Saharan African agriculture and bringing about 40% of economic damages to smallholders (World Bank, 2005; Burke et al., 2010; Tadesse et al., 2015). A single year of drought can set back years of social development for vulnerable members of society.

To reduce vulnerability to droughts, there is a need for a paradigm shift from a reactive, short-term posthazard management approach toward proactive, long-term ex ante drought management approach (Tsegai et al., 2015; Wilhite and Pulwarty, 2017). Related actions may include structural measures, such as planting drought-resistant crops, improved animal breeds, building infrastructure that supports water conservation and use regulation, or facilitating income generation that is less dependent on water availability. Nonstructural measures include drought-related knowledge development, risk and vulnerability assessments, insurance, capacity building on monitoring and forecasting, all embedded in comprehensive drought policies and plans (Duguma et al., 2017). These approaches have long-term outcomes to break intergenerational shock transfer and cycles of poverty (Barnett et al., 2008).

Drought insurance is one such instrument to protect farmers against uncertainties of crop yields in times of drought (Silvestre and Lansigan, 2015; Ward and Makhija, 2018). There have been earlier attempts to implement subsidized crop insurance programs against multiple perils, including drought, mostly unsuccessful (Hazel, 1992). The subsidy became too expensive for most governments to afford in the long run. Nevertheless, with the advancements in drought monitoring, forecasting, and early warning systems, along with assessments on drought vulnerability and impact, the potential for drought risk insurance cannot be undermined. Fig. 13.1 provides a nonexhaustive overview of farmer insurance schemes in developing regions.

Drought Challenges. **https://doi.org/10.1016/B978-0-12-814820-4.00013-4**

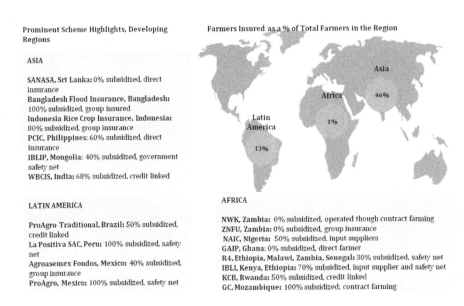

Prominent Scheme Highlights, Developing Regions

ASIA

SANASA, Sri Lanka: 0% subsidized, direct insurance
Bangladesh Flood Insurance, Bangladesh: 100% subsidized, group insured
Indonesia Rice Crop Insurance, Indonesia: 80% subsidized, group insurance
PCIC, Philippines: 60% subsidized, direct insurance
IBLIP, Mongolia: 40% subsidized, government safety net
WBCIS, India: 68% subsidized, credit linked

LATIN AMERICA

ProAgro Traditional, Brazil: 50% subsidized, credit linked
La Positiva SAC, Peru: 100% subsidized, safety net
Agroasemex Fondos, Mexico: 40% subsidized, group insurance
ProAgro, Mexico: 100% subsidized, safety net

Farmers Insured as a % of Total Farmers in the Region

Asia 46%

Africa 1%

Latin America 13%

AFRICA

NWK, Zambia: 0% subsidized, operated though contract farming
ZNFU, Zambia: 0% subsidized, group insurance
NAIC, Nigeria: 50% subsidized, input suppliers
GAIP, Ghana: 0% subsidized, direct farmer
R4, Ethiopia, Malawi, Zambia, Senegal: 30% subsidized, safety net
IBLI, Kenya, Ethiopia: 70% subsidized, input supplier and safety net
KCB, Rwanda: 50% subsidized, credit linked
GC, Mozambique: 100% subsidized; contract farming

FIGURE 13.1

Overview of farmer insurance schemes in developing regions.

Own data analysis from Lowder, S.K., Skoet, J., Raney, T., 2016. The number, size, and distribution of farms, smallholder farms, and family farms worldwide. World Development, 87, 16–29.

This chapter discusses the potential of an innovative insurance tool that could lead to a reduced subsidy by creating an incentive for farmers to work on their land for a reduced or temporarily free premium. This would help to achieve the twin goals of rehabilitating land and reducing vulnerability to drought that can revolutionize access to formal insurance for smallholder farmers and simultaneously increase the adoption of sustainable land management (SLM) practices.

Agricultural insurance schemes

Currently, two main drought insurance scheme types are prevalent: the traditional Multi-/Single Peril Crop Insurance (MPCI/SPCI) and Index-based Insurance. MPCI/SPCI protects beneficiaries against yield losses caused due to multiple perils or a single peril by allowing them to insure a certain percentage of historical crop production. While MPCI provides many advantages, especially damage-based indemnity system and the affordability of premiums, it is more complex and costlier to administer. This is because it requires preinspections and in-field measurement of crop yields to assess losses. Payout time is also another important factor. For MPCI, payout time could last months. In contrast, index insurance systems are easier and less costly to administer, and can have faster payouts because no field inspection is needed, payout is triggered by the trespassing of certain predefined parameters for a whole (small) region. In the Caribbean, Munich Climate Insurance Initiative (MCII) has created a 72-h payout guarantee in cooperation with the Caribbean Catastrophe Risk Insurance Facility (CCRIF). In addition, international experience throughout the world shows that MPCI can be subject to adverse selection and moral hazard (Mahul and Stutley, 2010).

So, index-based insurance schemes were developed to shorten payout time and to reduce transaction costs. Index insurance provides coverage against perils that are highly correlated to crop yields or revenue wherein insurance payouts are pegged to be easily measurable indicators such as weather or agroecological conditions. Index insurance can be categorized under two broad categories, direct and indirect insurance, as depicted in Fig. 13.2.

Direct insurance refers to insurance policies that cover yield and revenue losses for a given crop due to any meteorological event. The two prominent direct insurance products are those based on area yield and revenue. In area yield, compensation paid to the farmer depends on the statistical yield for a year in a predefined area, usually an administrative unit. Revenue insurance, on the other hand, combines yield and price insurance. The farmer is paid if the total value of the crop production falls below a predetermined threshold. The primary benefit of revenue index insurance is that it entails less basis risk. It is based on the premise that farmers, especially those who want to sell significant amount of their production in the market, are interested not only in the yield but in the income they receive from their output. For example, a farmer is no worse off when he experiences a 20% reduction in yield, if it is compensated by a simultaneous 20% increase in price that leaves his income unchanged (Mulangu, 2015). Revenue insurance, therefore, also provides livelihood protection to farmers instead of insuring only production cost or input loan.

However, in the case of indirect index insurance, instead of referring to the average yield in an area, it takes meteorological indicator or satellite images into account. Weather derivatives can be included in this category of insurances. Meteorological insurance can be designed based on rainfall and temperature. normalized difference vegetation index (NDVI), in certain cases, serves as better proxy for yield. Indirect insurance, therefore, can be based on crop stress due to drought such as soil moisture index insurance and vegetation-based index insurance constructed with remote-sensing data. However, for these products to work, they need to act as a reliable indicator for crop loss due to specific weather-related crop stress.

Cost-effectiveness is a well-recognized benefit of index insurance compared to MPCI/SPCI as the former avoids high administrative costs, moral hazard and individual loan assessment. Less transaction

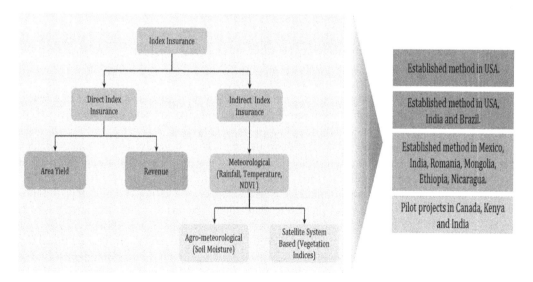

FIGURE 13.2

Index insurance categories.

Authors' own construct.

costs are incurred since the insurance contract is simple, independent of farmer's behavior, difficult to manipulate, transparent, and easy to monitor (Skees and Barnett, 1999; Miranda and Vedenov, 2001). Even though limited empirical evidence exists about the effectiveness of index insurance schemes in developing countries, a study conducted in Mexico (see Alan et al., 2011) revealed promising results. The authors analyzed the effects of rainfall index insurance on crop productivity, farmers' risk management strategies, as well as per capita income and expenditure. The analysis finds that the presence of insurance in the Mexican counties under study had significant and positive effects on maize productivity. Similarly, there was a positive association between the availability of insurance in the municipality and rural households' per capita expenditure and income.

While area yield insurance provides a promising means to farmers against a disaster event, these indices are typically developed for a single crop, as cropping mix within a village may significantly vary by farm, thus posing a challenge for development of the whole farm operations—based yield or revenue index. Also, data availability in developing countries is a prime challenge restricting the efficiency and utility of these products.

However, with the development of geospatial information systems (GIS)—based tools, data mapping in developing countries, especially for meteorological parameters has improved significantly. Crop yields may be subjected to many perils which may not be all parametric and covariate in nature such as quality of seeds planted and individual-level farm management practices. In such a case, developing an actuarially fair area yield index may pose a significant challenge.

Although other alternative parameters are being utilized for developing a robust insurance model such as NDVI, they are typically effective in homogenous terrain only, especially for designing an index. The presence of inhomogeneous, e.g., hilly terrain makes NDVI-based index products very uncertain (for predicting individual plot yields). Therefore, a major concern is the challenge in normalizing and adjusting underlying data differences in elevation and latitude, which results in coarse and imprecise data jeopardizing the design of the insurance product. Multiple index insurance, combining various parameters such as rainfall, temperature, and NDVI, can indeed provide better accuracy to the index insurance product. The consequence is that NDVI is generally relatively poorly correlated with actual plant growth even though the data are available for relatively small grids (Smith, 2009). However, more robust data analysis and standardization methods of reducing uncertainty and data errors must be devised prior to drafting multiple index insurance.

Weather index insurance has gained widespread attention in comparison to area yield index in developing countries. The advantage of weather insurance, relative to traditional crop insurance, is that it only covers weather-related losses instead of insuring yields in general. Weather is insured directly which is one of the strongest determinants of crop yields. It is advisable to assess the correlation of crop yields with weather (at least total rainfall and temperature), as analyzed through sufficiently long farm yield and rainfall time series data. However, time series data are often not available. In addition, in the case of orographic rainfall (rain produced from the lifting of moist air over a mountain), it can be difficult to insure topographically challenging regions as index insurance works best when terrain is flat.

Failure of a thorough assessment may significantly restrict the reliability of weather-based index insurance as an efficient tool to combat drought-related economic losses. In case of index-based insurance, the index variable value is also assessed over a larger area, comparatively farther from the individual farmer field which may exaccerbate the problem of *basis risk*.

Developing an index more highly correlated with farmer yields can be a good strategy to reduce basis risk. Innovative solutions in India, Kenya, and Ethiopia include development of temperature- and vegetation-based index systems where these variables have shown to be correlated with crop yields (Leblois and Quirion, 2011; Burke et al., 2010). Linking geospatial data with location-specific agronomic conditions can be used to support the aforementioned strategy. Few such potential solutions such as DSSAT (Decision Support System for Agro-technology Transfer) and CropSyst are currently being used in the United States to reduce the basis risk (Bobojonov and Sommer, 2011). These solutions combine existing knowledge of agro-ecology, crop agronomy, and plant physiology which can be consequently used to construct indices more consistent with crop yields (Fig. 13.3).

Certain organizations are trying to cover entire Africa under low resolution especially for evaporation and precipitation. Based on these data, index insurance can be designed for any location in Africa. One such organization is Environmental Analysis and Earth Monitoring (EARS) based in the Netherlands, specializing in satellite data for climate, water, and food.

The timing, not just the amount of rainfall during the various growth phases of a plant, is very important for satisfying the soil—water balance and, therefore, the ultimate yield (Fig. 13.4). This is what we call the temporal risk. Dry spells over the main phases of crop growth can cause yield loss, albeit cumulative seasonal rainfall being adequate. Thus, it is necessary to capture stress during critical crop growth phases to devise an effective insurance product.

As part of the Index Insurance Innovation Initiative (I4) of the Feed the Future Innovation Lab for Assets and Market Access at the University of California, Davis, a team of researchers designed an innovative multiscale index insurance product that uses a double trigger method for area yield—based instrument (Elabed et al., 2013). The insurance is triggered when a farmer group's area yield falls below a threshold which is then compared to a broader geographical area yield which serves as a second reference point. The multiscale design minimizes contract failure while limiting opportunities for

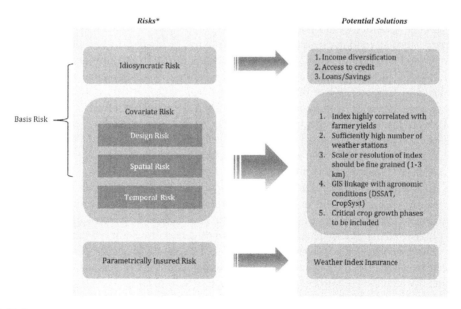

FIGURE 13.3

Sources of crop yield risks and potential solutions.

Adapted from Elabed, G., Bellemare, M.F., Carter, M.R., Guirkinger, C., 2013. Managing basis risk with multi-scale index. Agricultural Economics, 44 (4—5), 419—431.

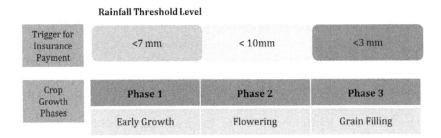

FIGURE 13.4

Phase-wise index insurance contract: An illustration.

Adapted from Ntukamazina, N., Onwonga, R.N., Sommer, R., Rubyogo, J.C., Mukankusi, C.M., Mburu, J., Kariuki, R., 2017. Index-based agricultural insurance products: challenges, opportunities and prospects for uptake in sub-Sahara Africa. Journal of Agriculture and Rural Development in the Tropics and Subtropics, 8 (2), 171–185.

moral hazard. Given small land acreage and fast degrading lands, farmers' ability to afford the premiums may be restricted where insurance is considered as a nontrivial demand by these households, which are often extremely poor (Schaefer and Waters, 2016).

Linking sustainable land management and drought insurance

Designing an innovative product which addresses not only ex post drought impact but also combats land degradation through adoption of appropriate ex ante strategies is crucial to forming holistic policies for reducing vulnerability of smallholders to drought. Combating land degradation is also vital for other purposes such as yield improvements, erosion control, or biodiversity losses. So, to achieve the twin goals of land sustainability and enhancing individual farmers' and entire countries' abilities to cope with drought, SLM practices can be used as a condition or a contractual obligation to benefit from a reduced or temporarily free premium insurance scheme to farmers who decide to participate in the insurance scheme. Community-level insurance is more practical for this purpose as community-level insurance is expected to reduce the need for assessment of individual farms for SLM practices compliance, as defined by contractual obligations, and some SLM practices are not feasible on individual plots or farms but take place on communal grounds, such as river bed stabilization or communal forests (source) protection.

During the audit of predetermined and agreed-upon observable characteristics, farms can be selected randomly for assessment and vouchers, stating compliance level which can be issued to the communities. These vouchers can then be utilized for receiving payouts as and when drought occurs. It is expected that the value and reduction in transaction cost achieved because of implementing index insurance may get diminished as a result of integrating insurance with SLM compliance audit model. However, in the long run, the benefit derived from continued use of SLM practices is expected to outweigh the costs associated with the integrated model.

After a few years of premium subsidized assistance, targeted farmers can be graduated from the integrated premium-free insurance scheme with improved agroecological conditions and infrastructure. Following the termination of this scheme, farmers can continue to sustain their livelihoods with lesser impact of climatic variations and standard index insurance schemes without premium subsidy.

It is important to note that premium subsidy schemes are effective in developing countries if they can be maintained over a long term without draining public resources. The scheme should also be able to promote wide-scale adoption of SLM practices leading to improved agricultural productivity and soil quality which continues even after the termination of the integrated and subsidized insurance scheme. On the other hand, governments and donors invest a lot for SLM projects to achieve public goods such as poverty alleviation, food security, erosion control and other agro-ecological benefits. So, a comparison of SLM through insurance and SLM through the other measures, with regards to costs and benefits, efficiency, and transaction costs should be considered (Fig. 13.5).

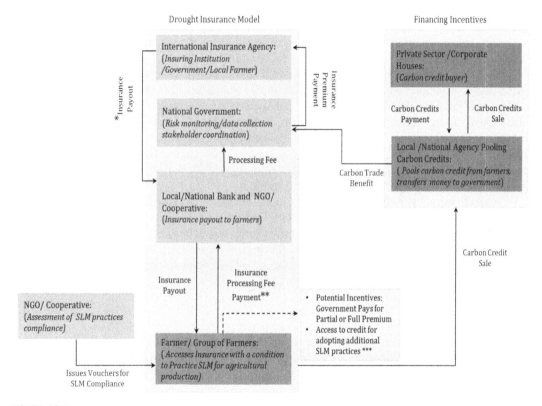

FIGURE 13.5

Hypothetical drought insurance—sustainable land management (SLM) integration model. * Maintaining a single channel for insurance payout is critical for timely payments. Payments can be disbursed to the local bank or an NGO, and can then be disbursed to farmers/farmer groups holding accounts with banks. In case farmers do not have a bank account, they can be issued checks. Access to formal bank accounts and resolving operational issues would be the necessary first step for timely payouts. Mobile banking system can also be harnessed where payments can be made directly to their accounts which can be accessed through mobile phones. ** For the premium subsidy (or premium-free insurance), making a small financial contribution (small processing fee) can prevent negative incentives that may arise out of free access to disaster relief assistance. *** In the case of "Access to Credit incentive" for buying agricultural inputs, the insurance payout would be applied to farmer's outstanding loans in times of drought. It can be paid directly to the financing institution, such as a bank, a cooperative or an input on credit supplier.

Authors' own design.

Institutional setup

Drawing from high adoption rates in Guatemala, India, Mexico, and Mongolia, subsidizing premium insurance helps in increasing smallholder farmers' demand for index insurance in the short run and to help them get accustomed to buying it (Cohen and Dupas, 2010; Cai et al., 2017). In case of access to premium-free drought insurance, individual farmers/farmer groups as a single unit can be insured free of premium. In addition to premium-free drought insurance, for covering against additional potential losses, they can be provided access to easy credit to procure agricultural inputs required for SLM where insurance serves as collateral for credit. Index insurance, therefore, can be used as a means of providing the microfinance institution (credit institution) or an input on credit (contract) supplier with the confidence to lend farmers with the capital necessary to purchase higher quality inputs, and in turn, increase productivity and sustainability which allows providers to hand out higher credit amounts and achieve better repayment rates for higher profits.

This model could target smallholder farmers cultivating primarily staple crops where yield-based weather index insurance provides a more suitable risk mitigation tool. In Africa, given the high correlation of crop production with weather parameters, weather-based index insurance is more suitable owing to the nature of their farms. Also, individual farm assessment for drought insurance involves higher transaction costs which can be subsided with weather-based index insurance. Commercial farmers on the other hand, are expected to benefit more from MPCI policies which can cover them with all kinds of agricultural risks such as price fluctuation of crops, droughts, floods, postharvest losses, disease and insect incidence, and fire.

To achieve operational efficiency, the index model would require cooperation of multiple stakeholders at different levels catering to government, farmers, and international donors and investors. In such a scenario, therefore, it would be important to clearly define the role of various stakeholders in supporting the development and implementation of agricultural insurance. Once the roles have been identified, it is imperative for insurers and governments to agree upon a predetermined index-based insurance model.

Insurance Payout Structure

As is to be observed with all cash incomes of smallholder farmers, it is likely that with insurance payout, target farmers may overwhelmingly spend money on essentials without saving for the next harvest season. Therefore, to promote judicious use of the cash received as part of the insurance payout scheme, farmers can alternatively be paid a part in cash and another in the form of vouchers to procure agricultural inputs. Drawing from Tanzania's *National Agricultural Input Voucher Scheme (NAIVS),* where a similar voucher-based smart input subsidy scheme was launched in 2008 with the aim to raise maize and rice production, targeted farmers can be offered two to three vouchers (for seed, pesticide, insecticide, or fertilizer) redeemable at a local agricultural retail outlet (Goyal et al., 2014). However, vouchers will only be appropriate in situations where agricultural inputs are available in local markets or can be relatively quickly supplied through market mechanisms. Therefore, developing a robust logistical and distribution structure would be essential. An existing distribution network or a tie-up with an international seed company operating in the country can also be harnessed where the input companies invest in the distribution network leveraging on gaining economies of scale because of bulk sale of inputs to large number of farmers (Table 13.1).

Table 13.1 Roles and responsibilities of various institutions in drought insurance and sustainable land management (SLM) integration.

Responsibilities	Government	Insurer	Agricultural Bank/ MFI as mediator	Private/Corporate sector	NGO/ Cooperative
Oversight of the scheme		–	–	–	–
Data collation/ surveys		–	–	–	–
Selection of insurers		–	–	–	–
Drafting insurance model			–	–	–
Outreach and marketing to beneficiaries				–	–
Collection of insurance processing fee		–		–	–
Processing of indemnity payment	–			–	–
Reinsurance (international capital markets)			–	–	–
Carbon credit stock taking	–	–	–		
Collection of carbon credit sale realization and transfer to government	–	–	–	–	
Carbon credit investment	–	–	–		–
SLM compliance assessment	–	–	–	–	
Monitoring and regulating SLM compliance assessment					

Authors' own design.

Noncompliance and farm SLM practice assessment procedure

If a farmer is assessed not to have followed the suggested practices as per the contract, a predetermined amount could be deducted from his insurance payment as a penalty. Farm audits can be conducted for assessing farmer's compliance to the adoption of SLM practices as directed under insurance contract. Drawing from similar setups in the United Kingdom, a farm audit would identify and define key

Table 13.2 Illustrative sustainable land management (SLM) Practices Based on Observed Characteristics Method (Processes Adhered to) for Compliance Audit.

Improvement	SLM practices	Benefit	Assessment unit
Improvement in agroecology	No tillage		Farm level
	Multicropping/ intercropping		Farm level
	Agroforestry	Reduced costs, improved resilience, stable yield, risk mitigation	Farm level
	Selection of suitable varieties		Farm level
Farm infrastructure	Watershed development/ ditch improvements		Farm level
	Postharvest infrastructure		Community level
Authors' own design.			

objectives of sustainable agricultural practices (Measures, n.d.). Primary indicators can then be selected for each criterion which can be measured and graded accordingly. Subsequently, farmers can be issued a voucher with their respective grades based on compliance to either processes adhered to or outcomes of the suggested farming system and practices. In the initial years, however, observed characteristics method (based on SLM processed adhered to) would be more suitable as SLM outcomes may not be visible in a short span of time. Despite the site-specific and individual nature of sustainable agriculture, several general practices and principles can be identified to help farmers select appropriate management practices (Table 13.2).

Improving operational efficiency through complementary strategies

Implementing *Farm Audit* may require creation of a standardized audit procedure and tools to register observations on use of SLM supporting agricultural practices. This feature would probably require investments in systems for collecting, standardizing, and auditing data for transparent and efficient assessment. In addition to identifying alternative financing mechanisms for this activity, potential synergies between *Farm Audit* and agriculture extension activities can also be explored. *Farm Audits* can be conducted by authorized extension workers trained according to auditing requirements. This can serve as a complementary strategy to improve both agricultural practices and risk mitigation activities.

Financing the incentives

a) *Carbon credits:* In developed countries, usually a significant portion of insurance premium is subsidized by the government. Even if it is beneficial from the social standpoint, such subsidies are more affordable for countries where farmers represent a small portion of the total population. This will be less true when agricultural activities make up a greater percentage of total economic activity in a country, as is typically the case for most developing countries. The funding for providing free insurance can be generated through carbon credit programs wherein an international investor or a private firm can invest in carbon credits equivalent to the tonnage of carbon expected to be sequestered through SLM practices. This would help in easing the financial burden on governments and local communities and is expected to achieve wider adoption among the farmers in developing countries. Relatively easy access to credit to invest in SLM practices in addition to premium-free drought insurance should give enough impetus for early adoption of this scheme. Carbon can be sold upfront in the international market mediated by the national government. Third-party verification by an independent body can confirm the amount of emissions reductions/carbon sequestered through the years. Funds generated from carbon sale can be deposited in a community fund or bank for carbon credits which can then be transferred to the government to compensate for premium subsidy of drought insurance being offered to farmers.

b) International donors: Payment can be received from international agencies and disbursed via government. However, this method will need to be strengthened further. Lack of sufficient funds for insurance and competing projects in developing countries hinder progress in this field.

c) Beneficiary pays: Businesses or communal authorities that regularly draws benefits from ecosystem services may pay the farmers directly in order to secure these services. This may prove to be an effective approach for cash crops such as coffee and cocoa, which hold vast export potential and have a reduced number of high-value market clients who are susceptible to the advantages and disadvantages of the sustainability of production.

d) Farmer pays: Farmers may want to invest in sustainable management practices convinced by government with a promise of consistent and increased income from agricultural production. However, this may prove to be a challenge for farmers surviving on a severely degraded land wherein their low income and stagnant yields may not provide sufficient funds to invest in their farms.

e) Catastrophe bonds issued by government: Catastrophe bonds, which act as a financial instrument to transfer risks to investors in the capital markets, are well suited to being redesigned to protect the vulnerable against increasingly severe weather extremes such as drought. However, for capital markets to invest in these bonds and design an instrument specifically tailored for a situation such as drought, high level of data accuracy is a prerequisite. African Risk Capacity (ARC) is currently operating the scheme under Extreme Climate Facility in Africa (Scott et al., 2017). Clear evidence of the success of this measure is yet to be seen.

Conclusions and recommendations

Insurance is one of the various measures to reduce drought risks. Whereas it is more beneficial to reduce short-term risk, it can also be used to reduce long-term risks. While individual farmer insurance has been successful in a few countries, community-level insurance may be the key to address low uptake issue and to further reduce transactional costs. Institutional insurance in developing countries also suffers from weak institutional capacity for combating disaster events, which often results in delayed insurance payouts. Institutional insurance, therefore, can be used to address large covariate catastrophic risks in the form of safety nets. However, to address intermediate risks, community-level insurance or individual farmer insurance should be considered (Fig. 13.6).

Even though index insurance is a promising instrument to address drought impacts, designing a comprehensive risk mitigation strategy should combine different instruments on a demand-driven

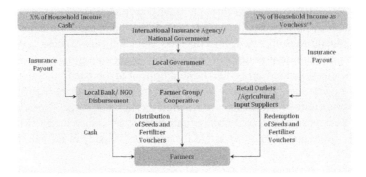

FIGURE 13.6

Insurance payout structure. * Based on target farmers' economic history, a certain percentage of their household income used for education, food, health and clothing can be paid in cash. ** Vouchers can be issued for agricultural inputs. They can be issued by farmer group/co operative and can be redeemed at local agricultural retail outlet.

Authors' own design.

basis. Insurance should typically serve as a complement to formal and informal risk mitigation activities such as savings, investment in sustainable cultivation practices, and easy access to agricultural credit and improved technology. Insurance could also be used for larger catastrophic risks and commercial risks, whereas credit and savings can be more suitable to smaller covariate risks.

However, there is a challenge in data. BI do not agree that credit and saving is useful for "covariate" risks only - why not also for idiosyncratic risks? uilding infrastructure and systems to provide reliable data can prove to be cost-effective in the future. Donor support should be focused on data development, whereas alternate financing tools, specifically market-based options involving capital markets and private sectors should be utilized for financing premium subsidies. Developing countries typically lack reasonable concentration of weather stations, a necessary requirement for developing a highly accurate index-based insurance product and data about historical damage. If it is not available, the premium's price will increase, and the product is less acceptable and affordable to farmers.

Roles and responsibilities of various stakeholders such as government, international insurers, cooperatives, NGOs, and local banks should be clearly defined. More importantly, clarity regarding contract design should be gained before enforcing contracts. Local regulations should be standardized to provide an enabling environment and achieve operational efficiency. Donor support can be leveraged for capacity building in order to govern insurance schemes effectively.

Integrated approaches toward drought insurance should be adopted. To realize the potential of this integrated approach with multiple layers of stakeholders, an enabling institutional framework should be put in place to avoid coordination discrepancies. Efforts should be made in educating farmers about benefits and costs of adopting SLM practices and losses incurred due to weather-related extreme events.

SLM practices, if effectively implemented, have the potential to address land degradation and mitigate the effects of drought. The key is to support the development of comprehensive land use policies, the identification of appropriate SLM technologies, and the respective needs for inputs and investments providing training to farmers and offering right incentives for wide-scale adoption. SLM practices are likely to be adopted where SLM has the potential to reduce drought impact, increase crop yields, and subsequently farm income. Therefore, policies which provide means to mitigating drought risk, tangible benefits with right mix of incentives, and a potential role for reducing the vulnerability to drought are likely to be effective in the long run. Such integrated policies are also better suited to achieve the twin goal of combating drought risk and land rehabilitation, rather than adopting isolated approaches, which are likely to be more expensive.

It is difficult to expect farmers in developing countries to pay for full costs associated with risk mitigation activities upfront, especially in the face of degrading lands and small land holding size, among others. Seen in this light, there may be a valid argument for providing incentives for wide-scale adoption of insurance products; however, in developing countries with significantly large number of farmers, arranging funds for premium subsidies for investment in SLM practices may prove to be challenging. As donor funds are largely directed toward risk mitigation and SLM implementation measures, lack of enough funds and competing projects in developing countries renders them ineffective.

To ease financial burden on governments and local communities, alternative financing sources should be explored. Carbon credit or carbon sequestration-based payment for ecosystem services is one such mechanism which has the potential to raise new sources of sustainable finances. However, instead of transferring the benefit derived from carbon credit directly to farmers, governments can utilize these funds to finance premium subsidie sand maybe also to fund the necessary infrastructure. Designing carefully drafted index insurance can avoid the issues of moral hazard and adverse selection; index insurance carrying premium subsidies offered to farmers with a condition to invest in the SLM practices are subsequently expected to promote judicious use of insurance payout. Additionally, insurance payout can be paid in part in cash and another in the form of vouchers to procure agricultural inputs for the next harvest season. This can further avoid overspending by farmers in essentials immediately after insurance payout is made.

Glossary

Adverse selection	Adverse selection generally refers to a situation where sellers have market or product information that buyers do not have, or vice versa. Because of adverse selection, insurers find that high-risk farmers are more willing to take insurance and pay greater premiums.
Moral hazard	Moral hazard occurs when someone increases their exposure to risk when insured, especially when a person takes more risks because someone else (mainly government) bears the cost of those risks.
Basis risk	This risk arises when the index measurements do not match an individual farmer's actual losses as insurance payout depends on an index and is not calculated on assessment of individual losses. There are two major sources of basis risk where one stems from poorly designed products and the other from geographical elements.
Retention risk	Risk where farmer's yield falls below long-term average, however, they can maintain their commercial viability for next harvest season.
Commercial risk	Risk where farmer's yield falls below long-term average, where they struggle to arrange working capital for next cropping season. For those under input debt, repayment is challenging without external support.

Catastrophic risk	Farmers need external support to avoid falling into irreversible poverty trap and long-term negative consequences on their livelihood.
Payment for ecosystem services	This refers to the incentives offered to farmers in exchange for managing their land sustainably and to provide some sort of ecological service.
Direct insurance	Direct insurance usually refers to insurance policies that cover yield and revenue losses for a given crop due to any meteorological event.
Area yield	In area yield insurance, compensation paid to the farmer depends on the statistical yield for a year in a predefined area, usually an administrative unit.
Revenue insurance	It combines yield and price insurance. The farmer is paid insurance amount if the total value of the crop production falls below a predetermined threshold. The primary benefit of revenue index insurance is that it entails less basis risk than typical yield index insurance. It is based on the premise that farmers, especially those who see significant amount of their production in the market, are interested not only in the yields but in the income they receive from their output. For example, a farmer is no worse off when he or she experiences a 20% reduction in yield, if it is compensated by a simultaneous 20% increase in price that leaves his income unchanged. Revenue insurance, therefore, also provides livelihood protection to farmers instead of insuring only production cost or input loan.
Indirect index insurance	Instead of referring to the average yield in an area, indirect index insurance takes meteorological indicator or satellite images into account. Weather derivatives can be included in this category of insurances (Agriculture and Fisheries Unit et al., 2006).
Meteorological insurance	Meteorological insurance can be designed based on rainfall and temperature or the newly devised NDVI-based index which, in certain cases, serves as a better proxy for yield.

Risks definitions

Spatial risk	Risk associated with variable topography and rainfall distribution pattern.
Temporal risk	Instrument does not reflect the actual growth stage that is sensitive to specific weather such as drought.
Parametrically insured risk	Risk transfer arrangement that does not indemnify the full loss for the protection of buyer. Broadly, it is an insurance program that is triggered based on an assessable index.
Idiosyncratic risk	The risk that a loss unique to an individual farmer may occur that is not observed by the instrument.
Covariate risk	Neighboring households suffer similar shocks.
Design risk	This risk reflects the imperfections of the index as a predictor of yield losses.

References

Alan, F., Hendrik, W., 2011. Concept and unintended consequences of weather index insurance: the case of Mexico. American Journal of Agricultural Economics 93 (2), 505−511.

Ali, S., Liu, Y., Ishaq, M., Shah, T., Ilyas, A.A., Ud Din, I., 2017. Climate change and its impact on the yield of major food crops: evidence from Pakistan. Food 6, 39.

Barnett, B.J., Barrett, C.B., Jerry, S., 2008. Poverty traps and index-based risk transfer products. World Development 36 (10), 1766−1785.

Bobojonov, I., Sommer, R., 2011. Alternative insurance indexes for drought risk in developing countries. In: International Congress of European Association of Agricultural Economists (EAAE), Zurich, August 30—September 2, 2011.

Burke, M., de Janvry, A., Juan, Q., 2010. Providing Index-Based Agricultural Insurance to Small Holders. Recent Progress and Future Promise. CEGA, University of California at Berkeley.

Cai, J., de Janvry, A., Sadoulet, E., 2017. Subsidy Policies and Insurance Demand. Working paper number 22702. The National Bureau of Economic Research, USA.

Chijioke, O.B., Haile, M., Waschkeit, C., 2011. Implication of Climate Change on Crop Yield and Food Accessibility in Sub-Saharan Africa. ZEF, Bonn, Germany.

Cohen, J., Dupas, P., 2010. Free distribution or cost-sharing? Evidence from a randomized malaria prevention experiment. The Quarterly Journal of Economics 125 (1), 1—4.

Dai, A., 2011. Drought under global warming: a review. WIREs Climate Change 2, 45—65.

Duguma, K.M., Brüntrup, M., Tsegai, D., 2017. Policy Options for Improving Drought Resilience and its Implication for Food Security: The Cases of Ethiopia and Kenya. Studies. DIE, Bonn, Germany.

Elabed, G., Bellemare, M.F., Carter, M.R., Guirkinger, C., 2013. Managing basis risk with multi-scale index. Agricultural Economics 44 (4—5), 419—431.

Goyal, A., Gine, X., Rohrbach, D., Biswalo, D., Mashindano, O., Mmari, D., Mink, S., 2014. Public Expenditure Review National Agricultural Input Voucher Scheme (NAIVS). World Bank, Washington DC.

Hazell, P.B.R., 1992. The appropriate role of agricultural insurance in developing countries. Journal of International Development 4, 567—581.

Leblois, A., Quirion, P., 2011. Agricultural Insurances Based on Meteorological Indices: Realizations, Methods and Research Challenges. Reading, Royal Meteorological Society.

Lowder, S.K., Skoet, J., Raney, T., 2016. The number, size, and distribution of farms, smallholder farms, and family farms worldwide. World Development 87, 16—29.

Mahul, O., Stutley, C.J., 2010. Government Support to Agricultural Insurance: Challenges and Options for Developing Countries. World Bank, Washington DC.

Measures, M., n.d. Farm auditing for sustainability. Available at: http://orgprints.org/10881/1/Farm_Auditing_for_Sustainability_Mark_Measures.pdf.

Miranda, M., Vedenov, D.V., 2001. Innovations in agricultural and natural disaster insurance. American Journal of Agricultural Economics 8 (3), 650—655.

Mulangu, F., 2015. How to use weather index insurances to address agricultural price volatility?. In: Seventh Multi-Year Expert Meeting on Commodities and Development, 15—16 April 2015. UNCTAD, Geneva.

Ntukamazina, N., Onwonga, R.N., Sommer, R., Rubyogo, J.C., Mukankusi, C.M., Mburu, J., Kariuki, R., 2017. Index-based agricultural insurance products: challenges, opportunities and prospects for uptake in Sub-Sahara Africa. Journal of Agriculture and Rural Development in the Tropics and Subtropics 8 (2), 171—185.

Pye-Smith, C., 2012. Increasing Rural Employment in Sub-Saharan Africa. CTA Policy Brief. No. 12.

Schaefer, L., Waters, E., 2016. Climate Risk Insurance for the Poor and Vulnerable: How to Effectively Implement the Pro-poor Focus of Insu-Resilience. http://www.climate-insurance.org/fileadmin/mcii/documents/MCII_2016_CRI_for_the_Poor_and_Vulnerable_full_study_lo-res.pdf.

Scott, Z., Simon, C., Mc Connell, J., Villanueva, P.S., 2017. Independent Evaluation of African Risk Capacity (ARC) Final Inception Report.

Silvestre, P.R., Lansigan, F.P., 2015. Drought risk management through rainfall-based insurance for rainfed rice production in Pangasinan, Philippines. Asia Pacific Journal of Multidisciplinary Research 3 (5), 111—120.

Skees, J.R., Barnett, B., 1999. Conceptual and practical considerations for sharing catastrophic and systemic risks. Review of Agricultural Economics 21 (2), 424—441.

Smith, V.H., Myles, W., 2009. Index Based Agricultural Insurance in Developing Countries: Feasibility, Scalability and Sustainability.

Tadesse, M.A., Bekele, S., Erenstein, O., 2015. Weather index insurance for managing drought risk in smallholder agriculture: lessons and policy implications for Sub-Saharan Africa. Agricultural and Food Economics 3 (26), 1–21.

Tsegai, D., Liebe, J., Ardakanian, R., 2015. Capacity Development to Support National Drought Management Policies: Synthesis. UNW-DPC, Bonn, Germany.

Ward, P.S., Makhija, S., 2018. New modalities for managing drought risk in rainfed agriculture: evidence from a discrete choice experiment in Odisha, India. World Development 107, 163–175.

Wilhite, D.A., Pulwarty, R.S., 2018. Drought and Water Crisis: Integrating Science, Management and Policy. Taylor and Francis Group, CRC Press, Boca Raton, London New York.

World Bank, 2015. New Insurance Model Protects Mongolian Herders from Losses.

Further reading

Akerlof, G., 1978. The market for 'Lemons': quality uncertainty and the market mechanism. Quarterly Journal of Economics 84 (3), 488–500.

Bardhan, P., 1994. Land, Labor, and Rural Poverty. Columbia University Press, New York.

Blanc, E., 2012. The impact of climate change on crop yields in sub-Saharan Africa. American Journal of Climate Change 1 (1), 1–13.

Boucher, S., Carter, M., Guirkinger, C., 2008. Risk rationing and wealth effects in credit markets: theory and implications for agricultural development. American Journal of Agricultural Economics 90 (2), 409–423.

Carter, M., de Janvry, Alain, Sadoulet, E., Sarris, A., 2015. Index-based weather insurance for developing countries: a review of evidence and a set of propositions for up-scaling. Revue d'Économie du Développement 23, 5–57.

Chalice, L., Coble, K.H., Barnett, B.J., Miller, C.J., 2017. Developing area-triggered whole-farm revenue insurance. Journal of Agricultural and Resource Economics 42 (1), 27–44.

Christiaensen, L., Dercon, S., 2007. Consumption risk, technology adoption, and poverty traps: evidence from Ethiopia. World Bank Policy Research Working 36 1 (1), 1–41.

Colea, J.B., Gibsonb, R., 2010. Analysis and feasibility of crop revenue insurance in China. Agriculture and Agricultural Science Procedia 1, 136–145.

Fuchs, A., Wolff, H., 2016. Drought and Retribution: Evidence from a Large-Scale Rainfall-Indexed Insurance Program in Mexico. Policy research working paper no. 7565. World Bank, Washington, DC.

Syroka, J., Reinecke, E.B., 2015. Weather index insurance and transforming agriculture in Africa: challenges and opportunities. In: Background Paper. United Nations Economic Commission for Africa. International Conference on "Feeding Africa", October 21–23, 2015. Dakar, Senegal.

An assessment of drought monitoring and early warning systems in Tanzania, Kenya, and Mali

Paschal Arsein Mugabe[a],*, **Fiona Mwaniki**[b], **Kane Abdoulah Mamary**[c], **H.M. Ngibuini**[d]

[a]*University of Dar es Salaam (UDSM), Dar es Salaam, Tanzania,*
[b]*Kilimo Media International, Nairobi, Kenya,*
[c]*Egerton University, Department of Agricultural Economics and Agribusiness Management,
Institut d'Economie Rurale-IER, Nakuru, Kenya,*
[d]*Independent Natural Resource Management consultant, Nairobi, Kenya*
*Corresponding author

Introduction

Climate change is expected to increase the frequency and intensity of droughts in many parts of the world including Africa (Santos et al., 2014). Drought causes widespread suffering and loss in rural communities. Recurrent severe droughts have resulted in hunger due to decline in food production, diseases related to nutrition, and outbreaks such as cholera and acute diarrhea. Other consequences are decline in school attendance and increase in dropout rates, conflicts between communities as resources such as pasture and water become scarce, and even deaths. For instance, the prolonged 2016/2017 drought in Kenya left about 3.4 million people food insecure and about half a million others with poor access to water (UNICEF, 2018). These impacts are felt more in the arid and semi-arid lands (ASALs) (Republic of Kenya, 2013) which represent 80% of Kenya's land mass and are home to 30% of the population and half of all livestock in Kenya (Uhe et al., 2017). Droughts are the major constraint to rain-fed agricultural production in the ASALs. The two major types of rain-fed subsistence agriculture practiced in the ASALs of Kenya are small-scale, mixed crop/livestock farming, and pastoralism.

Agriculture is the backbone of Africa's economy and contributes about 30% to the gross domestic product (GDP) annually. Additionally, the sector accounts for 65% of Africa's total exports and provides more than 80% of informal rural employment, making agricultural success a key to food security and reduction of poverty (Asafu-Adjaye, 2014). The 2008–11 drought in Kenya caused USD 2.1 billion in direct damages and losses, and a further USD 11.3 billion in lost income flows across all sectors of the economy, with the livestock sector accounting for 72% of the total damages and losses (Republic of Kenya, 2013).

Climate change–related risks in Africa from extreme events such as drought are already significant as a result of the current situation of 1°C additional warming over pre-industrial global average

Drought Challenges. https://doi.org/10.1016/B978-0-12-814820-4.00014-6

temperatures. In East Africa, the impact of changes in mean annual temperature is more pronounced and adverse to the economy than a similar change in annual precipitation (Seitz & Nyangena 2009). In Tanzania, an increase in temperature between 1.8 and 3.6°C in catchment areas of Pangani River, for example, will lead to a decrease of 6%−9% of the annual river flow, which will have adverse impacts on livelihoods and agricultural productivity (URT, 2007). Risks associated with some types of extreme events will increase at higher temperatures (IPCC, 2014). The implications are considerable. As an example, for Tanzania, models suggest that climate change will lead to losses of 1.5%−2% of annual GDP by 2030 (GCAP et al., 2011). This implies economic losses of at least $1.5 billion per year by 2030 (in 2006 prices). The cumulative effect of these losses is likely to reduce Tanzania's chances of achieving key economic and development targets, and to delay its plans for achieving middle-income country status (GCAP et al., 2011).

Given the significance of the effects of drought on Africa's agricultural sector, urgent action is needed to develop effective early warning systems (EWSs) to detect the probability of occurrence and the likely severity of drought. EWSs are defined as, "a set of capacities needed to generate and disseminate timely and meaningful warning information; to enable individuals, communities and organizations threatened by hazards to prepare and to act appropriately to [avert] the possibility of harm or loss" (Baudoin et al., 2014, p. 6). EWSs provide time for national and local governments in conjunction with humanitarian and development partners to take action (International Federation of Red Cross and Red Crescent Societies, 2014). Early warning information can reduce impacts of drought if delivered to decision makers in a timely and appropriate format and if mitigation measures and preparedness plans are in place (World Meteorological Organization, 2006).

In this chapter, we review the status and future needs of drought EWSs in Mali, Kenya, and Tanzania. Our review is based on a cross-comparison survey and analysis of government policy documents, research journal publications, project reports, and manuals, desk studies/reviews pertaining to existing drought EWSs in the three countries. What now follows is a short summary of our findings, written for a wide audience.

Drought early warning in Kenya

Following the 2011 drought that led to an unprecedented humanitarian crisis in Kenya and across the Horn of Africa, the Government of Kenya (GoK) established the National Drought Management Authority (NDMA). The NDMA is a statutory body gazetted in November 2011. The formation of the body, which now falls under the National Drought Management Authority Act 2016, is underpinned by Sessional Paper No. 8 of 2012 and the National Policy for the Sustainable Development of Northern Kenya and other Arid Lands. The NDMA is mandated to ensure that drought does not result in humanitarian emergencies and that the impacts of climate change are sufficiently mitigated. This means that Kenya now has a permanent and specialized institution tasked with providing leadership on drought management and coordinating the work of all stakeholders in implementing drought risk management activities. The government further endorsed its commitment to ending drought emergencies in Kenya by launching the second Medium Term Plan (MTP) for drought risk management for its Vision 2030 strategy in October 2013 (Republic of Kenya, 2013). The second MTP (2013−17), which is part of a long-term plan aimed at ending drought emergencies by the year 2022, is one of the key foundations to attaining the 10% GDP growth target envisaged in the Vision 2030 (https://

vision2030.go.ke/). In this plan, developed through extensive consultations between state and nonstate actors, it is recognized that drought emergencies have their roots in poverty and vulnerability, and that Kenya's drought-prone areas are also among those which have benefited least from investment in the past. The second MTP aims to strengthen Kenyans resilience to drought and to improve the monitoring of and response to drought conditions through programs and projects to be implemented under the leadership of the NDMA and other government sectors. Flagship projects under the leadership of the NDMA within the second MTP term included the following:

- the National Drought and Disaster Contingency Fund which reduces the response turnaround period;
- an integrated Drought Early Warning System (DEWS) which provides timely and accurate early warning information to all stakeholders and triggers for response;
- an Integrated Knowledge Management System for Drought; and
- the Hunger Safety Net Program which aims to reduce poverty and hunger among poor and vulnerable households through cash transfers via the NDMA and other development partners such as the Kenya Red Cross and World Food Program.

Some of the sectoral priority projects include the establishment of 23 County Early Warning and Response Hubs; the establishment of a food security; a drought information platform; investment in renewable energy; construction, upgrading and rehabilitation of roads; and construction and rehabilitation of water supply systems in the arid regions. The NDMA is also responsible for ensuring delivery of the Ending Drought Emergencies (EDE) strategy. The EDE strategy commits the GoK to reduce the impact of drought by 2022 using two approaches: strengthening the basic foundations for growth and development, such as security, infrastructure, and human capital and strengthening the institutional and financing framework for drought risk management (Second medium term plan, 2013). This strategy builds on the National Policy for the Sustainable Development of Northern Kenya and other Arid Lands as well as the constitution of Kenya (2010) both of which in part seek to address marginalization in the drylands that underpin current levels of vulnerability and risk.

Drought is difficult to forecast, but its impacts can be significantly mitigated through a proactive, risk-based management approach (FAO, 2016). In this regard, the President of Kenya in March 2017 launched various initiatives focused on reducing the impacts of drought on Kenyans, especially pastoralists. These initiatives include livestock insurance payouts, an enhanced livestock take-off exercise, and a cash transfer program. Pastoralist communities have lost many of their livestock due to droughts. For example, 70% of livestock was lost in the droughts of 1991—92 and 2004—06; and 72% in the 2010/2011 drought (Huho, Mugulavai, 2010). The government, through the Kenya Meat Commission, and other development partners such as the Kenya Red Cross Society (KRCS) have embarked on an exercise that aims at destocking weak cattle by buying cows, goats, and sheep from pastoralists at an agreed price. The KMC processes the meat from purchased animals to corned beef (a salt-cured beef product), while the KRCS distributes the meat from the purchased animals to vulnerable households. This exercise helps to remove animals before they become emaciated, lose their value, die, or pose a public health risk to communities. Through this exercise, pastoralists are able to salvage some capital from their at-risk cattle, which they can use to support their families during the drought period. Additionally, this exercise helps to relieve pressure on scarce water and pasture resources, protect their livelihoods, and strengthen the community's ability to recover from the short and long term effects of the drought (Uhe, 2017; International Federation of Red Cross and Red Crescent Societies, 2017).

The government and different insurance companies developed the Kenya Livestock Insurance Program (KLIP), which uses satellites to monitor forage available to livestock and triggers assistance for feed, veterinary medicines, and even water trucks when animal deaths are imminent. Pilot projects established payment levels linked to the state of grazing lands, with the goal of providing enough money to help pastoralists keep their animals alive until the return of the rains (APA, 2017). Under KLIP, payments to pastoralists and farmers are pegged to satellite measurements of available forage in a given area, so that payments are higher in areas where drought is particularly severe and modest in other areas. Through this program, an estimated Ksh 215 million in insurance payouts were made in six drought-stricken counties in early 2018 to avert future losses.

In addition to livestock insurance, the government launched the Kenya Agricultural Insurance and Risk Management Program in March 2016, with technical assistance from the World Bank, UNDP, Equity Kenya Commercial Bank, and insurance companies. This program is designed to address the challenges farmers face when there are large production shocks, such as droughts and floods, with a focus on maize and wheat producers. This insurance program uses an "area yield" approach, with large farming areas being subdivided into insurance units. All insured farmers in a given unit receive a payout if average production in the unit falls below a predetermined threshold. The program was initiated in Bungoma, Embu, and Nakuru counties, with plans to reach 33 other counties by 2020 (The World Bank, 2016).

The Kenya Agriculture and Livestock Research Organization (KALRO), with support from development partners (ICRISAT), has developed drought-resistant crops for the Eastern parts of Rift Valley, and North Eastern Kenya where famine is an ever-present risk. These crops include sorghum, millet, beans, cowpeas, pigeon peas, mung beans, cassava, and sweet potatoes. A challenge is that these crops are commonly perceived as poor people's foods, and it is thus difficult to promote them to Kenyan farmers as viable, commercially marketable foods. There is ongoing need for deliberate and concerted efforts by all stakeholders involved in food production to promote the production of these traditional food crops in the ASALs and their utilization nationally (Karanja et al., 2006).

There are four key challenges in achieving effective drought management, as is outlined in the MTP for Vision 2030 (Republic of Kenya, 2013). First, the most appropriate time to invest in drought resilience is when conditions are good; the reality, however, is that financial investments are more often triggered by crisis events, rather than in periods of normal climate conditions. Second, the allocation of funds by governments and development partners tends toward either emergency responses or conventional livelihoods activities; financing for critical foundational investments (such as roads, education, and peacebuilding initiatives) that ensure other activities achieve their maximum impact, is comparatively lower. Third, decision-making processes for drought early warning are slow and cumbersome, meaning that funds are often released too late in the drought cycle to effectively mitigate its impacts. Fourth, the significant resources that government does make available are typically channeled through each sector's normal financing and procurement channels, which are not nimble enough to take timely action when drought emergencies occur.

Drought early warning systems in Tanzania

Formal drought early warning and management systems are less well advanced in Tanzania as compared with Kenya. In Tanzania, a number of actors are involved in creating EWSs for natural hazards, including drought. Important government actors include the Tanzania Meteorological Agency (TMA); the Ministry of Energy and Minerals Seismology Unit; the Emergency Preparedness and Response Unit (EPRU) of the Ministry of Health and Social Welfare; and the Plant Protection Unit and Food Security Department of the Ministry of Agriculture, Livestock and Fishing. An important nongovernment actor is the Famine Early Warning System Network (FEWS NET, n.d.; Arenas, 2016) which was created by the USAID in 1985 with operations in 34 countries globally to provide early warning and analysis on food insecurity. In Africa, FEWS NET works in East, Central, and West African regions with a vast network of international partners, ranging from collaborators in data collection and analysis to consumers of the reports generated. Most of the early warning information is in the form of reports and maps posted on the FEWS NET website. The reports contain information in the form of monthly reports and maps detailing current and projected food insecurity; timely alerts on emerging or likely crises; specialized reports on weather and climate, markets and trade, agricultural production, livelihoods, nutrition, and food assistance (http://www.fews.net/about-us). Information on hazards is passed on to the community through government publications, the media, and meetings with local communities and their leaders. Tanzania has a National Disaster Management Policy and National Operational Guidelines for Disaster Management, which describe the roles and responsibilities of stakeholders and seek to improve coordination and cooperation among them. Use of mobile telephone technologies and community radio services has been cited as an important method for sharing and disseminating climate information for effective early warning and adaptation (Mutua, 2011).

Indigenous knowledge and person-to-person networks are important components of drought warning and preparation in Tanzania. For example, the CGIAR Research Program on Climate Change, Agriculture and Food Security (CCAFS), and the World Agroforestry Centre (ICRAF) have conducted monitoring and evaluation under the Global Framework for Climate Services Adaptation Programme in Africa (GFCSAPA) to assess the value and usefulness of climate services being provided to farmers and pastoralists in Tanzania and neighboring Malawi. Baseline surveys which were conducted in semi-arid districts of Northern Tanzania found that most farming households rely on indigenous knowledge on climate forecasts to inform their crop and livestock management decisions. For example, the occurrence of swallows and swans in large numbers, moving from south to the north in the months of September to November is regarded as an indicator of the onset of the short rains (Ayal et al., 2017). Farmers also trust information delivered by the well-established network of government agricultural extension agents.

In summary, there is no comprehensive DEWS in Tanzania. Existing EWSs for hazards are inefficient due to lack of enough skilled manpower, equipment, technology, and financial resources. These hinder the capacity for accurate and timely collection, process, and release of early warning data and information. A challenge is how to establish a National Emergency Operation Centre for generating appropriate warning systems and coordinating response arrangements for future disasters. For effective dissemination and use of early warning information, it is necessary to place greater emphasis on public education programs at both the national and local levels. Similarly, traditional/indigenous drought prediction mechanisms could be further developed to provide more reliable information.

Drought early warning systems in Mali

Rural communities in Mali primarily make use of DEWSs based on traditional environmental knowledge acquired through multigenerational experience in their home area (Shumba, 1999). There exist numerous strategies used by farmers in forecasting the weather (Zuma-Netshiukhwi, 2013). These strategies include observations of animal, insect, and plant behavior; assessment of clouds patterns, as well as analysis of humidity, temperature, and wind direction. A keen assessment of known historical trends in weather patterns is also a major strategy used as EWS. In Mali, as well as in many western African countries, rural populations use observations of birds, toads, and white ants to predict the onset of particular seasons as well as rains, temperature trends, and weather patterns. Activities of arthropods, such as fleas, cockroaches, houseflies, and spiders are indications for the arrival of the summer season in Mali, when they become more abundant. By contrast, cockroaches disappear with the onset of the winter season. Weather forecasting based on observations of flowering patterns of certain plants is common knowledge among rural communities in Mali. Although doubts are often raised by outsiders, these traditional weather and climate prediction methods are perceived by poor farmers to be both useful and dependable.

Modern science should be seen as complementing indigenous weather forecasting strategies; in some cases, modern science has benefited from valuable insights from indigenous knowledge (Mundy and Compton, 1991). Integrating the experience of modern science and indigenous knowledge in Mali is essential for achieving more rigorous weather forecasting. Accurate weather information is critical to agricultural decisions such as crop timing, management of herd sizes, and other aspects of rangeland management (Tekwa and Belel, 2009). The resilience of most rural farmers to climatic shocks is significantly influenced by their adoption and implementation of available weather forecasting systems (Ekitela, 2010; Doherty et al., 2010; Field, 2005). However, the increased unpredictability of weather patterns due to climate change means there is a need to ascertain the continued reliability of traditional environmental knowledge as an input for drought early warning.

Climate change can aggravate conflict, result in humanitarian crisis, negatively affect households' livelihoods, and reverse major economic gains (UNDP, 2013). Droughts and other extreme climate events over the past few decades have undermined agriculture and rural development in Mali (Kelly, 2008). EWSs help in monitoring, detecting, and forecasting risks and, when necessary, issuing of alerts about impending food security challenges. Strengthening and improving DEWSs will therefore be key in preventing the erosion of development progress in rural Mali. Mali faces several challenges in establishing and maintaining modern weather observation networks, including limited government budgets, lack of technical infrastructure, and insufficient expertise. There are also challenges due to political and socioeconomic issues, corruption, and poverty. There are, however, a number of approaches that may be used to overcome such challenges, including the following:

- Harnessing cellular telephone networks to gather and disseminate information; this will require developing novel public—private sector partnerships between government agencies, telecommunications companies, and commercial weather companies.
- Utilizing sensing systems that are designed specifically for the environment in which they are to be deployed, which can be installed on or colocated with cell towers, and, require minimal maintenance and calibration.
- Employing other complementary or emerging technologies such as obtaining soundings from instrumentation on commercial aircraft and deploying networks of small X-band radars where active sensing is needed.
- Making ongoing investments in training and professional development of operational staff a top priority.

Adaptation measures will need to be put into place to minimize the impacts of climate change in Mali. Rural Malians have traditionally dealt with a variable and unpredictable climate through coping strategies such as farm-level diversification, soil conservation, and venturing into nonfarming activities. Future adaptation strategies should build upon and enhance traditional coping mechanisms, and originate from the ideas and needs of the local communities, if they are to be successful. Examples of such adaptation options include agricultural improvements (adoption of crops with shorter growing seasons); agroforestry systems and seasonal forecasts; improved water harvesting and management techniques; and the strengthening of relevant government and nongovernmental institutions (for example, agricultural extension). Improved irrigation techniques, such as drip, sprinkling, and Californian systems, will also be an important adaptation option; these have been shown to be more cost-effective than traditional manual irrigation systems, especially for potatoes, shallots, and tomatoes. With the uncertainties associated with the impacts of climate change, efforts on strengthening the capacity of key institutions in Mali are also important. Government agencies, research organizations, and NGOs will need to integrate climate change into their activities and respond to the best available information. Increased harmonization and sharing of information and best practice between stakeholder groups should be encouraged.

Summary and additional recommendations

DEWSs vary considerably across the three countries surveyed, with Kenya's system being the most systematic and advanced and perhaps serving as a model for Mali, Tanzania, and other sub-Saharan African countries. In all the three countries, national partners should be working together with local stakeholders to improve drought responses. These include such activities as training agricultural extension workers in communicating climate services, synthesizing best practice tools, training meteorological services staff on tailored forecast packaging, and improving EWSs more generally. Climate services products demanded by farmers include forecasts of rainy season onset and end, frequency of extreme events, and distribution of dry and wet spells over the growing season. Given that farmers and pastoralists generally trust information provided by government extension agents, further training extension agents in understanding the value and use of climate forecasts, and relying on them to deliver information to farmers, will be critical to the success of DEWSs. Indigenous knowledge should be seen as complementing scientific climate and weather predictions, and vice versa. Extension agents, likely already familiar with indigenous forecasting techniques, must value and understand indigenous knowledge as an entry point for both introducing and complementing scientific forecasting.

The above section on Mali outlined a number of ways by which modern technology can support traditional DEWSs. Radio may also be an effective means of delivering climate information where it exists. In areas with poor access to information and communications technology infrastructure, the use of social networks and printed climate information bulletins may present a practical and scalable means to deliver climate information from the national meteorological service to village leaders and end users.

Monitoring and forecasting drought needs data inputs and analysis from agriculture and wider ecosystem specialists. This means creating interdisciplinary working groups with specialists from meteorological agencies, water authorities, and ministries of environment and agriculture in order to develop capacity to produce and analyze data and to communicate this information to those who need it in a timely and understandable manner.

References

APA, 2017. Government of Kenya Partners With Insurers to Make Record Payout for Herders. http://www. apainsurance.org/news/government-of-kenya-partners-with-insurers-to-make-##record-payout-for-herders/.

Arenas, A., 2016. Strengthening Climate Information and Early Warning Systems in Tanzania for Climate Resilient Development and Adaptation to Climate Change Project. UNDP.

Asafu-Adjaye, J., 2014. The economic impacts of climate change on agriculture in Africa. Journal of African Economies 23 (Suppl. 2), ii17−ii49. https://doi.org/10.1093/jae/eju011.

Ayal, y., Radeny, M., Mungai, C., 2017. Indigenous Knowledge in Weather Forecasting: Lessons to Build Climate Resilience in East Africa. https://ccafs.cgiar.org/news/indigenous-knowledge-weather-forecasting-lessons-build-climate-resilience-east-africa-0#.XBuCTlwzbIU.

Baudoin, M., et al., 2014. Early Warning Systems and Livelihood Resilience: Exploring Opportunities for Community Participation. http://www.riskaward.org/dms/MRS/Documents/ResilienceAcademy/2014_resilience_academy_wp1.pdf.

Doherty, R.M., Sitch, S., Smith, B., Lewis, S.L., Thorton, P.K., 2010. Implications of future climate and atmospheric CO^2 content for regional biogeochemistry. biogeography and ecosystem services across East Africa Global Change Biology 16, 617−640. https://doi.org/10.1111/j.1365-2486.2009.01997.

Ekitela, R., 2010. Adaptation to Weather Variability Among the Dry Land Population in Kenya: A Case Study of the Turkana Pastoralists. MSc thesis. Wageningen University and Research.

FAO, 2016. Easing the Impact of Drought in Kenya. http://www.fao.org/resilience/news-events/detail/en/c/1047550/.

Famine Early Warning Systems Network (FEWS NET), 2016. Remote Sensing Imagery. http://www.fewsnet/about-us. (Accessed 30 June 2016).

Field, C., 2005. Where There is No Development Agency. A Manual for Pastoralists and Their Promoters. NR International, Aylesford.

GCAP (Global Climate Adaptation Partnership), 2011. The Economics of Climate Change in the United Republic of Tanzania. GCAP.

Huho, J.M., Mugalavai, E.M., 2010. The effects of droughts on food security in Kenya. International Journal of Climate Change: Impacts and Responses 2 (2), 61−72.

International Federation of Red Cross and Red Crescent Societies, 2014. Early Warning Early Action Mechanisms for Rapid Decision Making; Drought Preparedness and Response in the Arid and Semi-arid Lands of Ethiopia, Kenya and Uganda, and in the East Africa Region. https://www.preventionweb.net/publications/view/42670.

International Federation of Red Cross and Red Crescent Societies, 2017. Kenya: Livestock Farmers Affected by Drought Receive a Helping Hand from the Red Cross, through a Destocking Exercise. http://www.ifrc.org/en/news-and-media/news-stories/africa/kenya/kenya-livestock-farmers-affected-by-drought-receive-a-helping-hand-from-the-red-cross-through-a-destocking-exercise/.

IPCC, 2014. Climate Change 2014: Synthesis Report. In: Pachauri, R.K., Meyer, L.A. (Eds.), Contribution of Working Groups I, II and III to the Fifth Assessment Report of the Intergovernmental Panel on Climate Change. IPCC, Geneva, Switzerland, p. 151.

Karanja, D.R., et al., 2006. Variety Characteristics and Production Guidelines of Traditional Food Crops. http://www.kalro.org/sites/default/files/Traditional-food-crops-manual.pdf.

Kelly, V., 2008. Agricultural Statistics in Mali: Institutional Organization and Performance. Background Paper for the World Bank (WB) on Agricultural. World Bank, Washington D.C.

Mundy, P., Compton, L., 1991. Indigenous communication and indigenous knowledge. Development Communication Report 74 (3), 1−3.

Mutua, 2011. Strengthening Drought Early Warning at the Community and District Levels: Analysis of Traditional Community Warning Systems in Wajir & Turkana Counties. http://www.fao.org/fileadmin/user_upload/drought/docs/Study%20on%20traditional%20drought%20early%20warning%20systems%20Oxfam%20GB%202011.pdf.

Republic of Kenya, 2013. Sector Plan for Drought Risk Management and Ending Drought Emergencies, Second Medium Term Plan 2013 − 2017. http://www.ndma.go.ke/index.php/resource-center/ede-reports/category/43-ending-drought-emergencies.

Santos, J.R., Pagsuyoin, S.T., Herrera, L.C., et al., 2014. Environment System Decisions 34, 492. https://doi.org/10.1007/s10669-014-9514-5.

Second Medium Term Plan, 2013. Transforming Kenya: Pathway to Devolution, Socio-Economic Development, Equity and National Unity. https://vision2030.go.ke/2013-2017/.

Seitz, J., Nyangena, W., 2009. Economic Impact of Climate Change in the East African Community. Global 21 Consulting and GIZ. http://www.global21.eu/download/Economic_ Impact_Climate_Change_EAC.pdf.

Shumba, O., 1999. Coping with Drought: Status of Integrating Contemporary and Indigenous Climate/Drought Forecasting in Communal Areas of Zimbabwe; Consultancy Report to the United Nations Development Program Office to Combat Desertification and Drought. UNSO/UNDP/WMO, Harare, Zimbabwe.

Tekwa, I., Belel, M., 2009. Impacts of traditional soil conservation practices in sustainable food production. Journal of Agriculture and Social Sciences 5, 128−130.

The World Bank, 2016. Kenyan Farmers to Benefit from Innovative Insurance Program. http://www.worldbank.org/en/news/press-release/2016/03/12/kenyan-farmers-to-benefit-from-innovative-insurance-program.

Uhe, P., et al., 2017. The Drought in Kenya 2016−2017. https://cdkn.org/wp-content/uploads/2017/06/The-drought-in-Kenya-2016-2017.pdf.

UNDP, 2013. Climate Risk Management for Sustainable Crop Production in Uganda: Rakai and Kapchorwa Districts.

UNICEF, 2018. UNICEF Kenya Humanitarian Situation Report, 1 February to 2 March 2018. https://reliefweb.int/report/kenya/unicef-kenya-humanitarian-situation-report-1-february-2-march-2018.

United Republic of Tanzania (URT), 2007. National Adaptation Plan of Action (NAPA). Vice President Office, Division of Environment. Dar es salaam, URT.

World Meteorological Organization, 2006. Drought Monitoring and Early Warning: Concepts, Progress and Future Challenges. https://public.wmo.int/en/resources/library/drought-monitoring-and-early-warning-concepts-progress-and-future-challenges.

Zuma-Netshiukhwi, G.N., 2013. The Use of Operational Weather and Climate Information in Farmer Decision Making, Exemplified for the South-Western Free State, South Africa. Ph.D. Thesis. University of the Free State, Bloemfontein, South Africa.

Further reading

Murungaru, C., 2003. Opening Statement by the Minister of State, Office of the President Republic of Kenya during the Second Conference on Early Warning Systems. Available: www.unisdr.org/ppew/info/Opening-Statement_Murungaru.doc.

Impact of drought-tolerant maize and maize—legume intercropping on the climate resilience of rural households in Northern Uganda

15

Kelvin Mashisia Shikuku*, Chris Miyinzi Mwungu, Caroline Mwongera

International Center for Tropical Agriculture (CIAT), Nairobi, Kenya
*Corresponding author

Introduction

Climatic shocks are increasingly threatening agricultural and economic development in sub-Saharan Africa (SSA). An estimated 70% of economic losses in SSA are attributed to droughts and floods (Bhavnani et al., 2008). Studies have indicated a fall in the annual growth in gross domestic product in SSA countries by 2%—4% (Brown et al., 2011); a substantial decline of about 22% in yields of maize, 17% each for sorghum and millet, and 18% for groundnuts by 2030 (Schlenker and Lobell, 2011); and a reduction in food security (Parry et al., 2005; Lybbert and Sumner, 2012; Wheeler and von Braun, 2013) as a consequence of climatic shocks. Identifying and promoting options that will help address the problems posed by climatic shocks is, therefore, a formidable challenge for policy in SSA.

Climate-smart agriculture (CSA)—an approach seeking to sustainably increase food security and enhance resilience of livelihoods while contributing to mitigation of greenhouse gas emissions—has received considerable attention (FAO, 2013). The popularity of CSA, globally and in Africa, is evident. The launch of a Global Alliance for Climate-Smart Agriculture (GACSA) in 2014, with the goal of helping 500 million smallholder farmers practice CSA, and the Alliance for Climate-Smart Agriculture in Africa (ACSAA) spearheaded by New Partnership for Africa's Development (NEPAD) was intended to help catalyze the scaling up of CSA to 25 million farm households across the continent by 2025.

Implementation of CSA is recognized to be effective in addressing climatic shocks and contributing to increased income required to eradicate poverty in SSA (AGRA, 2016). Empirical evidence suggests that there is a positive relationship between increased income and resilience of households (e.g., Gao and Mills, 2018). Recently, empirical evidence from cost—benefit analysis of CSA technologies has shown positive net present values and high internal rates of return, further suggesting that such technologies are worth investing in (Ng'ang'a et al., 2017; Sain et al., 2017). Some authors have found that CSA technologies not only contribute to increased yields but could also reduce probabilities of low yields (see, for example, Arslan et al., 2015).

Drought Challenges. https://doi.org/10.1016/B978-0-12-814820-4.00015-8

Although a few studies have assessed the effect of CSA on agricultural productivity (Arslan et al., 2015, 2017), empirical evidence about the impact on resilience is scanty. This study used a theory-based resilience measurement approach (Barrett and Constas, 2014; Cisse and Barrett, 2018; Upton et al., 2016) and assessed the impact of drought-tolerant (DT) varieties of maize and maize—legume (M-L) intercropping, in isolation and combination, on the climate resilience of farm households in northern Uganda.

Methodology
Data

The study uses a balanced panel dataset that comes from two waves of household surveys. The first survey was implemented in August to November 2015, whereas the second one was conducted in February to May 2017. The northern Ugandan context is particularly appropriate for our analysis, as poverty in the region is the highest in the country and population growth rate is much higher compared to other regions (Republic of Uganda, 2015), and households are incredibly vulnerable to weather-related shocks (Mwongera et al., 2014). Sampling followed a multistage procedure. In the first stage, Nwoya district in northern Uganda was purposively selected. All the four subcounties in Nwoya district were involved. In the subsequent steps, parishes were randomly sampled within subcounties, villages within parishes, subvillages within villages, and households within subvillages. The total sample size used in this study was 1310 observations. Both surveys covered a broad range of topics including household demographic characteristics, crops and livestock production and marketing activities, exposure to climatic shocks (drought, floods, unpredictable rainfall, less rainfall) and its associated impacts, access to credit and information, participation in farmers' groups, food security, off-farm income activities, and asset ownership.

Theory-based resilience measurement

Barrett and Constas (2014) advanced a theory that focuses explicitly on the stochastic dynamics of well-being and defined development resilience as "the capacity of an individual, household, or community, to avoid and escape from poverty over time and in the face of various shocks. If and only if that capacity is and remains high over time, then the unit is resilient." The moments-based methodological approach advocated by Barrett and Constas (2014) first estimates the conditional distribution of well-being. Based on the conditional distribution, the probability of meeting a threshold level of well-being can then be derived. Subsequently, one can examine the contextual factors that determine the probability of meeting a well-being threshold level in order to identify both targetable characteristics of the most vulnerable and actions that can enhance resilience. Furthermore, it is possible to aggregate resilience measures across individuals to construct group-level measures (Cisse and Barrett, 2018; Upton et al., 2016). This ability to aggregate is particularly important for the targeting of interventions, as it allows one to identify the most vulnerable groups based on readily observable indicators, such as gender and education (Upton et al., 2016).

Employing the approach discussed above, we use annual household income per adult equivalent as our well-being indicator of interest. Several previous studies have used a similar indicator to proxy well-being (see, for example, Adato et al., 2006; Giesbert and Schindler, 2012; Shikuku et al.,

2017). First, we determine an appropriate minimum threshold level of the household income per adult equivalent. We follow previous studies and use a minimum threshold of one US dollar as the threshold for northern Uganda (Kassie et al., 2011). Resilience is then a function of the estimated probability that the household will meet or surpass the income threshold.

The estimation began by econometrically estimating conditional mean income regression equations and then using the residuals from the conditional mean equations to similarly estimate conditional variance as a function of household characteristics and climatic shocks (please refer to Table A15.1 in the Appendix for the parameter estimates). Our dependent variable in the conditional mean equation was the natural log of household income per adult equivalent. The explanatory variables included sex, education, and age of the household head; dependency ratio; access to agricultural and climate-related information; access to credit; membership to farmers' associations; size of kinship network; experience with climatic shocks; loss of crop or livestock as a result of climatic shocks; household and agricultural assets ownership; number of months of food deficit; average distance to the nearest water source in wet and dry seasons. The mean income equation was estimated using ordinary least squares regression with robust standard errors clustered at subvillage level. Residuals from the conditional mean income equation were obtained, squared, and regressed on the same covariates to estimate conditional variance. Using the conditional mean and variance estimates, the probability of a household attaining household income per adult equivalent threshold of one US dollar per day was estimated as that household's resilience score.

Estimation of the impact of climate-smart agriculture on resilience

Our objective was to estimate the causal effect of adopting CSA technologies—in this case, use of DT maize varieties and M-L intercropping (in isolation and in combination)—on the resilience of households to weather-related shocks. Our dependent variable for this analysis was the household-specific resilience measure, which falls in the [0,1] interval. Specifically, the reduced form equation can be written as:

$$y_{it} = \beta_0 + \beta_1 csa_{it} + \beta_2 \mathbf{X}_{it} + \varepsilon_{it} \quad i = 1, \ldots, N; t = 1, 2, \tag{15.1}$$

where y_{it} represents the resilience outcome variable for household i at time t; csa_{it} is a dummy variable equal to one if the household grew a DT maize variety and zero otherwise; $\mathbf{X}$ is a vector of exogenous explanatory variables similar to those used in the estimation of conditional mean and variance equations; and ε is the random error term.

It is possible that the adoption variable, csa_i, is endogenous due to measurement error, simultaneity, or selection bias. In other words, the unobserved factors that predict the outcomes of interest, y might be correlated with the household's decision to adopt a CSA technology. Estimating Eq. (15.1) without adequately controlling for endogeneity would result in biased estimates, overstating the impacts of the CSA technology on resilience.

Eq. (15.1) was, therefore, estimated via two-stage least squares (2SLS) regression with fixed effects in panel data, with the CSA adoption in the first stage even though the assumed model is linear. Angrist and Krueger (2001) stated that even in the case of a dichotomous variable in the first of the two equations, a 2SLS equation produces consistent estimators that are less sensitive to assumptions about functional form.

The study was implemented as part of a randomized controlled trial (RCT) that examined different incentives for agricultural knowledge diffusion from peer farmers to their neighbors.[1] Randomly selected peer farmers, therefore, received training about growing of DT maize, including how to intercrop the DT maize with grain legumes, and were required to help their neighbors to learn about the varieties. We, therefore, used the RCT arms interacted with the endline dummy variable as our instrumental variables in the regressions testing the endogeneity of the binary variables for CSA adoption in outcomes of interest. The first instrument is constructed from a dummy variable equal to one if the farm household comes from a subvillage in which a peer farmer point was assigned to receive a training only and zero otherwise. The second instrument is constructed from a dummy variable equal to one if the household belongs to a subvillage in which a trained peer farmer was assigned to receive social recognition for training his or her covillagers and zero otherwise.

The Hansen J test cannot reject the hypothesis that our instruments are uncorrelated with the error terms for all the technologies considered. We, therefore, accept the orthogonality conditions required for our instruments to be valid. To evaluate the strength of our instruments, we used the Kleibergen—Paap test of underidentification and the Cragg—Donald F-statistic of our first stage regressions. We found that our instruments were jointly significant throughout. Furthermore, the Cragg—Donald F-statistic mostly exceeds the critical 15% value for weak instruments proposed by Stock and Yogo (2005) that was 5.33 for our specifications. The test for exogeneity of the use of DT maize could not, however, reject the null hypothesis that the variable was exogenous. Since under the null the standard fixed effects estimator is more efficient than the fixed effects instrumental variables estimator, we also estimate a standard fixed effects model for the DT maize equation.

Results
Summary statistics

In this study, the adoption of a CSA technology is defined as binary variable taking a value of one for a household which implemented the technology on at least one of its plots (irrespective of the area covered) and zero if the technology was not implemented. Table 15.1 presents descriptive statistics on the use of the CSA technologies. About 7% of the sample households grew a DT variety of maize at baseline compared to 18.5% at endline. Although M-L intercropping was more common than growing of a DT maize variety, about 11% of the households that used M-L intercropping at baseline had stopped practicing it at endline. This further reinforces the need to understand the impacts of CSA technologies. Use of both DT maize and M-L intercropping was very low both at baseline although descriptive results showed a considerable increase between the baseline and the endline. We, therefore, also examined determinants of adoption of the technologies.

Table 15.2 presents summary statistics of the outcome variable (resilience), at baseline (2015) and endline (2017), for adopters and nonadopters of CSA technologies. Panel A presents statistics for DT

---

[1]A total of 126 peer farmers were randomized into one of the three RCT groups: training only, training plus a material reward, and training plus social recognition. In the training only group, selected peer farmers only received training about climate-smart technologies. Peer farmers in the training plus material reward were awarded a weighing scale in case of sufficient knowledge diffusion, whereas those in the training plus social recognition received a reputational reward that involved public announcement of their good performance in knowledge diffusion.

Table 15.1 Summary statistics of adoption of recommended climate-smart agriculture technologies at baseline (2015) and endline (2017).

Variable	2015		2017	
	Mean	SD	Mean	SD
Drought-tolerant (DT) maize (1 = household grew a DT variety of maize, 0 = otherwise)	0.070	0.256	0.185	0.388
Maize–legume (M-L) intercrop (1 = household practiced M-L intercropping, 0 = otherwise)	0.304	0.460	0.208	0.406
DT maize plus M-L intercrop (1 = household grew a DT variety of maize and practiced M-L intercropping, 0 = otherwise)	0.026	0.159	0.061	0.240
Number of observations	655		655	

varieties of maize. As shown, both at baseline and endline, farmers who grew a DT variety of maize were more resilient to weather shocks compared with their nonadopting counterparts. For example, the resilience index was 21% higher at baseline and 26% higher at endline for adopters of a DT maize variety than for nonadopters. Whereas the resilience index rose by about 3% for farmers who grew a DT variety, the index fell by the same magnitude for nonadopters, between the baseline and the endline.

Panels B and C of Table 15.2 show that similar to DT maize varieties, the resilience index was higher for adopters of M-L intercropping and a combination of DT maize with M-L intercropping compared with nonadopters. Comparing across time periods, however, descriptive analysis revealed that the increase in resilience among adopters was lower compared to the corresponding increase among nonadopters of M-L intercropping. Looking at combined DT maize and M-L intercropping, the increase in resilience among adopters outweighed that for nonadopters.

Our choice of explanatory variables that would influence the decision to adopt the CSA technologies and resilience was informed by economic theory, empirical literature, and availability of data. The explanatory variables are mainly drawn from studies on the adoption of agricultural technologies (Kassie et al., 2011, 2013; Asfaw et al., 2012; Shiferaw et al., 2014) and those that focus on adoption of technologies that may potentially enable rural households to adapt to weather shocks (Di Falco and Bulte, 2012; Di Falco and Veronesi, 2013; Arslan et al., 2015, 2017). Further, there are the ones that estimate resilience of households and its determinants (Barrett and Constas, 2014; Cisse and Barrett, 2018; Upton et al., 2016), and those that examine the effect of agricultural technologies on food security (Smale et al., 2015; Kabunga et al., 2014).

Table 15.2 Summary statistics of the outcome variables, by adopter category at baseline (2015) and endline (2017).

| | 2015 | | | | 2017 | | | |
| | Adopters | | Nonadopters | | Adopters | | Nonadopters | |
Variables	Mean	SD	Mean	SD	Mean	SD	Mean	SD
Panel A: Drought-tolerant (DT) varieties								
Resilience	0.578***	0.199	0.457	0.008	0.596***	0.190	0.441	0.195
Observations	46		609		121		534	
Panel B: Maize–legume (M-L) intercropping								
Resilience	0.502***	0.184	0.450	0.200	0.525***	0.194	0.455	0.203
Observations	199		456		136		519	
Panel C: Combined DT maize with M-L intercropping								
Resilience	0.668***	0.168	0.460	0.195	0.648***	0.183	0.458	0.199
Observations	17		638		40		615	

*Notes: *, **, *** means statistically significant difference between adopters and nonadopters at 10%, 5%, and 1% level, respectively.*

Variables commonly considered to influence adoption behavior of rural households include households' human capital (sex, age, and education of the household head, and household's dependency ratio); productive capital (household assets–based wealth index); access to information, credit, and markets; social networks; and exposure to climatic shocks. The next section provides descriptive statistics for the variables used in the analysis. Table 15.3 provides summary descriptive statistics for the whole sample at baseline and endline.

Households are predominantly male-headed with an average of 43 years of age and 6 years of completed formal education. The dependency ratio is 54%. Most households report to have received agricultural and climatic information as well as finance, although the statistics are lower at endline compared to baseline. On average, a household has two other households in the same village with which they are related by blood or marriage. Ownership of both agricultural and nonagricultural assets is very low and characterized by high variance at baseline and endline. Participation in farmers' associations, however, increased at endline compared with baseline. Households walk about 12 min to the nearest main road on average, and about the same distance to the nearest source of water. Distance to markets is, however, further; it takes about 45 walking minutes to the nearest main market. Exposure to weather shocks may influence the decision to implement CSA technologies. We asked respondents whether they had experienced droughts, floods, or unpredictable rainfall. Using these reported data, a dummy variable equal to one if a household reported to have observed the weather event and zero

Table 15.3 Summary statistics for explanatory variables at baseline (2015) and endline (2017).

Variable	2015		2017	
	Mean	SD	Mean	SD
Panel A: Household level				
HHH is male[a]	0.831	0.375	0.821	0.383
Age of HHH (years)	42.530	14.301	42.974	14.121
Education of HHH (years)	6.176	3.917	5.756	3.317
Dependency ratio	0.539	0.212	0.547	0.199
HH received agricultural information[a]	0.649	0.478	0.490	0.500
HH received weather information[a]	0.747	0.435	0.515	0.500
HH received credit[a]	0.705	0.456	0.576	0.495
HH has a member participating in a farmers' association[a]	0.756	0.430	0.841	0.366
Kinship network (number of relatives in same village)	1.795	1.070	1.889	1.231
Ownership of agricultural assets (index)	0.133	3.916	0.113	4.116
Ownership of nonagricultural assets (index)	0.054	3.737	0.143	3.581
Panel B: Weather shocks				
HH observed drought, floods, or unpredictable rainfall[a]	0.985	0.123	0.644	0.479
Household experienced loss of crops or livestock due to a weather shock[a]	0.980	0.140	0.887	0.317
Number of observations	655		655	

Notes: HH, household; HHH, household head. Statistics for continuous variables.
[a]variable is a dummy equal to 1 if true and 0 if otherwise.

if otherwise was defined and used in our analysis. Exposure to weather shocks was high although more farmers reported having experienced the events in 2015 (98%) than in 2017 (64%). Together, these variables may not only influence adoption behavior but also resilience of households. We test this explicitly using the estimation strategy already described above.

Determinants of use of the CSA technologies

Results of first stage regression of the CSA technologies are presented in Table A15.2 (Appendix). Having more relatives in the same village correlated positively with the likelihood to grow DT maize varieties and practice M-L intercropping in isolation and combination. Furthermore, households which experienced droughts and unpredictable rainfall had a higher likelihood to adopt a combination of DT maize and M-L intercropping. This latter finding is consistent with Arslan et al. (2017) and suggests that farmers perceived the combined use of DT maize and M-L intercropping as a strategy to mitigate the effects of unpredictable climatic conditions. The probability of adopting combined DT maize with M-L intercropping, however, decreased when households experienced a longer period of food shortage. This finding supports Kristjanson et al. (2012) who argued that extended periods of hunger tend to reduce the innovative capacity of farmers, hence hindering them from implementing improved technologies.

Impact of drought-tolerant maize and maize—legume intercropping on resilience

The results of 2SLS fixed effects regression to assess the effect of CSA technology on resilience are presented in Table 15.4. For the DT maize varieties, we also present results of the standard fixed effects regression. According to 2SLS estimates, the probability of having income per adult equivalent above the region-specific poverty line increased by 9% points on average with adoption of DT maize, suggesting that growing a DT variety of maize enhanced resilience to weather shocks. Estimates from the standard fixed effects regression showed a similar positive impact, although the magnitude is lower. Varieties that are tolerant to drought reduce the probability of yield decline when drought occurs. Households in northern Uganda are experiencing frequent occurrence of prolonged dry spells both within and between seasons (Mwongera et al., 2014). Our findings, therefore, suggest that adopting DT varieties may contribute toward reducing the risks of households obtaining income below the subsistence level.

Consistent with what we would expect, there was a positive relationship between resilience of households and education of the household head, access to agricultural information and credit, membership to farmers' associations, size of kinship network, and assets ownership. Exposure to weather-related shocks reduced the resilience of households. Similarly, households with a high dependency ratio and those which experience many months of food shortage were less resilient.

The practice of M-L intercropping in isolation reduced resilience of households to climatic shocks. Specifically, resilience decreased by about 10% points with M-L intercropping. This is puzzling because one of the main reasons why farmers reported to practice M-L intercropping was to reduce the risk, that is, if one crop failed then they could rely on the other crop. Our finding suggests that reliance on M-L intercropping alone would make households worse off in terms of resilience. We speculate several reasons that may help to explain this. First, if both the main crop and the intercrop are equally intolerant to the climatic shock, the objective of minimizing risks in the event that a shock occurs will be compromised. Second, most farmers in the study site perceive the soil to be very fertile (Mwongera et al., 2014; Shikuku et al., 2015). Farmers may be, therefore, not interested in the soil fertility and productivity enhancing properties of M-L intercropping. Although it could be argued that farmers have always been practicing intercropping, in a postconflict context, such as northern Uganda, knowledge about intercropping may not be sufficient. As already mentioned, most farmers

Table 15.4 Two-stage fixed effects (unweighted) least squares estimates of the impact of recommended climate-smart technologies on resilience and food security.

	DT maize 2SLS-FE	DT maize Standard FE	M-L intercropping 2SLS-FE	DT maize with M-L intercropping 2SLS-FE
CSA technology	0.089** (0.042)	0.023** (0.010)	−0.095* (0.055)	0.277* (0.151)
HHH is male	−0.082*** (0.016)	−0.074*** (0.014)	−0.071*** (0.016)	−0.085*** (0.018)
Age of HHH (years)	−0.003***(0.001)	−0.003***(0.001)	−0.003***(0.001)	−0.004***(0.001)
Education of HHH (years)	0.013***(0.002)	0.013***(0.002)	0.014***(0.002)	0.012***(0.002)
Dependency ratio	−0.257***(0.017)	−0.261***(0.017)	−0.248***(0.023)	−0.250***(0.021)
Received agricultural information	0.029***(0.007)	0.032***(0.007)	0.034***(0.008)	0.027***(0.009)
Received climate information	0.002(0.007)	0.001(0.007)	−0.005(0.008)	0.001(0.009)
Received credit	0.090***(0.007)	0.092***(0.007)	0.091***(0.008)	0.088***(0.008)
Belongs to a farmer group	0.022***(0.008)	0.018***(0.008)	0.023***(0.009)	0.025***(0.009)
Size of kinship network	0.034***(0.003)	0.034***(0.002)	0.040***(0.003)	0.032***(0.004)
Experienced floods, droughts, or unpredictable rainfall	−0.022***(0.008)	−0.013***(0.008)	−0.018*(0.010)	−0.031***(0.008)
Months of food shortage	−0.032***(0.002)	−0.033***(0.002)	−0.035***(0.003)	−0.029***(0.003)
HH assets index	0.027***(0.001)	0.027***(0.001)	0.028***(0.002)	0.026***(0.002)
Endline dummy		0.018***(0.005)		
Diagnostics				
H_0: Exogeneity of practices	2.340		6.744***	4.068**
Kleibergen–Paap rk LM statistic	20.217***		11.252***	7.665**
Hansen J	0.015		0.001	0.043
Cragg–Donald Wald F-statistic	17.473		7.858	4.315
R^2 (within)	0.7538	0.773		
Fraction of variance due to fixed effects		0.481		
Observations	1310	1310	1310	1310

Notes: HHH, household head; DT, drought-tolerant; M-L, maize–legume; 2SLS, two-stage least squares; FE, fixed effects. *P < .1, **P < .05, ***P < .001. Robust standard errors clustered at subvillage level are reported in parentheses.

did not know benefits of M-L intercropping beyond diversified production. Similarly, lack of knowledge about right crop combination and spacing in intercropping systems is cited as a barrier to use of the practice (Shikuku et al., 2015) and may lead to lower water, radiation and nutrient use efficiency with negative effects on climate resilience. Without good knowledge, studies have shown that payoffs with improved technologies may substantially decline (Ben Yishay and Mobarak, 2018; Kondylis et al., 2015) rendering households less resilient.

Households which combined DT maize with M-L intercropping experienced enhanced resilience to climatic shocks. Several studies have also documented greater benefits when CSA technologies are implemented as packages as opposed to single technologies (see for example, Teklewold et al., 2013; Di Falco and Veronesi, 2013; Arslan et al., 2017). Our finding suggests that including stress-tolerant varieties in the M-L intercropping would outweigh the negative effect of the latter practice and enhance resilience more compared with growing of DT maize varieties in isolation.

Conclusion

We studied the impact of DT maize varieties and M-L intercropping on resilience of rural households in northern Uganda. Based on a recently recommended theory-based approach to resilience measurement that combines insights from the vulnerability literature and studies on poverty dynamics, this study used income per adult as an indicator of rural household resilience to drought and precipitation variability. Using this estimation, we found that only 10% of rural households could be considered resilient. Our study found that adoption of DT maize in isolation increased resilience while the practice of M-L intercropping reduced resilience. Combining both technologies, however, had greater positive impact on resilience. We further found that kinship networks play an important role to enhance adoption of the technologies both in isolation and combination. Exposure to climatic shocks increased the likelihood to practice combined DT maize and M-L intercropping. However, households which suffered extended periods of hunger were less likely to adopt such a combination.

Our findings suggest several important implications for policymaking. As the majority of households are vulnerable to weather shocks, finding ways to enhance their resilience is imperative. Efforts targeting to promote M-L intercropping as a single technology may not succeed to enhance resilience. Encouraging the growth of adapted varieties of crops within M-L intercropping may, however, yield higher payoffs in terms of increased resilience. The success of such efforts will depend on the ability to leverage existing kinship ties and reducing the prevalence of acute food shortage. In the immediate term, addressing the problem of acute food shortage may require provision of social safety nets to the most vulnerable households.

Acknowledgements

This work was implemented as part of the CGIAR Research Program on Climate Change, Agriculture and Food Security (CCAFS), which is carried out with support from the CGIAR Trust Fund and through bilateral funding agreements. For details please visit https://ccafs.cgiar.org/donors. The views expressed in this document cannot be taken to reflect the official opinions of these organizations.

Appendix

Table A15.1 Ordinary least squares estimates of household income per adult equivalent (per AEHH income), variance (per AEHH income).

Variable	Per AEHH income		Variance (per AEHH income)	
	Coefficient	Robust SE	Coefficient	Robust SE
Household head is male	−0.186**	0.090	0.106	0.156
Formal education of household head (years)	0.023**	0.009	−0.018	0.015
Age of the household head (years)	−0.008***	0.003	0.005	0.004
Dependency ratio	−0.627***	0.157	−0.204	0.349
Household assets index	0.041***	0.009	0.041**	0.020
Agricultural assets index	0.050***	0.008	−0.051***	0.018
Household received credit	0.221***	0.067	−0.162	0.134
Household has at least one member participating in a farmers' group	0.012	0.085	0.035	0.173
Household received agricultural information	0.114*	0.067	−0.267**	0.118
Household received climate-related information	−0.064	0.066	0.284**	0.122
Distance to nearest source of water (walking minutes)	−0.009***	0.002	−0.004	0.004
Months of food shortage experienced by the household	−0.071***	0.025	0.012	0.036
Kinship network (number of relatives)	0.067**	0.027	−0.009	0.047
Household experienced droughts, floods, or unpredictable rainfall	−0.050	0.080	−0.181	0.140
Number of observations	1310		1310	

Source: Household surveys, 2015 and 2017. Robust standard errors (SE) are clustered at subvillage level. *, **, *** show significance at 10%, 5%, and 1% level, respectively.

Table A15.2 First-stage regression results of determinants of adoption of climate-smart agricultural technologies.

	DT maize	M-L intercropping	DT maize with M-L intercropping
	(1)	(2)	(3)
HHH is male	0.125**(0.049)	−0.002(0.083)	0.050(0.034)
Age of HHH (years)	0.001(0.002)	−0.001(0.003)	0.002*(0.001)
Education of HHH (years)	0.004(0.006)	0.011(0.011)	0.003(0.004)
Dependency ratio	−0.041(0.056)	0.135(0.097)	−0.042(0.042)
Received agricultural information	0.037(0.026)	0.029(0.036)	0.019(0.018)
Received climate information	−0.012(0.029)	−0.026(0.038)	−0.011(0.016)
Received credit	−0.009(0.030)	0.019(0.038)	0.005(0.017)
Belongs to a farmer group	−0.012(0.035)	0.024(0.051)	−0.012(0.017)
Size of kinship network	0.016*(0.010)	0.048***(0.014)	0.014*(0.007)
Experienced floods, droughts, or unpredictable rainfall	0.005(0.029)	0.044(0.043)	0.033*(0.019)
Months of food shortage	−0.011(0.008)	−0.025**(0.012)	−0.012**(0.006)
HH assets index	−0.001(0.005)	0.012(0.006)	0.002(0.003)
Observations	1310	1310	1310

Notes: HHH, household head. DT, drought-tolerant; M-L, maize–legume; *P < .1, **P < .05, ***P < .001. Robust standard errors clustered at subvillage level are reported in parentheses.

References

Adato, M., Carter, M.R., May, J., 2006. Exploring poverty traps and social exclusion in South Africa using qualitative and quantitative data. Journal of Development Studies 42, 226–247.

Alliance for Green Revolution in Africa (AGRA), 2016. Africa Agriculture Status Report: Progress Towards Agriculture Transformation in Sub-Saharan Africa. Nairobi, Kenya. Issue No. 4.

Angrist, J.D., Krueger, A.B., 2001. Instrumental variables and the search for identification: from supply and demand to natural experiments. Journal of Economic Perspectives 15 (4), 69–85.

Arslan, A., McCarthy, N., Lipper, L., Asfaw, S., Cattaneo, A., Kokwe, M., 2015. Climate smart agriculture? Assessing the adaptation implications in Zambia. Journal of Agricultural Economics 66 (3), 753–780.

Arslan, A., Belotti, F., Lipper, L., 2017. Smallholder productivity and weather shocks: adoption and impact of widely promoted agricultural practices in Tanzania. Food Policy 69, 68–81.

Asfaw, S., Shiferaw, B., Simtowe, F., Lipper, L., 2012. Impact of modern agricultural technologies on smallholder welfare: evidence from Tanzania and Ethiopia. Food Policy 37, 283–295.

Barrett, C.B., Constas, M.A., 2014. Toward a theory of development resilience for international development applications. Proceedings of the National Academy of Sciences 111 (40), 14625–14630.

BenYishay, A., Mobarak, A.M., 2018. Social learning and incentives for experimentation and communication. Review of Economic Studies 1–34.

Bhavnani, R., Vordzorgbe, S., Owor, M., Bousquet, F., 2008. Report on the Status of Disaster Risk Reduction I the Sub-Saharan Africa Region. Commission of the African Union. United Nations and the World Bank. Available from: http://www.unisdr.org/files/2229DRRinSUBSaharan AfricaRegion.pdf.

Brown, C., Meeks, R., Hunu, K., Yu, W., 2011. Hydroclimate risk to economic growth in sub-Saharan Africa. Climatic Change 106, 621−647.

Cissé, J.D., Barrett, C.B., 2018. Estimating development resilience: A conditional moments-based approach. Journal of Development Economics 135, 272−284.

Di Falco, S., Bulte, E., 2012. The impact of kinship networks on the adoption of risk-mitigating strategies in Ethiopia. World Development 43, 100−110.

Di Falco, S., Veronesi, M., 2013. How can African agriculture adapt to climate change? A counterfactual analysis from Ethiopia. Land Economics 89 (4), 743−766.

FAO, 2013. Sourcebook on Climate Smart Agriculture, Forestry and Fisheries (Rome, Italy: Food and Agriculture Organization of the United Nations (FAO). Available at: http://www.fao.org/climatechange/37491-0c425f2caa2f5e6f3b9162d39c8507fa3.pdf.

Gao, J., Mills, B.F., 2018. Weather shocks, coping strategies, and consumption dynamics in rural Ethiopia. World Development 101, 268−283.

Giesbert, L., Schindler, K., 2012. Assets, shocks, and poverty traps in rural Mozambique. World Development 40 (8), 1594−1609.

Kabunga, N.S., Dubois, T., Qaim, M., 2014. Impact of tissue culture banana technology on farm household income and food security in Kenya. Food Policy 45, 25−34.

Kassie, M., Shiferaw, B., Muricho, G., 2011. Agricultural technology, crop income, and poverty alleviation in Uganda. World Development 39 (10), 1784−1795.

Kassie, M., Jaleta, M., Shiferaw, B., Mmbando, F., Mekuria, M., 2013. Adoption of interrelated sustainable agricultural practices in smallholder systems: evidence from rural Tanzania. Technological Forecasting and Social Change 80 (3), 525−540.

Kristjanson, P., Neufeldt, H., Gassner, A., Mango, J., Kyazze, F.B., Desta, S., Sayula, G., Thiede, B., Förch, W., Thornton, P.K., Coe, R., 2012. Are food insecure smallholder households making changes in their farming practices? Evidence from East Africa. Food Security 4 (3), 318−397.

Kondylis, F., Mueller, V., Zhu, S., 2015. Measuring agricultural knowledge and adoption. Agricultural Economics 46, 449−462.

Lybbert, T.J., Sumner, D.A., 2012. Agricultural technologies for climate change in developing countries: policy options for innovation and technology diffusion. Food Policy 37, 114−123.

Mwongera, C., Shikuku, K.M., Twyman, J., Winowiecki, L., Ampaire, A., Koningstein, M., Twomlow, S., 2014. Rapid Rural Appraisal Report of Northern Uganda. International Center for Tropical Agriculture (CIAT), CGIAR Research Program on Climate Change, Agriculture and Food Security (CCAFS).

Ng'ang'a, S.K., Notenbaert, A., Mwungu, C.M., Mwongera, C., Girvetz, E., 2017. Cost and Benefit Analysis for Climate-Smart Soil Practices in Western Kenya. Working Paper. CIAT Publication No. 439. International Center for Tropical Agriculture (CIAT), Kampala, Uganda, 37 pp.

Parry, M., Rosenzweig, C., Livermore, M., 2005. Climate change, global food supply and risk of hunger. Philosophical Transactions of the Royal Society B 360 (1463), 2125−2138.

Republic of Uganda, Poverty status report 2014: Structural change and poverty reduction in Uganda, 2015, Kampala, Uganda

Sain, G., Loboguerrero, A.M., Corner-Dolloff, C., Lizarazo, M., Nowak, A., Martinez-Baron, D., Andrieu, N., 2017. Costs and benefits of climate-smart agriculture: the case of the dry corridor in Guatemala. Agricultural Systems 151, 163−173.

Schlenker, W., Lobell, D.B., 2011. Robust negative impacts of climate change on African agriculture. Environmental Research Letters 5 (1), 1−8.

Shiferaw, B., Kassie, M., Jaleta, M., Yirga, C., 2014. Adoption of improved wheat varieties and impacts on household food security in Ethiopia. Food Policy 44, 272—284.

Shikuku, K.M., Mwongera, C., Winowiecki, L., Twyman, J., Atibo, C., Läderach, P., 2015. Understanding Farmers' Indicators in Climate-Smart Agriculture Prioritization in Nwoya District, Northern Uganda. Centro Internacional de Agricultura Tropical (CIAT), Cali, CO, 56 pp. (Publicación CIAT No. 412).

Shikuku, K.M., Winowiecki, L., Twyman, J., Eitzinger, A., Perez, J.G., Mwongera, C., Läderach, P., 2017. Smallholder farmers' attitudes and determinants of adaptation to climate risks in East Africa. Climate Risk Management 16, 234—245.

Smale, M., Moursi, M., Birol, E., 2015. How does adopting hybrid maize affect dietary diversity on family farms? Micro-evidence from Zambia. Food Policy 52, 44—53.

Stock, J.H., Yogo, M., 2005. Testing for weak instruments in linear IV regression. In: Andrews, D.W.K., Stock, J.H. (Eds.), Identification and Inference for Econometric Models: Essays in Honor of Thomas Rothenberg. Cambridge U y Press, Cambridge, pp. 80—108. Working paper version: NBER Technical Working Paper 284. http://www.nber.org/papers/T0284.

Teklewold, H., Kassie, M., Shiferaw, B., 2013. Adoption of multiple sustainable agricultural practices in rural Ethiopia. Journal of Agricultural Economics 64 (3), 597—623.

Upton, J.B., Cisse, J.D., Barrett, C.B., 2016. Food security as resilience: reconciling definition and measurement. Agricultural Economics 47, 135—147.

Wheeler, T., von Braun, J., 2013. Climate change impacts on global food security. Science 341, 508—513.

Further reading

Angrist, J., 2001. Estimation of limited-dependent variable models with binary endogeneous regressors: simple strategies for empirical practice. Journal of Business & Economic Statistics 19 (1), 2—16.

Can social protection schemes contribute toward drought resilience? Evidence from rural Ethiopia

16

Mesay Kebede Duguma

Consultant, PSNP Donor Coordination Team/World Bank, Addis Ababa, Ethiopia

Introduction

Over the last two decades, the government of Ethiopia has introduced a number of policies and strategies toward tackling food insecurity, combating drought emergencies, and assisting vulnerable rural households maintaining their livelihoods in the face of frequent drought disasters. Most prominently since 2005, the country has pursued a country-wide social protection scheme, the Productive Safety Net Program (PSNP), with substantial donor support (Gilligan et al., 2009; Porter and Dornan, 2010). The aim of the program is to assist chronically food-insecure rural households bridge their food gaps through transfer of food and cash while providing long-term solutions through the creation of community and (to a lesser degree) household assets. In relation to this, the public work component of the program seeks to support sustainable livelihoods by promoting natural resource management interventions and building local infrastructure (MoA, 2014a).

Besides the standard components (direct cash/food transfer and food for work), the PSNP includes contingency funds at the federal district levels that are used to expand coverage in the case of mainly drought and other small scale shocks. In other words, contingency budget provides timely resources for transitory food insecurity in response to shocks within the existing program areas. Such scalable safety net uses early warning systems and contingency planning to tackle an impending drought in PSNP districts. Early response through contingency budget has therefore been considered to have a potential to avoid a shock from becoming an emergency since its benefit lies in the fact that it is early and preventive, rather than late and reactive (Ashley, 2009).

In light of the above stated components, PSNP, both through its public work component and contingency funding mechanisms, seeks to provide a platform for drought risk management practices and resilience building at household and community levels. Furthermore, effective and permanent system for delivering elements of social protection and disaster risk management (DRM) was particularly strengthened during the current (fourth) phase of PSNP which has started implementation in January 2016 (MoA, 2014a). According to the 2014 Program Implementation Manual (PIM), the estimated maximum annual program caseload till 2020 will be 10 million clients/beneficiaries, consisting of 8.3 million chronically food-insecure individuals and with the capacity to support an additional 1.7 million transitory beneficiaries if need exists (MOA, 2014b, p. 2–6).

Drought Challenges. https://doi.org/10.1016/B978-0-12-814820-4.00016-X

State-of-the-Art literature

A wide range of literature exists regarding the role of social protection in reducing chronic poverty and vulnerability to disasters as well as in facilitating long-term investment in human and physical capital (Arnold et al., 2011; Barrientos, 2010; Dercon, 2011; Devereux, 2010; Ellis et al., 2008). Ethiopia is one of the countries whose experience on the one hand suggests that productive safety nets can make a valuable contribution to protecting assets against "distress sales" for food and nonfood needs, improving household food security, raising household incomes, and enhancing resilience (Devereux et al., 2008; Headey et al., 2012; Jones et al., 2010). On the other hand, some studies have shown less optimism with regard to the role of the program in protecting households from the negative impacts of livelihood shocks such as droughts and in ultimately building the resilience of its beneficiaries. For instance, using a panel survey conducted in four regions (Tigray, Amhara, Oromia, and SNNP), Béné et al. (2012) found that the positive achievements of the program were rather shallow as regards guaranteeing complete protection of its beneficiaries from the impacts of severe shocks. Similarly, Anderson et al. (2011) did not find evidence that PSNP protected households' livestock in times of climate or economic difficulties/shock, while Gilligan et al. (2009) documented that PSNP had little impact on participants on average, due in part to transfer levels that were far below program targets.

Reports regarding the destructive impact of frequent droughts on the lives and livelihoods of people, including the 2015/16 drought, also show that the program has not yet fully succeeded in protecting households from the effects of catastrophic crises or in terms of building resilience. During 2015/2016, Ethiopia experienced one of the worst droughts in decades. The two main rainy seasons[1]—that support over 80% of Ethiopia's agricultural yield and employ 83% of the workforce—failed in 2015. The 2015 spring/*belg* rains failed and crop production in *belg* growing areas fully collapsed, and the pastoral areas in southeastern and south were hit hard by the drought. This affected smallholder farmers and pastoralists in the north eastern rangelands of Afar and the northern Somali regions. Subsequently, the summer rains were weak and erratic due to El Niño, which negatively affected *meher*-dependent farmers and pushed pastoralists into severe food insecurity. According to the government-led multiagency assessments, 10.2 million people in 2016 and 7.8 million people in 2017 faced severe food shortages and required emergency food assistance. The financial resources required for the former is estimated at US dollars 1.4 billions, and more than 948 million for the latter (NDRMCC, 2016, 2017).

In conclusion, a number of studies have documented the achievements of the program over the last 10 years. However, very few of these studies had an explicit focus on the impact of the program on building long-term resilience. Motivated by some of PSNP's success stories and the existing knowledge gap with respect to the potential contribution of social protection schemes to drought resilience, this chapter primarily aims to assess Ethiopia's PSNP against some of the key lessons that can be learned from the 2015/16 drought in the country. This chapter also focuses on three specific areas for its analysis. First, it assesses the performance of PSNP using selected examples/case studies of regions implementing the program. Second, it explores the (technological/institutional/policy) bottlenecks that weaken the impact of PSNP on drought resilience. Third, it advises policy for enhanced drought resilience through social protection schemes. The next section highlights the methodology used for the study.

---

[1]*Meher* (major rainy season) and *Belg (small rainy season)* are the two main rainy seasons in Ethiopia.

Methodology

This chapter has substantially benefited from study missions in 2015 and 2016 in Ethiopia. Accordingly, qualitative data were gathered through case studies of two specific regional states (Oromia and Afar) and stakeholder interviews with 26 high-level government and nongovernment officials and experts both at federal and subnational level. Furthermore, desk review of secondary sources has further substantiated the field level primary data.

The data collection for conducting the case studies was assisted by checklists and guidelines that are structured into key thematic areas, namely agro-climatic characteristics of study sites, prevalence of drought, scale of the impact of drought, status of food security, the scale of drought impact on people's livelihood, and the strengths and weakness of PSNP in contributing to drought resilience. The data gathered through field case studies and key informant interviews were thematically organized, synthesized, and presented in a form of qualitative study report (Duguma et al., 2017).

Case studies

Oromia and Afar regional states shared common environmental challenges of frequent droughts over the years which included the 2015/16 drought. However, PSNP has different life span in these two regions with longer presence in Oromia than Afar region. The state of infrastructural and socioeconomic development (Oromia region being relatively more developed than Afar), and the livelihood and cultural setup (crop growing Oromia versus pastoral and semipastoral Afar) are also attributes that differentiate one from the other. Against the above backdrop, it is assumed that the results from these two case study regions, presented below, can shade light on how and to what extent PSNP is contributing with respect to building resilience to droughts in two geographical regions which are at different stages of economic development, infrastructural, and socio-cultural constellation. However, it has to be stressed that whether the results are transferrable to other districts remains an open research question.

Oromia region—Dodota *woreda*[2]

In Oromia region, PSNP has made notable progress in terms of filling food gaps and reducing the depletion of households' assets due to disasters. There was also wide consensus among regional stakeholders that through its public works (rehabilitating degraded land and the rehabilitation of destroyed social infrastructures), the PSNP had succeeded in bridging relief and development efforts as compared to one-dimensional relief efforts in the past.

Despite the above-stated achievements, most of the *woreda* and regional level experts interviewed agreed that, with frequent droughts, the program would need additional tools to ensure sustainability and to build long-term resilience at household level. For instance, the 6 months of support/food/cash transfers made under the program were perceived by the *woreda* experts as insufficient to sustain the food needs of chronically food-insecure households throughout the year. Lack of integration of livelihood supporting schemes into the program has also hindered asset-building among drought-

---

[2]The local term for district.

vulnerable communities. Weak linkage with the DRM approaches (up until the fourth phase of PSNP) has undermined the potential of the response operation made under the contingency budget. As a result, the contingency budget has predominately focused on saving lives through provision of relief assistance *after* the occurrence of droughts. Because of this, the system was not able to support recovery and rehabilitation interventions identified on the basis of the local context. Neither was it implemented in an integrated manner with development plans and programmes to rehabilitate affected people and reduce future risk and vulnerability.

Besides the above, the poor quality of public works and flaws in the technical design had a negative effect on the environmental and technical sustainability of the public assets produced. This has been visible particularly in the construction of roads and water infrastructures. Key informants also expressed their concern regarding the pressure on *woredas* to force clients/households to graduate from the program prematurely. It was reported that nearly 50% of PSNP clients have graduated from the program over the last 10 years. The study has documented the experience of Dodota *woreda* of the Arsi Zone of Oromia regional state, located at a distance of 125 km from the capital city, Addis Ababa (Fig. 16.1).

Box 16.1 below presents the profile of Dodota *woreda*, one of the most affected *woredas* of Oromia region, during the 2015/16 drought.

Afar region

The Afar National Regional State is characterized by an arid and semi-arid climate with low and erratic rainfall. It is increasingly drought prone. The production system of the Afar region is dominated by

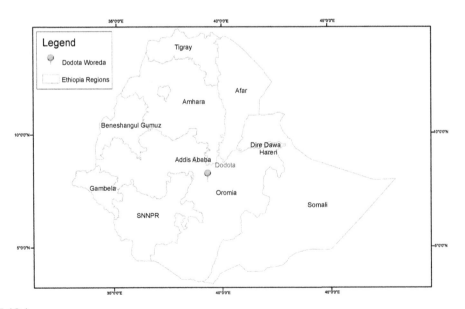

FIGURE 16.1

Location of Dodota woreda on the map of Ethiopia.

BOX 16.1 Case study 1: Dodota *woreda* in Oromia region and the impact of the 2015/2016 drought

Dodota woreda, one of the drought-prone and food-insecure *woredas,* consists of 15 *kebeles,* of which 12 are rural. The rural *kebeles* are home to a little more than 10,000 households. Located in the dry (*kola*) climatic zone of the country, the *woreda* has been experiencing frequent droughts over the last decade. It was reported that nearly 50% of PSNP clients have been able to graduate from the program over the last 10 years. According to the focus group discussion held with the *woreda* officials, the 2015 El Niño had a widespread impact in the *woreda,* affecting all of the 12 rural *kebeles* and putting the *woreda* on the emergency recipient list. In fact, by the time of the field visit, PSNP was operational in all of the 12 *kebeles.* The impact of the drought went to the extent of weakening the coping capacity of even the well-performing and relatively wealthy households in the *woreda.* Some graduated PSNP clients were forced to return to their beneficiary status after losing most of their assets. As of 2016, the total number of PSNP clients in the 12 rural *kebeles* stood only at 66,565. Yet, given the situation during the field data gathering period, an additional 13,587 clients would require support through the PSNP.

---

pastoralism (90%) from which agro-pastoralism (10%) is now emerging following some permanent and temporary rivers on which small-scale irrigation is developed (EPA, 2010).

Afar region has experienced severe droughts in the years 2002, 2006, 2010, 2011, and 2015. The region is also vulnerable to severe erosion, environmental degradation, bush encroachment, and invasive (alien) species among which *Prosopis juliflora* is the dominant one. Frequent destructive floods from the neighboring regions, particularly from the highlands of neighboring Wollo region have been a constant threat to the lives of people and livestock. The situation is creating huge pressure on land and water resources, exacerbating the vulnerability of the pastoralists and agro-pastoralists and instigating conflicts over the dwindling resources. Box 16.2 presents the profile of Afar region in the midst of the 2015/16 drought.

From the evidence in Oromia and Afar region, it is clear that, under the existing implementation mode and operational capacity, PSNP is still unable to make a fundamental positive impact on long-term household-level drought resilience although it is able to save lives and provide short-term support against asset depletion. This assertion also resonates with some of the earlier empirical findings (Anderson et al., 2011; Béné et al., 2012). Against the above well-established facts, many respondents emphasized that any "stand-alone" approach to drought resilience is less likely to achieve success in the face of frequent droughts and their associated negative impacts on nutrition and food security which calls for a more integrated approach.

Structural bottlenecks

Since 2005, PSNP has gone through a number of progressive steps to improve its performance in terms of meeting its program objectives such as filling food gaps, reducing asset depletion, and creating assets both at household and community levels thereby contributing to livelihood improvement and resilience building at household level. At aggregate level, the advances seen within the program are to be acknowledged. Nevertheless, certain factors have also been hindering its progress with respect to its role in drought resilience over the years. Key informants identified the following six limiting factors.

BOX 16.2 Case study 2: PSNP in Afar region

By early April 2016, the seasonal rains in Afar region had failed for almost two consecutive years. Severe drought in the region caused large-scale livestock deaths and a severe shortage of food, weakening the coping capacities of the pastoralist and the semi-pastoralist communities. As one of the long-term rural development programs, the PSNP has been implemented in all of the rural *woredas* of the Afar region since 2008. Since then, the program has helped chronically food-insecure households bridge their food gaps and protect the depletion of assets at household level. A number of community asset-building works have also been carried out through the public works. Nevertheless, as one PSNP expert pointed out, one cannot strictly assume that the program has made a substantial impact in terms of creating a drought-resilient community. It was further stressed that the constraints seen over the years have both nature- and capacity-related dimensions. As regards the former, it was pointed out that mobilizing the community for public works during longer periods of drought has become a difficult undertaking since shortage of water has frequently forced communities to move to new settlement areas including neighboring regions of Amhara and Tigray. Ultimately, the respondent made a statement saying: "There is sometimes a wrong perception out there that the government alone can change the existing bad situation; we are beginning to see that this hardly is possible unless nature cooperates.[3]"

The program also faces certain capacity gaps, including the lack of skilled manpower to function at its full capacity. For instance, despite the availability of a much wider range of technological options for the public works projects, the question of finding proper support for design and supervision of soil and water conservation structures is still unresolved. For this reason, some of the technologies used for land rehabilitation under the public works are of poor quality. To address this gap, the government is seeking technical collaboration, for instance, with a program of the German development cooperation agency GIZ (Deutsche Gesellschaft für Internationale Zusammenarbeit) to strengthen its "cash-for-work" program. Furthermore, shortage of other resources (such as equipment and vehicles) and poor public infrastructure remain a serious problem in the region. At the time of the interviews (mid of April 2016), it was reported that the region owned only two trucks to distribute forage (obtained through aid) to all the districts in the region and that field experts were unable to reach remote districts in time.

[3]Captured on April 15, 2016/Afar, Semera.

Lack of common understanding on the concept of "drought resilience"

Even though "drought resilience" is one of the most frequently used terminologies among DRM stakeholders, there is **no clear and uniform understanding of drought resilience** among government stakeholders including those at subnational level. It was noted that some stakeholders lacked clarity in distinguishing between the contributions of short-term responses and long-term development measures with respect to its relevance in building up drought resilience. For instance, the government's response to the 2015/16 drought derives mainly from its use of national reserves, and thus the successes made in terms of preventing loss of human lives per se was frequently mentioned as an indication of resilience by some stakeholders. Against such facts, it has to be noted that leaving people dependent on central government (and donor) support is a risky strategy because it leaves resilience to the vagaries of political decisions and availability of funding. For instance, it took the government several months to accept the early warnings of the 2015 El Nino drought and declare drought emergency (according to some interviewees). This, by itself, not only resulted in delaying early action, inflicting damages to livelihood assets such as animals, and increasing costs of responding, but also demonstrated that resilience building on discretional political will is a dangerous strategy. In addition, while this time the government could use own funds to scale up PSNP and other emergency activities, which only came once it accepted the facts, there is no guarantee that the government continues to do so. On top of this, it has to be understood that donor funding may also become scarce if priorities shift toward other issues.

Therefore, in the absence of proactive measures that aim at preventing such losses as well as lack of focus on building household internal resilience, preparedness through short-term response to drought alone apparently undermines the role of social protection schemes in drought resilience. Therefore, capacity development should follow a subsidiary approach in order to enhance the internal capacity of the very people vulnerable to droughts. Here, PSNP can serve an important platform for such bottom-up approach to long-term drought resilience building at household level.

Inadequate coordination and harmonization of activities among government institutions, development partners, and NGOs

The mandate (role and responsibility) of stakeholders was not clear enough in the first three implementation phases of PSNP. Thus, the implementation of the program suffered from **weak coordination** among government stakeholders at federal, regional, and lower levels. For instance, key informants unanimously emphasized that the linkage between agriculture and health sectors has suffered from lack of coordination particularly at regional and lower level of administration which created gaps to effectively address nutrition and health needs of beneficiaries in PSNP operational districts. Weak coordination with the water sector has also negatively impacted the quality of water infrastructures leading to dysfunctional water ponds and pipelines in some PSNP-implemented *woredas*. This is because most infrastructures often demand support from qualified experts which could be readily available from the water sector yet with the absence of cooperation between the agriculture and water sectors, the situation often led to a lost opportunity. In addition to these, the contingency fund/risk financing of PSNP has been poorly integrated in the overall DRM framework. As a result of this, there was weak harmonization of PSNP activities with early warning information to ensure early action which depends on fast and timely/early utilization of contingency funds. Such kind of coordination gaps at program level tends to undermine the contribution of PSNP interventions in addressing long-term drought effects.

Decentralization and capacity gap

Kena (2016) argues that decentralization in Ethiopia has, on many grounds, weakened the power of local authorities as regional states remain instruments of political control by the central government. The author further claims that this and other related factors contributed to lack of local autonomy and good governance fueling a wide public protest for the past many years. In another study, decentralization in Ethiopia was reported for achieving positive economic outcomes in some respects. For instance, it was documented that initially worse performing cities were able to catch up with the better-performing counterparts in their region after decentralization, leading to lower spatial inequality (Chaurey and Mukim, 2015). This study argues that strong implementing capacity at all levels is of vital relevance to the success of drought risk reduction efforts in Ethiopia. In this regard, the study identified institutional, organizational, technological, and financial **capacity gaps** at multiple levels.

This has been more pronounced in pastoral regions/emerging regions of the country (including Afar, Somali) in which years of neglect by previous governments has caused sharp development imbalance with the rest of the country.[4] These regions were not only the last to implement decentralization[5] from region to *woredas* but also have suffered from slow progress mainly attributed to acute capacity gaps mentioned earlier. Organizational, technological, and human capacity limitations at *woreda* level are almost a common problem across the country. Stakeholders emphasized that *woredas* lack adequate office space, equipment, communications, and IT facilities including computers to transmit early warning and risk and vulnerability assessment information. Shortage of skilled man power due to high staff turnover has a huge impact on both the number and quality of staff in government offices. It was stressed that measures are urgently needed to attract and retain qualified civil servants. Without higher staff stability, much of the capacity development will be done in vain. Such problems have to be addressed to unlock the potential of social protection schemes for long-term drought resilience.

Poor quality of public works under PSNP

Weak infrastructures built under public works disrupt the sustainability of the program, put pressure on the available budgets, and create resource gaps for other development activities. In relation to this, field visits to Chifra *woreda* of Afar region confirmed that poor-quality land rehabilitation structures built under the public works have further exacerbated land degradation and slowed down regeneration of vegetation. Lack of technical expert advice prior to designing and building structures has caused much of the problem. Generally, such quality gaps undermine the contribution of public works in improving people's livelihoods on long-term basis.

Shortage of funding for complementary livelihood components

PSNP is predominantly **a donor-funded program** since its inception in 2005. While donor's contribution to PSNP's core program components has been quite substantial, complementary interventions such as HABP (livelihoods support) had generally received less donor interest to finance its implementation. As a result, the program was unable to operate in its full capacity over the last 10 years. In light of such facts, it has to be stressed that relying solely on regular transfers made through PSNP may not exceed beyond fulfilling the immediate food needs of households for short-term survival. Linked to this, it is well acknowledged by most stakeholders that complementary livelihood components boost the programs' positive impact. Funding issues such as the one explained require better attention if

---

[4]Literacy levels and school enrollment beyond primary cycle education are very low particularly in the pastoral regions and not much different in the agro-pastoral regions as well. Fore instance, Afar and Somali have one of the lowest enrollment rates in secondary education, which is 8.62% and 11.19% compared to Tigray (45.45%), Amhara (33%), and Addis Ababa (82%) (MoE, 2017). The emerging regions are characterized by small, scattered and nomadic populations making it more challenging to provide public services. Most of the areas are inaccessible with poor or no roads and few social services (EPA, 2010).

[5]Increased support to the decentralization of power to these regions and then to the *woredas* has been a centerpiece of the development strategy for tackling the high vulnerability of these regions to droughts and other disasters. This has been strongly advocated for fast realization of improved accountability, responsibility, and flexibility in service delivery and increased local participation in democratic decision-making on factors affecting the livelihood of the grassroots population.

PSNP has to grow its capacity from protection of assets to building long-term resilience of households to droughts.

"Silo thinking" and limited knowledge and limited political will for "multisectoral" approaches

Multisectoral approaches are new concepts for many program implementers at federal, regional, and lower levels of administration. **Limited knowledge of multisectoral approaches** among district- and lower-level program implementing stakeholders is noted by all interviewed government stakeholders. In addition to this, lack of political will among implementers has been slowing the process. Therefore, breaking the "silo thinking" among stakeholders has been a challenging task over the years. Due to this, PSNP within agriculture sector has enjoyed weak linkage with the other sectors including the Health Labour and Social Affairs and Disaster Management. Lack of synergetic relationship among sectors and program implementers has therefore hampered the progress in terms of strengthening linkages with other drought resilience initiatives, which are under the responsibility of various line ministries.

Weak monitoring and follow-up and knowledge management

Poor monitoring of development activities not only compromises the quality of the services provided through social protection schemes but also reduces the accountability for all sectors involved. It has been reported by some stakeholders that, some promising results of pilot drought resilience projects by NGOs were not adequately documented and never scaled up. Drought resilience projects implemented by some of the long-standing NGOs in the country (e.g., Oxfam, Mercy Corps) have achieved enormous success stories in terms of strengthening the coping capacity of vulnerable population in some of the remote low land pastoralist regions of the country.[6] However, despite the impressive achievements made through such resilience initiatives, weak follow-up and complementarity with government development programs as well as poor uptake and integration of some of the best practices into long-term development programs by the government has undermined the sustainability of programs by NGOs. This limits progress that could have been accelerated through the sharing of knowledge and skills among stakeholders.

Conclusion and policy implications

Poverty and inequality are two of the root causes of vulnerability to the impacts of droughts. This is why many of the actions needed to mitigate these impacts require long-term and proactive

---

[6]Mercy Corps has been leading the implementation of USAID funded project Pastoralist Areas Resilience Improvement through Market Expansion (PRIME). This 5-year project aims to improve the lives of chronically food-insecure and vulnerable population in pastoralist communities in dry lands with a special focus on pro-poor market development. Oxfam America and the United Nations World Food Program (WFP) have been implementing a Rural Resilience Initiative called R4 which builds on the initial success of HARITA (Horn of Africa Risk Transfer for Adaptation), an integrated risk management framework developed by Oxfam America, the Relief Society of Tigray (REST), Ethiopian farmers, and several other national and global partners. The initiative combines improved resource management (risk reduction), insurance (risk transfer), livelihoods diversification and microcredit (prudent risk taking), and savings (risk reserves).

development interventions. PSNP, on one hand, has made a number of positive impacts in preventing the depletion of household assets among chronically food-insecure households. The rehabilitation of degraded environments and the creation of public asset through the expansion of social infrastructures and environmental rehabilitation of degraded environment through its public works are good examples of such achievements. However, given the frequency of droughts and their destructive impacts on the lives and livelihoods of people, one cannot overstate the need to advance and introduce serious adjustments to boost the effect of the program in building long-term resilience to droughts. Thus, in a more comprehensive sense of drought resilience, the program faces certain factors that limit its impact. Some of these shortcomings are expected to be solved within the program and are indeed addressed in the fourth phase of the PSNP. Others go beyond the mandate of a social protection scheme, yet PSNP could play a major role if carefully linked to other long-term development and resilience building programs. Taking into account the special role that can be played by social protection schemes and the above mentioned experiences, the following policy conclusions are drawn to make Ethiopia's PSNP—and safety net programs in general—work better for drought resilience.

First, policy makers should build awareness on drought risk management and the role of PSNP for enhanced drought resilience at community, subnational, national, regional, and global levels. These may include use of mass media to create awareness on drought, its multisectoral impact as well as its wider implication for national and regional peace and stability. Gatherings for payments could be used to sensitize beneficiaries on drought issues. The linkages of PSNP with other sectors could also be further communicated to develop new, location-specific ideas about raising drought resilience beyond the standard program.

Second, policy makers should improve communication among donors/NGOs and government institutions, which is decisive for efficient and properly functioning of social protection schemes, drought early warning system, and tailored long-term drought resilience programs. There is a need of establishing a regional/national independent platform that consolidates the early warning information on droughts from various sources. This can be in a form of a consortium of various governments, NGOs, research institutions with high-profile expertise and reputation.

Third, policy makers should improve the capacity of individuals, institutions, and organizations in order to use and mobilize resources. Especially, strengthening the skill and technology transfer for local PSNP implementers and enhance internal capacity of PSNP districts. For instance, expanding banking options and complementary business trainings for farmers so that they are able to invest in various sectors (also outside of agriculture) in their community. This could be used to create sources of employment and buffer in disaster periods.

Fourth, if social protection schemes such as PSNP are to serve their purpose as a long-term development approach in building drought resilience, then it is important to create and maintain quality infrastructures. Therefore, it is imperative to ensure active participation of the most vulnerable group. In other words, adequate grassroots level community participation from planning to implementation and evaluation should be strengthened. Furthermore, both technical and local human capacity development should be enhanced through learning and experience sharing platforms with assistance of development partners.

Fifth, the impacts of drought are multifaceted and its management requires strong multisectoral collaboration. Therefore, a robust and comprehensive central institution is essential to enhance coordination among governments, development partners, and nongovernment organization in carrying out long-term activities toward drought resilience. Thus, it is necessary to establish a strong coordination unit with solid authority and clear accountability to oversee the coordination of drought resilience activities among sectors. This coordination unit can be placed within the Food Security Coordination Directorate.

Sixth, strong monitoring and knowledge management is vital for effective follow-up, reporting, and documentation of drought resilience efforts and achievements. Thus, it is important to facilitate the exchange of information among PSNP stakeholders and those in the NGO sectors who are implementing/implemented drought resilience initiatives. This has to be accompanied by documentation of lessons learned and scale-up of best practices.

References

Anderson, C., Mekonnen, A., Stage, J., 2011. Impacts of the Productive Safety Net Program in Ethiopia on livestock and tree holdings of rural households. Journal of Development Economics 94 (1), 119–126.

Arnold, C., Conway, T., &Greenslade, M., 2011. DFID Cash Transfers Literature Review. Department for International Development (DFID), London.

Ashley, S., 2009. Guidelines for the PSNP Risk Financing Mechanism in Ethiopia. The IDL group, Bristol.

Barrientos, A., 2010. Social protection and poverty. International Journal of Social Welfare 20 (3), 240–249.

Béné, C., Devereux, S., &Sabates-Wheeler, R., 2012. Shocks and Social Protection in the Horn of Africa: Analysis from the Productive Safety Net Programme in Ethiopia (IDS Working Paper 395). Institute of Development Studies (IDS), Brighton.

Chaurey, R., Mukim, M., 2015. Decentralization in Ethiopia–Who Benefits? World Bank, Washington, DC. © World Bank. https://openknowledge.worldbank.org/handle/10986/23574. License: CC BY 3.0 IGO.

Dercon, S., 2011. Social Protection, Efficiency and Growth (CSAE Working Paper WPS/2011-17). University of Oxford, Centre for the Study of African Economies (CSAE), Oxford.

Devereux, S., 2010. Seasonal food crisis and social protection in Africa. In: Harriss-White, B., Heyer, J. (Eds.), The Comparative Political Economy of Development: Africa and South Asia. Routledge, Taylor & Francis Group, London, pp. 111–135.

Devereux, S., Sabates-Wheeler, R., Slater, R., Tefera, M., Brown, T., &Teshome, A., 2008. Ethiopia's Productive Safety Net Programme (PSNP): 2008 Assessment Report. Institute of Development Studies (IDS), Brighton.

Duguma, M.K., Brüntrup, M., Tsegai, D., 2017. Policy Options for Improving Drought Resilience and its Implication for Food Security. Studies 98. Deutsches Institut für Entwicklungspolitik. https://www.die-gdi.de/uploads/media/Study_98.pdf.

Ellis, F., White, P., Lloyd-Sherlock, P., Chhotray, V., Seeley, J., 2008. Social Protection Research Scoping Study. University of East Anglia, Governance and Social Development Resource Centre, Overseas Development Group, Norwich.

EPA, 2010. Afar National Regional State Programme of Plan on Adaptation to Climate Change. Semera: FDRE.

Gilligan, D.O., Hoddinott, J., &Taffesse, A.S., 2009. The impact of Ethiopia's productive safety net programme and its linkages. Journal of Development Studies 45 (10), 1684–1706.

Headey, D.D., Taffesse, A.S., You, L., 2012. Enhancing Resilience in the Horn of Africa: An Exploration into Alternative Investment Options (IFPRI Discussion Paper 01176). International Food Policy Research Institute (IFPRI), Washington, DC.

Jones, N., Tafere, Y., &Woldehanna, T., 2010. Gendered Risks, Poverty and Vulnerability in Ethiopia: To what Extent Is the Productive Safety Net Program (PSNP) Making a Difference? Overseas Development Institute (ODI), London.

Kena, D.J., 2016. Decentralization of power and local good governance in Ethiopian Federal System: a Look at two decades experiment. Urban and Regional Planning 1 (3), 45–58.

MOA, 2014a. Productive Safety Net Programme 4 [Design Document]. Addis Ababa: Federal Democratic Republic of Ethiopia, Disaster Risk Management and Food Security Sector (DRMFSS). Food Security Coordination Directorate.

MOA, 2014b. Productive Safety Net Programme, Phase 4. Programme Implementation Manual. Federal Democratic Republic of Ethiopia, Addis Ababa.

MoE, 2017. Education Statistics Annual Abstract. 2008 E.C. (2015/16. FDRE, Addis Ababa.

NDRMCC, 2016. Joint FDRE and Humanitarian Partners' Appeal. Humanitarian Requirement for Ethiopia, Addis Ababa.

NDRMCC, 2017. Joint FDRE and Humanitarian Partners' Appeal. Humanitarian Requirement for Ethiopia, Addis Ababa.

Porter, C., Dornan, P., 2010. Social Protection and Children: A Synthesis of Evidence from Young Lives Longitudinal Research in Ethiopia, India and Peru. Policy Paper no.1, Young Lives. Department of International Development, University of Oxford, Oxford, UK.

Drought preparedness and livestock management strategies by pastoralists in semi-arid lands: Laikipia North, Kenya

S. Wagura Ndiritu

Strathmore University Business School, Nairobi, Kenya

Introduction

The impacts of drought[1] in the developing world are wide-ranging: they affect almost all sectors of development, but the most vulnerable in this regard are agriculture, water, and health. The loss of income due to the loss of crops, the loss of livestock, and/or the loss of employment in these sectors cause great stress on people's livelihoods. Kenya is a highly drought-prone country because of its peculiar ecoclimatic conditions, as only about 20% of the territory receives adequate rainfall. The remaining 80% of the territory is made up of arid and semi-arid lands (ASALs) where periodic droughts are part of the climate system. Rainfall in arid areas ranges between 150 and 550 mm a year, while semi-arid areas receive between 550 and 850 mm (GoK, 2012a). In recent years, the chance of drought occurring in Kenya's ASALs has increased to as much as once every 3 years, while severe droughts have been forecast to occur there every 5 to 10 years (GoK, 2012a). Droughts directly impact on the household food security of over 10 million people who live in drought-prone areas. Moreover, although drought affects the entire nation, it has the direst consequences for the livestock-based economies and livelihoods in Kenya's ASALs (Zwaagstra et al., 2010): it adversely affects the already fragile food and livestock situation, and seriously impairs ASAL economies and sociocultural structures. Droughts erode the assets of poor communities and undermine their livelihood strategies, culminating in a downward spiral of increasing poverty and food insecurity.

Kenya has experienced two severe droughts in recent history, in 2008/09 and 2011 droughts (Kioko, 2013; World Bank, 2011). These severe droughts brought extreme hardship to livestock-based communities in the ASALs (Zwaagstra et al., 2010; World Bank, 2011): pastoralists' livestock production and associated incomes were negatively impacted by an abrupt and drastic decline in grazing resources, thereby exposing pastoralists to dire transitory food insecurity. Suffering and emaciated livestock translates into food insecurity by reducing the pastoralist's purchasing power through (1) lower incomes from animal sales and (2) from higher food prices, especially maize (World

---

[1]Drought is a naturally occurring phenomenon that exists when precipitation recorded is significantly below normal recorded levels causing a serious hydrological imbalance that affects land resource production systems (UNEP and GOK, 2000).

Drought Challenges. https://doi.org/10.1016/B978-0-12-814820-4.00017-1

Bank, 2011). Indeed, the 2008/09 and 2011 droughts were so severe that pastoralist households responded with new livestock management strategies to help reduce the negative impacts of future droughts.

ASALs support over 70% of all livestock in Kenya, and livestock is the main economic activity in ASALs (GoK, 2012b). In Kenya, the pastoral groups such as the Maasai pastoralists inhabit ASALs and in total account for 36% of the total population which is around 14 million people (GoK, 2012b). Some of the key features of ASALs are climate variability and extremes. However, climate change is assumed to be responsible for the likelihood of a rise in the frequency and intensity of droughts (Nkedianye et al., 2011; Speranza, 2010). If one considers that the ASAL context indicates fragile pasture conditions and variable access to pastures and water, the already intense pressure being exerted on these stressed and degraded resources by increasing human and livestock populations is likely to be exacerbated by climate change (Speranza, 2010).

Pastoralism is a common source of livelihoods and a traditional response to the harsh climate of Kenya's ASALs. Pastoral production systems are defined as those in which at least 50% of the household gross revenue comes from livestock or livestock-related activities (GoK, 2008; Catley at al., 2016). Pastoralism is the extensive production of livestock in rangeland environments. It takes many forms, but its principal defining features are livestock mobility and the communal management of natural resources. These are regulated by sophisticated governance systems within pastoral societies. There are two major categories of pastoralists: some are very mobile (in the driest areas), while others are more sedentary. Pastoralists use dryland natural resources sustainably where no other land-use systems can thrive. In doing so, pastoralists make use of water and fodder, the availability of which varies widely with respect to time and space (Catley et al., 2013; Krätli et al., 2015). To cope with the erratic conditions that characterize ASALs, the pastoral Maasai use a number of different strategies, drought refuges, disease-free areas, grass banks, and carefully chosen settlement sites, to enable them to exploit the landscape and weather conditions throughout the year (Lind et al., 2016). The basic economics of pastoralism combines the need to manage risk and to increase financial assets in a context of uncertain rainfall and therefore uncertain access to pasture and water for livestock (Catley at al., 2016). The primary policy challenge is how to protect and promote mobility and support the customary institutions which underpin pastoralism in a society which is otherwise sedentary and tending toward more individualized modes of organization and production (Catley at al., 2016). Historical herding institutional changes in pastoral lands have constrained adaptation to drought and markets (Unks et al., 2019).

Ranching is a form of beef production system practiced within a defined unit of land. In a ranch, it is possible to maintain optimal stocking rates, conserve and preserve pasture, and develop livestock support facilities such as dips and water points.[2] This system is practiced in both arid and semi-arid areas either as private or pastoralist group ranch. A pastoralist group ranch is a livestock production system or enterprise where a group of people jointly own freehold title to land, maintain agreed stocking levels, and collectively share herding responsibilities for livestock that are individually owned (Government of Kenya, 1968; Davis, 1970). It is noteworthy that selection of members to a particular group ranch was based on kinship and traditional land rights. Group ranches in Kenya were designed by the Kenya government in consultation with interested parties—Maasai elders and financiers—to increase productivity of pastoral lands, improve the earning capacity of pastoralists, avert landlessness,

---

[2]Source: http://www.infonet-biovision.org/node/5288.

prevent environmental degradation, establish a livestock production system that allows the modernization or modification of livestock husbandry and preserve the traditional way of life without causing social friction. The formation of group ranches was seen as a way to develop Maasailand in a manner that would not make the Maasai landless, and as the ranches were already stocked with cattle, it seemed to be a cost-effective solution (Lind et al., 2016). However, recent subdivision of some group ranches (e.g., Kajiado county) led to increased settlement and loss of mobility for the herds (Mwangi, 2005). The result was increased land degradation and increasing the pastoralists' vulnerability to drought (Western and Nightingale, 2004).

In the literature, the livestock drought coping strategies includes herd splitting, changing species composition, destocking, hired/purchased pasture, livestock mobility, and grazing based on rotation between dry and wet season (Butt et al., 2009; Huho et al., 2011; Mengistu and Haji, 2015; Oba, 2001; Opiyo et al., 2015; Speranza, 2010; Ng and Yap, 2011; Western and Nightingale, 2004). Some of the traditional drought recovery strategies included livelihood diversification, pastoral movement, restocking by purchase of breeding stock, diversifying livestock species, relief assistance, and herd splitting (Ahmed et al., 2002; Lesorogol, 2009; Ouma et al., 2012). Some of the coping strategies are also carried forward into the recovery phase, thus making the coping strategies and the recovery strategies interlinked and interrelated (Ouma et al., 2012).

The long-term livestock adaptation strategies used include destocking, livestock mobility, effective early warning system (EWS), water harvesting, diversifying livestock feeds, diversification of herd composition to benefit from the varied drought and disease tolerance, as well as fecundity of diverse livestock species (Bobadoye et al., 2016; Mude et al., 2007; Opiyo et al., 2015; Speranza, 2010; Silvestri et al., 2012). However, there remains a significant dearth of data on drought preparedness, especially in relation to the adoption of modern livestock management strategies, a gap this study aims to fill.

Farmers in arid and semi-arid drought-stricken areas of Kenya have given little attention to drought preparedness, mitigation, or early warning actions—i.e., risk management—that could reduce these impacts (Wilhite et al., 2000). The following strategies are deemed essential for drought preparedness in the developing world in general (Solh and Ginkel, 2014): (1) geographical shifts of agricultural systems; (2) climate proofing rainfall-based systems; (3) making irrigated systems more efficient; and (4) expanding the intermediate rain-fed irrigation systems.

Since these strategies largely apply to crop farming systems, however, for our study, I relied on previous research on livestock management and proposed the following strategies for drought preparedness and/or longer adaptation process (induced by droughts, shrinking space, and new options) relating to livestock farming systems in particular in respect of Kenya's semi-arid Laikipia North region: (1) early selling or destocking; (2) improved water management; (3) climate proofing in livestock systems including improved vaccination; (4) adoption of more drought-tolerant species or breeds; and (5) improved disease surveillance (e.g., Aklilu and Wekesa, 2002; Morton and Barton, 2002; Federal Democratic Republic of Ethiopia, 2008; Zwaagstra et al., 2010; Macon et al., 2016). The purpose of this study, therefore, was to investigate the determinants of the choice behind the various livestock management strategies adopted as a proxy for understanding pastoralists' levels of preparedness for future droughts in Laikipia North.

By around 2013, the livestock subsector contributed about 10% toward Kenya's gross domestic product (GDP) (KIPPRA, 2013). This contribution amounted to 30% of the total agricultural contribution to GDP. In that year, the livestock subsector employed about 50% of the national agricultural workforce and about 90% of the ASAL workforce. Close to 95% of all ASAL household income was derived from this subsector. It follows, therefore, that development of the pastoralist economy is critical to poverty reduction in Kenyan ASALs.

When the Kenyan government issued its Sessional Paper No. 8 of 2012 on National Policy for the Sustainable Development of Northern Kenya and other Arid Lands (GoK, 2012b), it aimed to increase investment in those regions and ensure that the use of resources there was fully reconciled with the realities of people's lives. These policy provisions were consistent with the African Union's Policy Framework for Pastoralism in Africa approved in January 2011 (African Union, 2010). In addition, Kenya's National Policy provided the framework to establish the National Drought Management Authority (NDMA) for coordinating all matters relating to drought management in the country.

One important element of the policy for ASAL and in the NDMA is an effective drought EWS. EWSs have high potential in contributing toward tackling the cycle of droughts because they provide timely, relevant, and comprehensible information on impending drought events. This information could be used to mitigate the effect of such calamities and thereby reduce their negative impact on fauna and flora (Masinde, 2014). However, an EWS is much more than a forecast: it is a communication system which links risk information—including information on people's perception of risk—and actively engages communities involved in preparedness (Pulwarty and Sivakumar 2014).

By 2001, Kenya had already initiated drought EWSs at district level as part of a national policy to alleviate the effects of drought and decrease the threat of famine and food insecurity in its ASALs. Surprisingly, Kenya is one of the few countries in the world to have designed and put into practice EWSs that specifically target drought in the pastoral livestock sector (Barton et al., 2001). These systems offer information on when it is going to rain and how much rain is predicted, which is critical for vulnerable households in semi-arid economies—especially with the effects of climate change. Moreover, Kenya's EWSs are efficient and effective in terms of identifying the various stages leading to an emergency. However, they are expensive to run unless funds are immediately available to enact contingency plans (Barton et al., 2001; Swift, 2001).

This chapter's contribution to the current literature is threefold. First, unlike previous studies on short-term drought coping mechanisms for agro-pastoralists in Kenya (e.g., Butt et al., 2009; Huho et al., 2011; Mengistu and Haji, 2015; Oba, 2001; Opiyo et al., 2015; Speranza, 2010; Ng and Yap, 2011; Western and Nightingale, 2004), this paper investigates long-term drought preparedness strategies of pastoralists in respect to livestock management in the country. A second novelty of this study is that it considers multiple livestock management strategies, unlike the usual approach to study a single strategy, and thus complementarity and substitutability among the various livestock management strategies are studied. In practice, pastoralists commonly combine several different livestock management mechanisms when they prepare for future droughts. This multipronged approach enables them to obtain the synergic benefits of such combinations. Finally, I investigate the role of access to early warning information in the adoption of improved livestock management strategies, since EWSs targeting ASALs are currently being championed by the government through the NDMA.

The rest of the chapter is organized as follows: Section 2 describes the study area, Section 3 discusses the variables and sets out the sampling procedures, and methodology employed, while Section 4 discusses the descriptive statistics and the analytical results. Section 5 concludes the chapter.

Study area

The data used in this study are part of the multicountry Pathways to Resilience in Semi-arid Economies (PRISE) Project.[3] In Kenya, the Project targets residents in Laikipia North. The main motivation for selecting this semi-arid area was its unique combination of livestock production systems (pastoralism and ranching). The sites targeted in the area were considered to have good potential for pastoralists and ranching livestock production systems.

The climate in Laikipia North is mainly semi-arid, with an annual average rainfall in the 400−750 mm range. The study area is situated where extremes of climate variation such as drought and unpredictable rainfall patterns are regular phenomena. Consequently, the area experiences cyclic drought, with the more recent drought events having been recorded in 2000, 2005, 2009, 2014, and 2017. Famines usually follow these droughts. The stakeholders interviewed generally regarded the increasingly arid conditions in the study area as being due to the impacts of more intense climate variability.

The 2009 drought was the worst in 60 years in Laikipia North where high livestock mortality rates were the order of the day, particularly for cattle (64%) and sheep (62%) (Mwangi, 2012). These livestock losses were extremely high when compared with similar losses during other recent droughts and the one in 1976. Thus, the 2009 drought was not only extreme from a meteorological and rangeland production perspective, but it also had a devastating impact on livestock resources (Zwaagstra et al., 2010).

Methodology

Before implementing the intended survey of households in Laikipia North in July 2016, a reconnaissance visit to all the study sites was conducted to collect secondary data. This comprised data reflecting the basic socioeconomic profiles of the respondent households, comprehensive data on livestock production, and information on marketing livestock from government officials at the Laikipia County livestock offices. Informal discussions with pastoralists and key informants were also conducted. Based on all the information collected, the sampling strategy was developed.

In order to have good understanding of respondents' perceptions of climate change, their vulnerability to climate extremes such as drought, and their employment of long-term drought preparedness strategies for livestock management, interviews were conducted with representatives of 440 households from eight group ranches by way of a pretested structured questionnaire. The population distribution within a group ranch was considered in order to stratify the ranch and determine distribution for the sample. Ranches where our security could not be guaranteed were excluded from the survey, as was the ranch used for pretesting the questionnaire. Enumerators who were familiar with the sampling area were selected from each group ranch.

---

[3]The PRISE project was a 5-year, multicountry (i.e., Senegal, Burkina Faso, Pakistan, Tajikistan, and Kenya) research project "harnessing opportunities for climate-resilient economic development in semi-arid lands: adaptation options in the cotton and beef value chains" and follows the three-step Value Chain Analysis for Resilience in Drylands (VC-ARID) methodology. For Laikipia, Kenya, VC-ARID enabled the identification of climate risk, adaptation options, and opportunities for development of actors involved in the beef value chain.

Given that I investigate several livestock management strategies, I will allow for interdependence of the strategies since pastoralists may adopt several of them simultaneously to complement or substitute any existing strategy. Because such management measures are chosen either simultaneously or sequentially and, thus, the error terms associated with them may be correlated, I use a multivariate probit (MVP) specification. MVP allows for systematic correlations among choices for the various strategies. A positive correlation of the error terms will indicate that the livestock management strategies involved are likely to be complementary, while negative correlations of these terms imply that the strategies are substitutes. A positive correlation may also be consequence of the fact that some households are more aware in general and thus chose more options.

The basic model is characterized by a set of latent dependent variables (L_i^*) and binary dichotomous choice variables (L_i) specified as follows:

$$L_i^* = \beta_{ij} X_j + \varepsilon_i \tag{17.1}$$

$$L_i = \begin{cases} 1 & \text{if } L_i^* > 0 \\ 0 & \text{otherwise} \end{cases}, \tag{17.2}$$

where $i = 1 \ldots k$ denotes the type of livestock management strategy adopted by the pastoralist. I construct dummy variables for the following livestock management strategies: vaccination before drought, disease surveillance, water management, early selling or destocking, and adoption of new species or breeds. Morton and Barton (2002) argue that selling strategies when animals are in good shape are analogues of the sort of destocking practiced by ranchers and commercial farmers as their long-term strategy to drought mitigation. X_j are the control variables by the jth household, which are the same for the various strategies. β_{ij} is a vector of parameters to be estimated. ε_i are error terms that may be correlated; if they are not, I estimate the univariate probit model (Greene, 2008). Following my sampling procedure, ε_i are multivariates normally distributed with zero means, unitary variance, and an $n \times n$ contemporaneous correlation matrix $[Q = \rho_{ij}]$.

Following the literature on coping strategies for climate extremes, the variables hypothesized to influence adaptation and other coping strategies include *temperature, rainfall, human capital* (proxied by *education* and *age*), *gender, credit facilities, income, household size, ownership of properties* (e.g., land and household assets), *infrastructure,* and *livestock population size* (e.g., Speranza, 2010; Berhanu and Beyene, 2015; Getachew et al., 2014; Mengistu and Haji, 2015). Long-term mean rainfall and temperature from 1950 to 2014 are obtained from the Kenya Meteorological Department. It is hypothesized that access to private ranch grazing and the receipt of early warning messages influence the choice of long-term drought preparedness livestock management strategies in these semi-arid economies.

Empirical results

Descriptive statistics

Table 17.1 provides descriptive statistics of the climate variables in the study area and the socioeconomic characteristics of the study group.

Table 17.1 Definition and descriptive statistics of model variables.

Variable	Description	Mean	Standard deviation
Avg_rainfall	Average annual rainfall	649.584	80.240
Avg_temp	Average annual temperature	28.009	0.633
Earlywarning	Access to early warning information (Yes = 1, No = 0)	0.389	
Grazingpranch	Access to grazing provided by private ranches (Yes = 1, No = 0)	0.689	
Wealthscore	Wealth index	0.000	1.627
Lvstksize	Livestock size in a standardized unit (total livestock units/TLU)	19.463	21.048
Age	Age of the household head (years)	44.186	12.974
Male	Male dummy (Male = 1, Female = 0)	0.923	
Higheduc	Highest level of education in the household (years of schooling)	9.566	3.822
Hhsize	Household size	6.423	2.575
Dist2manmkt	Distance to the main market (km)	7.956	5.213
Credit	Access to credit (Yes = 1, No = 0)	0.189	
Pastoralist	Pastoralism the main activity of this household (Yes = 1, No = 0)	0.816	

Source: Field survey, 2016.

As Table 17.1 reveals, the majority of the interviewed households were male-headed (92%) and regarded pastoralism as their main livelihood activity (82%). This was to be expected, given the climatic conditions of semi-arid lands, where well-managed rangelands can offer good livestock ranching. The highest level of education across households averaged out at the equivalent of 9.5 years. However, the household average in respect of education was higher than that for household heads, which was a low primary level, equivalent to 5.5 years. This shows that household heads invest in the school enrollment of their children. As regards respondents' access to credit, the data display a very low figure, namely 19%. The average age of a household head in the study area was 44 years.

In terms of the NDMA's efforts to promote early warnings in the ASALs about impending climate risks, only 39% of the households interviewed had ever received such messages. Of the households that had received early warning information, some 71% reported having adjusted their livestock management accordingly.

All the households reported to have lost livestock due to severe droughts that have become very frequent. Therefore, drought was ranked as the most important climate shock in Laikipia North. The main consequences of drought were loss of and reduced quality in livestock, both of which led to reduced sales and household incomes.

As an additional coping mechanism, pastoralists in Laikipia North were granted access to grazing on private ranches at a fee. About 70% of the respondents accessed such private grazing. Given that pastoralists could be learning from the ranches how to properly manage rangelands, this could influence their own livestock management strategies. This is further investigated using an econometric model.

Table 17.2 sets out the various livestock management strategies employed by pastoralists in Laikipia North to cope with drought.

Droughts are known to have short- and long-term effects on pastoralists' livelihoods. The short-term effects are the shocks caused by heavy losses of their livestock due to a drastic and abrupt decline of grazing resources, and their consequent exposure to severe transitory food insecurity. The long-term effects entail that pastoralists who experience chronic food insecurity due to the lack of recovery of their herds and/or long-term deterioration of grazing resources (grass, bushes, and trees) are forced to consider giving up their lifestyles and face utter impoverishment. Thus, pastoralists are obliged to manage their livestock properly during nondrought years, since effective management entails a prolonged food source.

Pastoralists adopt several strategies to prepare for future droughts and other related hazards. Table 17.2 reports the livestock management strategies considered in this study. Following past studies (e.g., Aklilu and Wekesa, 2002; Federal Democratic Republic of Ethiopia, 2008; Macon et al., 2016; Morton and Barton, 2002; and Zwaagstra et al., 2010), the following long-term drought preparation

Table 17.2 Livestock management strategies (N = 440).

Variable	Definition	Respondents who adopted the strategy (%)
Vaccination	Adopted vaccination (Yes = 1, No = 0)	16.4
Surveillance	Disease surveillance (Yes = 1, No = 0)	11.0
Watermgt	Change in water management (Yes = 1, No = 0)	38.4
Destocking	Early selling/destocking (Yes = 1, No = 0)	52.0
Newbreeds	New species/breeds adoption (Yes = 1, No = 0)	15.5

Source: Field survey, 2016.

livestock management strategies were specifically considered: vaccination against diseases, disease surveillance, water management, early selling or destocking, and adoption of new species or breeds.

Animal health interventions such as vaccination campaigns are perceived to be effective in reducing livestock mortality. The impact of drought can be reduced when a livestock vaccination or treatment plan is included in district-led drought preparedness strategies (Aklilu and Wekesa, 2002). Although vaccinating during a drought constituted poor timing, it was found that a lack of funding for routine vaccination during nondrought periods caused it to be carried out during drought times, that is, when emergency funds became available (Zwaagstra et al., 2010). Only about 16.4% of the respondents reported that they had adopted vaccination against disease before drought as a livestock management strategy.

Disease surveillance was low, namely only 11%. In 2017, FAO trained veterinary officers in Laikipia to conduct disease surveillance and reporting using mobile phone technology.[4] The target was to improve disease reporting coverage from the grassroots level along the whole reporting chain including the private sector to at least 50% of cases reported. In addition, the training equipped the officers with the skills to conduct outbreak investigations rapidly.

During drought times, dams usually dry up and the only source of water for livestock and households is boreholes. During the interviews, the targeted community expressed the need for more boreholes and dams as well as proper distribution of other available water resources. When respondents were probed as to whether their group ranch had been engaged in new water management strategies in preparation for future droughts, 38.4% of the respondents reported to having adopted such changes.

Livestock populations should be monitored during nondrought periods and, if necessary, reduced by either being consumed or sold before the harshest effects of drought impact them. Early selling or destocking is a livestock management strategy that not only helps pastoralists manage droughts and dry spells but also enables them to participate in the market, keep livestock as a business, and reduce losses due to drought. Thus, pastoralists need to formulate a selling policy to help them decide what classes of animals to sell and the rate at which they should be put on the market (Mcdougald et al., 2001). In addition, pastoralists are advised to check all heifers and cows for pregnancy and cull those that are open, saving the most desirable and younger individuals (McDougald et al., 2001). By carrying these selected animals on minimal rations, pastoralists will be able to save valuable breeding stock and replenish the herd when conditions improve. About 52% of the respondents had been engaged in early selling in Laikipia North.

Diversification of herd composition and species is a key strategy that has enabled pastoralism to thrive in a harsh environment for centuries (Speranza, 2010). Today, pastoralists are changing their herd composition by adopting new species or breeds that can adapt to the harsh climate, are good for beef production, and are marketable. Households involved in diversification of herd composition and species have a higher offtake and thereby improve access to food during drought (Opiyo et al., 2015). However, only 15.5% of the pastoralists in the study area were diversifying their herd with new species.

The regression results from the MVP model are presented in Table 17.3.

A likelihood ratio test was carried out, using the null hypothesis that the correlation coefficients (ρ statistics) are jointly equal to 0 versus the alternative hypothesis that ρ are not jointly equal to 0. The hypothesis of independence between the error terms is strongly rejected; hence, the use of MVP is

---

[4]http://www.fao.org/kenya/news/detail-events/en/c/1073996/.

Table 17.3 Multivariate probit results.

Variable	(1) Vaccination	(2) Surveillance	(3) Watermgt	(4) Destocking	(5) Newbreeds
Avg_rainfall	0.010***	0.004***	0.012***	0.003***	−0.009***
	(0.003)	(0.001)	(0.002)	(0.001)	(0.001)
Avg_temp	1.069**	0.493**	−0.264	−0.577***	0.115
	(0.523)	(0.201)	(0.207)	(0.163)	(0.175)
Earlywarning	0.029	−0.191	0.471**	0.374**	0.599**
	(0.267)	(0.236)	(0.221)	(0.171)	(0.259)
Grazingpranch	0.543**	0.808***	0.621***	0.357**	0.492**
	(0.228)	(0.242)	(0.201)	(0.170)	(0.229)
Wealthscore	−0.108*	−0.087	−0.025	0.098*	0.209***
	(0.065)	(0.064)	(0.055)	(0.051)	(0.062)
Lvstksize	−0.024***	0.006	0.011***	0.006*	0.003
	(0.007)	(0.004)	(0.004)	(0.003)	(0.004)
Age	−0.004	−0.009	0.009	0.003	0.004
	(0.008)	(0.009)	(0.007)	(0.006)	(0.008)
Male	0.253	0.121	0.148	−0.406*	0.018
	(0.304)	(0.327)	(0.280)	(0.241)	(0.362)
Higheduc	0.037	0.036	−0.007	−0.008	0.024
	(0.025)	(0.023)	(0.019)	(0.018)	(0.023)
Hhsize	−0.003	0.084**	0.022	−0.053	−0.037
	(0.044)	(0.039)	(0.041)	(0.034)	(0.038)
Dist2manmkt	−0.178***	−0.011	−0.151***	−0.040**	0.078***
	(0.036)	(0.020)	(0.037)	(0.016)	(0.017)
Credit	−0.287	0.076	0.589**	0.217	0.139
	(0.255)	(0.264)	(0.234)	(0.203)	(0.222)
Pastoralist	0.496*	−0.381	−0.040	0.101	−0.084
	(0.270)	(0.232)	(0.207)	(0.189)	(0.234)
Constant	−37.103**	−18.747***	−1.213	14.438***	0.204
	(16.180)	(5.935)	(6.126)	(4.733)	(5.102)
Observations	435	435	435	435	435
Model chi-square	552.3	552.3	552.3	552.3	552.3

Binary correlation	Correlation coefficient	Robust standard error	P-value
rho21	−0.196	0.083	0.018
rho31	−0.286	0.095	0.003
rho41	−0.156	0.095	0.102
rho51	−0.082	0.107	0.443

Table 17.3 Multivariate probit results.—cont'd

Binary correlation	Correlation coefficient	Robust standard error	P-value
rho32	−0.232	0.109	0.034
rho42	−0.195	0.088	0.026
rho52	0.176	0.102	0.084
rho43	0.301	0.096	0.002
rho53	0.034	0.123	0.78
rho54	−0.480	0.084	0.000

Robust standard errors in parentheses; ***P < .01, **P < .05, *P < .1.
Likelihood ratio test of rho21 = rho31 = rho41 = rho51 = rho32 = rho42 = rho52 = rho43 = rho53 = rho54 = 0; chi^2(10) = 59.9434; prob > chi^2 = 0.0000.

supported by this test. For the livestock management strategies under consideration, some had positive correlations. This indicated that they complemented each other. It could also mean that some decision makers are more aware than others. Others had negative correlations, revealing that they were substitute strategies.

As expected, higher rainfall correlated with an increase in water management strategies adopted by the respondents. This outcome could be explained by the fact that higher rainfall brought increased water supplies to be harvested from community ranch dams.

The study also revealed that an increase in average rainfall concomitantly increased the likelihood of adopting the livestock management strategies of destocking and early selling, as well as those of increased vaccination and disease surveillance. Although it was generally better for pastoralists to sell animals during nondrought periods, the fact that respondents engaged in commercial emergency destocking when climate extremes begin to hit nonetheless helped them salvage as much stock as possible before their animals begin to die (Ayal et al., 2018). When pastoralists employ the destocking/selling strategy during non-drought periods, this indicates that they have adopted good livestock management measures. It reduces overstocking around villages and ensures that their animals have higher weight and, therefore, get better prices at market. This ensures that their animals have higher weight values and, therefore, get better prices in the market, besides reducing overstocking around village settlements. Thus, the finding that respondents employed destocking—albeit only during drought periods—support the contention put forward by others (e.g., Derner and Augustine, 2016; Macon et al., 2016; Shrum et al., 2018) that the most common method of long-term drought adaptation is conservative stocking,[5] preferably employed even during nondrought periods.

Although it was an unexpected finding, the study also revealed that an increase in rainfall reduced the adoption of improved breeds, while an increase in average temperatures reduced early selling but increased disease surveillance and vaccination. One obvious reason why pastoralists might be reducing

---

[5]Conservative grazing involves routinely stocking rangelands 10%−30% below grazing capacity (Holechek et al., 1998). Lower animal performance, lower vegetation productivity, and the risk of financial crisis during drought all make routinely stocking at capacity an unsound strategy. Generally, partial destocking due to drought is only needed in about 1−2 years out of 10 with conservative stocking compared to half the years when rangelands are fully stocked (Holechek et al., 1998).

their adoption of improved breeds during nondrought times is that the pressure to possess breeds that adapt to unfavorable environments is minimal.

The study found that access to early warning information had positive and significant effects not only on decisions to engage in early selling and destocking but also on decisions to improve or invest water management on the group ranch and/or adopt improved breeds that were more productive and resistant to extreme weather conditions. An EWS was seen to empower communities by informing households to prepare for severe droughts so that they could reduce animal losses and enhance their drought resilience and food security.

As mentioned above, private ranches in Laikipia North allow pastoralists to graze at a fee. Results showed that access to grazing offered by such private ranches had a significant and positive effect on the adoption of all the livestock management strategies under study. This could be explained by the fact that pastoralists grazing on private ranches could observe the very sophisticated livestock management strategies in operation there which increased private ranchers' resilience in times of drought. Pastoralists could then adopt these measures themselves later.

Furthermore, the study tested the hypothesis that the further away a pastoralist was from the nearest market, the probability that s/he would adopt livestock management strategies that required education, exposure to innovations, contacts to extension officers and government services would increase, since markets are a source of such information. The findings also showed that, the further pastoralists lived away from the main market, the less likely it was that they employed the strategies of improved water management, vaccination, and early selling. This could be explained by the fact that pastoralists live in very remote areas: as distances to market increased, access to vaccination drugs and opportunities to sell livestock decreased. Thus, households that were furthest away from the market had no access to advice on current trends on livestock management strategies, and they could not sell their livestock when they needed to do so.

Another unexpected finding was that the greater the household's distance to market, the greater the likelihood was that such a household would adopt an improved breed. Those households have less access to other livestock management technologies, and often live in harsher conditions, so adopting breeds that are more drought-resistant is the simplest option.

The study also showed that wealthier households were more likely to adopt improved livestock breeds, sell some of their livestock before a drought hit, and engage in water management. In addition, access to credit increased the likelihood that pastoralists would respond to drought by employing a long-term water management strategy to enable them to cope with future droughts. Previous study in Ethiopia on coping strategies due to extreme events show that wealthy households (proxied by livestock size) are more likely to engage in destocking (Mengistu and Haji, 2015).

Conclusions and policy implications

Drought preparedness matters in respect of livestock management in many semi-arid lands mainly because drought has become very frequent and severe in such regions and threatens the food security and livelihoods of the communities who live there. However, there remains a significant dearth of data on drought preparedness, especially in relation to the adoption of modern livestock management strategies. The current study builds on the emerging literature on drought coping strategies (Butt et al., 2009; Huho et al., 2011; Mengistu and Haji, 2015; Oba, 2001; Opiyo et al., 2015; Speranza, 2010; Ng and Yap, 2011; Western and Nightingale, 2004; Morton and Barton, 2002) on addressing the gaps in long-term drought preparedness strategies and the effect that household characteristics, climate variables, access to markets, early warning messages, and private ranch grazing have on the adoption of multiple livestock management strategies.

The econometric results show that the livestock management strategies adopted by the respondents either substituted or complemented those they had been using. The study further revealed that access to markets was a positive driver when it comes to adopting livestock management strategies. There is, therefore, a need to improve access to markets in semi-arid economies since they offer the principal means of being able to take advantage of financial resources and other necessities on offer there. Increased distances to markets concomitantly decreased market participation by pastoralist households, driving them to fall back on their traditional practices and exposing them further to climate shocks. Thus, improving access to markets, which include all weather tarmac or gravel roads, improved livestock holding facilities in the markets, access to reliable information on prices which can be broadcast weekly on local radio networks in local languages (Bailey et al., 1999) could play a significant role not only in improving pastoral livelihoods per se but also in enhancing the traditional livelihood system within the framework of climate change. These interventions of pastoralism toward more commercialized systems increase their share of important livestock markets (Catley at al., 2016).

Moreover, since access to private ranch grazing increased the adoption of modern livestock management strategies, it is recommended that private ranches should be encouraged to offer relevant livestock management information when pastoralists use their grazing so that these vulnerable communities can be duly empowered. The group ranches should also be encouraged to initiate learning from the best practice livestock management techniques employed by the private ranches and implement such enhancements in their own communal group ranches.

Access to early warning information has significant positive effects not only on the decision to engage in early selling/destocking but also on the choice to increase water management on the group ranch and to adopt improved breeds that are more productive and resistant to extreme weather conditions. An EWS empowers communities and informs households in advance so that they are better prepared to cope with severe droughts; this in turn reduces the stock losses that pastoralists encounter and enhance their food security and resilience to drought in general. Although the NDMA has already implemented an EWS, the agency needs to increase its visibility in Laikipia North through an early warning information center and by educating group ranch leaders on how to communicate the information in good time to enable the communities concerned to adjust their livestock management. The main source of early warning information is the government, after which community leaders are the next best source. NDMA action on building climate-proofed infrastructure in ASALs needs to be fast-tracked.

The government ending drought emergencies initiative on the climate proofed infrastructure pillar needs to be fast-tracked by NDMA given that ASALs in Kenya continue to experience deficit of climate-proofed productive infrastructure. The framework for this pillar presents prioritization of national infrastructure projects in ASAL areas; standard guidelines for climate-proofed design infrastructure; county capacity to plan, contract and implement infrastructure. An important tool in overcoming food insecurity in case of drought is the strengthening and development of permanent water points. Permanent water points will enable communities to keep the remaining part of their livestock alive while the excess herd is being destocked and also enable the communities to better manage common natural pasture areas and trekking routes, thereby contributing to conflict reduction.

A rigorous analysis of the losses and impacts of droughts has not been addressed in this chapter. In addition, there are several other issues that have not been considered in the present work, including an analysis of the impact of drought preparedness livestock management strategies on reduced livestock losses. These are issues that warrant further research.

Acknowledgments

The data used in this study were supported by DFID and IDRC and led by ODI for the PRISE project is highly appreciated. I would also like to thank the anonymous reviewers and editors of this book for great comments which helped to improve the quality of the paper considerably. In addition, I gratefully thank Jesper Stage and Sandie Fitchat for copyediting and the Jan Wallander and Tom Hedelius Foundation and Strathmore University Business School for the financial support. The views expressed here are those of the authors and do not necessarily reflect the views of the donor or the authors' institution. The usual disclaimer applies.

References

African Union, 2010. Policy Framework for Pastoralism in Africa: Securing, Protecting and Improving the Lives, Livelihoods and Rights of Pastoralist Communities. Department of Rural Economy and Agriculture, Addis Ababa. Available at: https://au.int/sites/default/files/documents/30240-doc-policy_framework_for_pastoralism.pdf.

Ahmed, A.G.M., Azeze, A., Babiker, M., Tsegaye, D., 2002. Post Drought Recovery Strategies Among the Pastoral Households in the Horn of Africa: A Review, Organization for Social Science Research in Eastern and Southern Africa (OSSREA), Ethiopia.

Aklilu, Y., Wekesa, M., 2002. Drought, Livestock and Livelihoods: Lessons from the 1999–2001 Emergency Response in the Pastoral Sector in Kenya. Overseas Development Institute, London.

Ayal, D.Y., Radeny, M., Desta, S., Gebru, G., 2018. Climate variability, perceptions of pastoralists and their adaptation strategies: implications for livestock system and diseases in Borana zone. International Journal of Climate Change Strategies and Management 10 (4), 596–615. https://doi.org/10.1108/IJCCSM-06-2017-0143.

Bailey, D., Barrett, C.B., Little, P.D., Chabari, F., 1999. Livestock Markets and Risk Management Among East African Pastoralists: A Review and Research Agenda. Available at: SSRN: https://ssrn.com/abstract=258370 or https://doi.org/10.2139/ssrn.258370.

Barton, D., Morton, J., Hendy, C., 2001. Drought Contingency Planning for Pastoral Livelihoods, Policy Series15. Natural Resources Institute, Chatham, UK.

Berhanu, W., Beyene, F., 2015. Climate variability and household adaptation strategies in Southern Ethiopia. Sustainability journal 7, 6353–6375.

Bobadoye, A.O., Ogara, W.O., Ouma, G.O., Onono, J.O., 2016. Assessing climate change adaptation strategies among rural Maasai pastoralist in Kenya. American Journal of Rural Development 4 (6), 120–128. https://doi.org/10.12691/ajrd-4-6-1.

Butt, B., Shortridge, A., WinklerPrins, A.M.G.A., 2009. Pastoral herd management, drought coping strategies, and cattle mobility in Southern Kenya. Annals of the Association of American Geographers 99 (2), 309–334. https://doi.org/10.1080/00045600802685895.

Catley, A., Lind, J., Scoones, I., 2016. The futures of pastoralism in the Horn of Africa: pathways of growth and change. Revue Scientifique et Technique International Office of Epizootics 35 (2), 389–403.

Catley, A., Lind, J., Scoones, I., 2013. Development at the margins: pastoralism in the horn of Africa. In: Catley, A., Lind, J., Scoones, I. (Eds.), Pastoralism and Development in Africa: Dynamic Change at the Margins. Routledge, New York, USA.

Davis, R.K., 1970. Some Issues in the Evolution, Organization and Operation of Group Ranches in Kenya. Discussion Paper No. 93. Institute for Development Studies, University College, Nairobi, Kenya.

Derner, J.D., Augustine, D.J., 2016. Adaptive management for drought on rangelands. Rangelands 38 (4), 211–215.

Federal Democratic Republic of Ethiopia, 2008. National Guidelines for Livestock Relief Interventions in Pastoralist Areas of Ethiopia Ministry of Agriculture and Rural Development. Addis Ababa Ethiopia).

Greene, W.H., 2008. Econometric Analysis, 6th (International) ed. Prentice-Hall, Inc, Upper Saddle River, N.J., USA.

Getachew, S., Tilahun, T., Teshager, M., 2014. Determinants of agro-pastoralist climate change adaptation strategies: case of rayitu Woredas, Oromiya region, Ethiopia. Research Journal of Environmental Sciences 8, 300–317.

Government of Kenya (GoK), 1968. Group (Land) Representatives Act. Chapter 287 of the Laws of Kenya (Nairobi, Kenya).

Government of Kenya (GoK), 2008. National Livestock Policy. Ministry of livestock development Nairobi Kenya.

Government of the Republic of Kenya (GoK), 2012a. Development Strategy for Northern Kenya and Other Arid Lands.

Government of Kenya (GoK), 2012b. Sessional Paper No. 8 of 2012 on National Policy for the Sustainable Development of Northern Kenya and Other Arid Lands. Ministry of State for Development of Northern Kenya and Other Arid Lands, Nairobi. Available at: http://www.adaconsortium.org/images/publications/Sessional-Paper-on-National-policy-for-development-of-ASALs.pdf.

Holechek, J.L., Gomes, H.S., Molinar, F., Galt, D., 1998. Grazing intensity: critique and approach. Rangelands 20, 15–18.

Huho, J.M., Ngaira, J.K.W., Ogindo, H.O., 2011. Living with drought: the case of the Maasai pastoralists of Northern Kenya. Educational Research 2 (1), 779–789.

Kenya Institute for Public Policy Research and Analysis (KIPPRA), 2013. Kenya Economic Report 2013 KIPPRA.

Kioko, M.J.B., 2013. Who stole the rain? The case of recent severe droughts in Kenya. European Scientific Journal 9 (5), 29–30.

Krätli, S., 2015. Valuing Variability: New Perspectives on Climate Resilient Drylands Development. International Institute of Environment and Development, London.

Lesorogol, C.K., 2009. Asset building through community participation: restocking pastoralists following drought in Northern Kenya. Social Work in Public Health 24 (1–2), 178–186. https://doi.org/10.1080/19371910802569740.

Lind, J., Sabates-Wheeler, R., Kohnstamm, S., Caravani, M., Eid, A., Nightingale, D.M., Oringa, C., 2016. Changes in the Drylands of Eastern Africa: Case Studies of Pastoralist Systems in the Region. Institute of Development Studies, University of Sussex.

Macon, D.K., Barry, S., Becchetti, T., Davy, J.S., Doran, M.P., Finzel, J.A., George, H., Harper, J.M., Huntsinger, L., Ingram, R.S., Lancaster, D.E., Larsen, R.E., Lewis, D.J., Lile, D.F., McDougald, N.K., Mashiri, F.E., Nader, G., Oneto, S.R., Stackhouse, J.W., Roche, L.M., 2016. Coping with drought on California rangelands. Rangelands 38 (4), 222–228. https://doi.org/10.1016/j.rala.2016.06.005.

Masinde, M., 2014. In: An Effective Drought Early Warning System for Sub- Saharan Africa: Integrating Modern and Indigenous Approaches Paper Presented at Conference: SAICSIT, At Pretoria, South Africa. https://doi.org/10.1145/2664591.2664629. September 2014.

Mcdougald, K.N., Frost, E.W., Phillips, L.R., 2001. Livestock Management during Drought Rangeland Management Series. University of California ANR Publication 8034. Accessed at: http://www.ncrcd.org/files/7513/8091/4244/General_-_Livestock_Management_During_Drought.pdf. on 19th December 2017.

Mengistu, D., Haji, J., 2015. Factors affecting the choices of coping strategies for climate extremes: the case of Yabello district, Borana zone, Oromia national regional state, Ethiopia. Science Research 3 (4), 129–136. https://doi.org/10.11648/j.sr.20150304.11.

Morton, J., Barton, D., 2002. Destocking as a drought mitigation strategy: clarifying rationales and answering critiques. Disasters 26 (3), 213–228.

Mude, A., Ouma, R., Steeg, J. van de, Kaiuki, J., Opiyo, D., Tipilda, A., 2007. Kenya Adaptation to Climate Change in the Arid Lands: Anticipating, Adapting to and Coping with Climate Risks in Kenya - Operational Recommendations for KACCAL. ILRI Research Report, No. 18. ILRI, Nairobi (Kenya), 135pp.

Mwangi, E., 2005. The Transformation of Property Rights in Kenya's Maasailand: Triggers and Motivations" CAPRi Working Paper No. 35, CGIAR System wide Program on Collective Action and Property Rights. International Food Policy Research Institute IFPRI, Washington D.C.

Mwangi, E., 2012. Droughts in the GHA: A Case Study of the 2010/2011 drought in Kenya.

Nkedianye, D., de Leeuw, J., Ogutu, J.O., Said, M.Y., Saidimu, T.L., Kifugo, S.C., Kaelo, D.S., Reid, R.S., 2011. Mobility and livestock mortality in communally used pastoral areas: the impact of the 2005–2006 droughts on livestock mortality in Maasailand. Pastoralism: Research, Policy and Practice 1 (17), 1–17.

Ng, A., Yap, N.T., 2011. Drought preparedness and response as if development matters: case studies from Kenya. Journal of Rural and Community Development 6 (2), 15–34.

Oba, G., 2001. The importance of pastoralists' indigenous coping strategies for planning drought management in the arid zone of Kenya. Nomadic Peoples 5 (1), 89–119.

Opiyo, F., Wasonga, O., Nyangito, M., Schilling, J., Munang, R., 2015. Drought adaptation and coping strategies among the Turkana pastoralists of Northern Kenya. International Journal of Disaster Risk Science 6 (3), 295–309. https://doi.org/10.1007/s13753-015-0063-4.

Ouma, C., Obando, J., Koech, M., 2012. Post drought recovery strategies among the Turkana pastoralists in Northern Kenya. Scholarly Journal of Biotechnology 1 (5), 90–100.

Pulwarty, R.S., Sivakumar, M.V.K., 2014. Information systems in a changing climate: early warnings and drought risk management. Weather and Climate Extremes 3, 14–21.

Shrum, T.R., Travis, W.R., Williams, T.M., Lih, E., 2018. Managing climate risks on the ranch with limited drought information. Climate Risk Management 20, 11–26.

Silvestri, S., Bryan, E., Ringler, C., Herrero, M., Okoba, B., 2012. Climate change perception and adaptation of agro-pastoral communities in Kenya. Regional Environmental Change 12 (4), 791–802.

Solh, M., Ginkel, M., 2014. Drought preparedness and drought mitigation in the developing world's drylands. Weather and Climate Extremes 3, 62–66.

Speranza, I.C., 2010. Drought coping and adaptation strategies: understanding adaptations to climate change in agro-pastoral livestock production in Makueni district, Kenya. The European Journal of Development Research 22 (5), 623–642.

Swift, J., 2001. District-level drought contingency planning in arid districts of Kenya. In: Pastoralism, Drought and Planning: Lessons from Northern Kenya and Elsewhere. Natural Resources Institute, Chatham, UK, pp. 40–84.

UNEP and GOK, 2000. Devastating Drought in Kenya. Environmental Impacts and Responses, Nairobi, Kenya.

Unks, R.R., King, E.G., German, L.A., Wachira, N.P., Nelson, D.R., 2019. Unevenness in scale mismatches: institutional change, pastoralist livelihoods, and herding ecology in Laikipia, Kenya. Geoforum 99, 74–87.

Western, D., Nightingale, D.L.M., 2004. Environmental change and the vulnerability of pastoralists to drought: the Maasai in amboseli, Kenya. In: Africa Environment Outlook Case Studies: Human Vulnerability to Environmental Change. UNEP, Nairobi, pp. 35–50.

Early warning systems for drought preparedness and drought management. In: Wilhite, D.A., Sivakumar, M.V.K., Wood, D.A. (Eds.), 2000. Proceedings of an Expert Group Meeting Held in Lisbon, Portugal, 5–7 September 2000. World Meteorological Organization, Geneva, Switzerland.

World Bank, 2011. The Drought and Food Crisis in the Horn of Africa. The World Bank, Nairobi Kenya.

Zwaagstra, L., Sharif, Z., Wambile, A., de Leeuw, J., Said, M.Y., Johnson, N., Njuki, J., Ericksen, P., Herrero, M., 2010. An Assessment of the Response to the 2008-2009 Drought in Kenya. A Report to the European Union Delegation to the Republic of Kenya. ILRI, (International Livestock Research Institute), Nairobi, Kenya, 108 pp.

Building resilience to drought among small-scale farmers in Eastern African drylands through rainwater harvesting: technological options and governance from a food–energy–water nexus perspective

Nicholas O. Oguge

Centre for Advanced Studies in Environmental Law & Policy (CASELAP), University of Nairobi, Nairobi, Kenya

Introduction

It is well established (AfDB, 2011) that Africa is among the world's most vulnerable regions to climate change despite being among the smallest emitters of greenhouse gases (NASCA, 2015). Modeling of aggregate damages as a percentage of GDP shows that, with an average global temperature increase of 4°C, Africa would experience impacts from climate change greater than in any other region of the world (AfDB, 2011). The negative impacts of climate change will have a large effect on water availability and food production in Africa (IPCC, 2013), and Eastern Africa in particular will experience significant declines in yields of staple crops of maize, rice, and wheat (Parry et al., 2007; AfDB, 2011). Eastern Africa will also experience an acceleration of the hydrologic cycle, leading to increased variance in rainfall (WWF, 2006). Such changes in precipitation patterns will increase the likelihood of short-run crop failures and long-run production declines (Bazzaz and Sombroek, 1996).

As a whole, Africa may experience a slight increase in water availability under climate change scenarios, but conditions may worsen in Eastern Africa. Due to a gap between demand and supply growth from an increase in population and development, acute water shortages could arise as a result (AfDB, 2011). Eastern Africa is also projected to experience an increase in annual run-off (particularly in southern Somalia, Kenya, and southern Ethiopia), creating further challenges for water resource management and food production in an already drought-prone region (AfDB, 2011; Daron, 2014).

Within agricultural production systems in Eastern Africa, dryland environments are particularly vulnerable to the acceleration of the hydrologic cycle given their general climate sensitivity and the low adaptive capacity of the region. Large segments of the rural population are unable to adjust to and rebound from impacts of extreme climatic events. Rain-fed agriculture, commonly practiced in

Drought Challenges. https://doi.org/10.1016/B978-0-12-814820-4.00018-3

these areas, is particularly vulnerable to increasing climate variability, as any reduction of soil moisture adversely affects plant growth and pasture availability. Any changes in patterns of seasonal rainfall and extreme weather events will have devastating socioeconomic and environmental impacts in Eastern Africa (NASCA, 2015).

Extreme climate events are also of public health, socioeconomic, environmental, and development importance in Eastern Africa (Easterling et al., 2000). The extent to which these impacts are felt depends on local and national adaptive capacities. The region's vulnerability to climate events is a function of inadequate technological capacity and sociopolitical and institutional factors. The most vulnerable groups and communities are typically poor (Shepherd et al., 2013), have low investment in health and educational systems, and are politically disenfranchised and marginalized. These communities are among the first to experience the impacts of adverse climate and are typically the least equipped to diversify their livelihoods.

There is a strong case, therefore, for building resilience toward climate change in these drylands by addressing adaptive capacity and base level development. Although there is no single prescriptive adaptation solution to these challenges, a series of adaptive responses that are region- or area-specific will be necessary to reduce vulnerability of affected communities (Chaudhury, 2017). Such adaptive solutions will require, among others, the building of institutional capacity in technology and governance, particularly in the food—energy—water (FEW) nexus.

Regional and national policies, legal and institutional frameworks related to production and the use of food and energy, as well as the management of water resources, constitute a fundamental basis for resilience building. These need to be accompanied by appropriate technologies such as rainwater harvesting (RWH) systems that can be applied for supplemental irrigation, and in conjunction with innovative agricultural technologies (De Trincheria et al., 2017a) to ensure food, energy, and water security at the household level among small hold farmers. It is necessary that newly introduced technologies be acceptable to farmers and are supported by farmer awareness campaigns and capacity development (De Trincheria et al., 2017b).

Existing strategy documents focusing on food security in Eastern Africa take particular note of the low capacity for RWH as a constraint to attaining rural development goals in the region (IWA, 2015). Action plans on the use of water for agricultural production tend to focus on the promotion of integrated water resources management at the macro-level, advocating for appropriate technologies that are adaptive to climate change impacts (IWA, 2015). The focus is on sectoral regulatory frameworks that have a strong but narrow focus (e.g., environment, water, land, forestry, energy, transport, agriculture, livestock, fisheries, health, disaster risk management, gender) and lack cross-sectoral integration. Meanwhile, such plans often fail to provide strategies directly targeted at supporting small hold farmers, beyond advocating for a greater availability of agricultural inputs at an affordable price (Mdee, 2017). Consequently, there is a lack of strategic focus on technological options and governance approaches to address food security of small hold farmers through the FEW nexus perspective. The remainder of this chapter focuses on needs and opportunities for building climate resilience among small hold farmers in Eastern Africa, taking an approach centered on the FEW nexus.

The food—energy—water nexus

Globally, agriculture accounts for 70% of total water withdrawal, while food production and the supply chain accounts for about 30% of total global energy consumption (Fabiola and De Rosa Dalila, 2016).

Food, energy, and water demands increase proportionally with population increase (Yumkella and Yillia, 2015), meaning that appropriate technological solutions and institutional arrangements for efficient production and consumption are urgently needed in Africa, where the continent's population is expected to double from 1.2 to 2.5 billion by 2050 (IPBES, 2018). This is especially the case given that food production systems are also linked to climate change and biodiversity loss (IPBES, 2018; Yumkella and Yillia, 2015). Moreover, the high reliance on wood fuel, accounting for 85% of energy use in the region (Othieno and Awange, 2016), necessitates a long-term strategy to sustain supply. This demand is anticipated to rise due to population growth, strong urbanization dynamics, and relative price changes of alternative fuels (Africa Renewable Energy Access Program, 2011). Hence, there is demand and opportunity to grow wood fuel in an integrated program that involves food production and water conservation.

The concept of FEW nexus provides a starting point for addressing challenges surrounding the need to balance competing demands on food production, energy and water access, and security under a changing climate (Pardoe et al., 2017). It was developed on the pedestal of integrated planning and decision-making across sectors (Yumkella and Yillia, 2015; Schreiner and Baleta, 2015). Its emphasis is on technical solutions to environmental problems, achievement of efficiency gains, and a preference for technocratic forms of environmental managerialism (Cairns and Anna, 2016) to support equitable and sustainable development. Although the concept was developed to support attainment of the UN sustainable development goals (SDGs), it is also critical for the climate change agenda, as changes in water availability due to climate change will affect food and energy production (Yumkella and Yillia, 2015). Nexus-type approaches have been employed elsewhere as a strategy to adapt to the challenges posed by population growth, climate change, and variability, such as in the Southern African Development Community Region (SADC) (http://www.sadc.int/files/9914/6823/9107/SADC_Water_4th_Regional_Strategic_Action_Plan_English_version.pdf). Effective implementation of the FEW nexus needs to be supported by robust science; however, robust science alone does not automatically lead to effective implementation (Kurian, 2017) and needs to be accompanied by appropriate governance structures.

Example: technological options in rainwater harvesting for domestic use

RWH as a technique has been used for irrigation since 4500 BC in southern Mesopotamia (Oweis et al., 2001) and continues to be practiced worldwide for obtaining both drinking water and water for agricultural purposes (De Trincheria et al., 2017c). My focus here is the microcatchment approach that involves the collection of runoff from roofs and ground surface, its storage, and its productive uses (Trincheria et al., 2016) (Table 18.1). A number of technology-focused approaches to RWH have the potential to increase soil moisture, reduce soil temperature, and recharge groundwater and other water resources in order to bolster rain-fed crop production of small hold farmers and enhance their food security and economic development (https://wocatpedia.net/images/3/3b/Table_1%2C_Examples_of_in_situ_rainwater_harvesting_systems_with_description.jpg) (Table 18.1). RWH therefore provides a low-cost adaptation mechanism at the farm level to build resilience among small-scale farmers who are continually exposed to the risk of climate-induced moisture variations in the vast drylands of Eastern Africa. Technological approaches must, however, be adapted to local environmental, social, and economic conditions (https://wocatpedia.net/wiki/Rainwater_harvesting) to be sustainable (Fig. 18.1).

Table 18.1 Uses and importance of microcatchment techniques in rainwater harvesting.

Potential	Impacts
Uses	• Crop production • Fruit tree growing • Livestock watering • Woodlot • Crops, forest/rangeland restoration • Rehabilitation of degraded, denuded, and hardened land for crop growing, grazing, or forestry • Erosion prevention

Climate change adaptation potential	• Increase in amount of available water
	   ◦ Crop survival during dry period
	   ◦ Increased rate of tree survival
	• Slowing down of runoff
	• Protection of land during heavy rainfall
	• Erosion protection
	• Lowering of soil temperature
	• Long-term wind erosion control

Environmental and socioeconomic impacts	• Developing woodlot for biomass energy supply • Restoration of vegetation cover • Reduction of risk for gulley erosion and siltation downstream • Regreening of the environment and promotion of biodiversity • Lessen pressure for extensification of area under cultivation • Reduce vulnerability of plants during droughts • Retention of water, trapping of fertile sediment and fertilizer, improving soil fertility and structure • Reduction of flood water • Increased livelihood options • Increased farmer earnings • Access to water and sanitation

Source: Modified from Oweis et al. 2001 https://wocatpedia.net/images/3/3b/Table_1%2C_Examples_of_in_situ_rainwater_harvesting_systems_with_description.jpg.

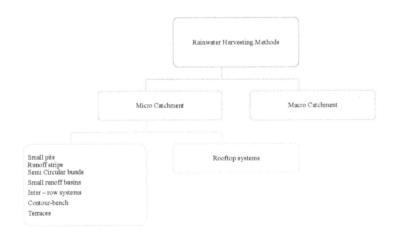

FIGURE 18.1

Microcatchment techniques in rainwater harvesting.

Modified from Oweis, T., Prinz, D., Hachum, A., 2001. Water harvesting: indigenous knowledge for the future of the drier environments.
ICARDA, Aleppo, Syria. 40 p.

Benefits of these RWH technologies include the reduction of negative impacts on ecosystems services, such as extensive soil erosion which is typical of areas with large-scale surface or groundwater withdrawals. Farm-level technologies can protect against water losses through seepage and evaporation, and consequently yield a higher aggregated amount of collected water than is captured through, for example, construction of large dams (AfDB, 2011). Recent studies (Reddy, 2016) further show increased benefits of enriched soil nutrients, increased biomass production, and subsequent higher yields with RWH, leading to higher food security, increased income, and delivery of positive impacts on other ecosystems. The additional social benefits include access to water and sanitation (https://www.unece.org/fileadmin/DAM/env/water/meetings/TF_EWE/2nd_meeting/power_points/Water_Harvesting.pdf). RWH is important for biodiversity conservation (Pascual et al., 2017) by augmenting the Provisioning and Regulating Services of agro-ecosystems as described in the Millennium Ecosystem Assessment (Millennium Ecosystem Assessment, 2005).

The limitations of RWH technologies are largely due to technical, financial, and knowledge gap challenges that face farmers and promoters of local initiatives (De Trincheria et al., 2017c). These include (1) inadequate construction guidelines for tanks, gutters, filters, etc.; (2) inadequate technological transfer to the beneficiaries; (3) lack of suitable training programs; (4) poor technical selection and usage of local materials; (5) inappropriate sizing of storage tanks, with respect to rainfall data and costs; (6) lack of water quality improvement structures and control; and (7) inappropriate design, construction (support), and maintenance (https://www.google.com/search?client=safari&rls=en&q=challenges+in+rainwater+harvesting+adoption&ie=UTF-8&oe=UTF-8). Consequences include abandonment of use of RWH systems by farmers, particularly for in situ types such as pits, trenches, and contour ridges. Another reason these technologies may not succeed is inappropriate agricultural advisory approaches that are supply driven (i.e., priority given to increasing food supply in the fight against hunger), rather than demand driven (i.e., services driven by farmer needs and demands) (https://wocatpedia.net/wiki/Rainwater_harvesting). Institutional capacity may also be a challenge, as installation of some of these technologies requires heavy machinery that is not affordable for subsistence farmers. Farmers may also find that some RWH measures are not profitable given fluctuating market conditions for their produce.

Although an enabling environment at the policy level is essential for promoting the concept and implementation of RWH systems on a large scale, rainwater captured by crops is often not included in integrated water resource management (IWRM) plans. IWRM plans focus primarily on stream flow or groundwater resources (Barron et al., 2009). Given the multiple benefits of RWH technologies for development and sustainability, expanding its use will necessitate mainstreaming it in policy agendas, awareness raising, capacity building, and technical exchanges.

Regional governance frameworks in Eastern Africa

Policymaking drives regional, national, and sectoral priorities, making it an important point of emphasis for integrating issues of climate change adaptation. It is also important in establishing the frameworks through which cross-sectoral collaborations may be facilitated (Pardoe et al., 2017). The FEW nexus perspective encourages efficient resource use and policy coherence. In addition, it creates a shift from a sectoral to a transdisciplinary approach that avoids competing and counterproductive actions in addressing the climate change agenda. The literature suggests that, "the overarching governance problem is that policies are fragmented across the water, energy and food sectors, which lead to unintended consequences; and the goal is to achieve policy coherence by identifying synergies and trade-offs, optimizing policy options, and adapting governance arrangements" (Weitz et al., 2017) (p. 166).

Achieving policy coherence in Eastern Africa is complicated given the large governance landscape (Fig. 18.2A−D). The Eastern African subregion, as per United Nations categorization of the African continent, enjoins 20 countries with an estimated population of 430.6 million people that is growing at a rate of 2.7% per annum (Table 18.2) (http://www.worldometers.info/world-population/eastern-africa-population). This population is projected to reach 888 million by 2050, hence posing major development challenges. Regional governance consists of three economic communities: The East African Community (EAC), the Intergovernmental Authority on Development in Eastern Africa (IGAD), and the Southern Africa Development Community (SADC). While EAC is nested within the subregion, IGAD includes the Sudan, which falls in the Northern African subregion, while SADC comprises 14 countries of which only six (Madagascar, Malawi, Mauritius, Tanzania, Zambia,

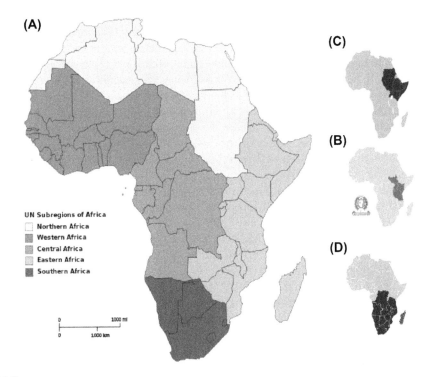

FIGURE 18.2

(A) United Nations Subregions of Africa. (B) The member States of EAC are: Burundi, Kenya, Rwanda, South Sudan, Uganda, and United Republic of Tanzania. (C) Intergovernmental Authority on Development (IGAD). (D) Southern Africa Development Community (SADC).

Reproduced from (A) From http://pitt.libguides.com/c.php?g=12378&p=65814; (B) From EAC − East African Community, ©United Nations. Reprinted with the permission of the United Nations; (C) Source: https://en.wikipedia.org/wiki/Intergovernmental_Authority_on_Development; (D) From https://de.wikipedia.org/wiki/Datei:SADC-Staaten.png.

Table 18.2 Population of Eastern Africa (2018 and historical).

Year	Population	Yearly % change	Yearly change	Migrants (net)	Median age	Fertility rate	Density (P/km²)	Urban population %	Urban population	Eastern Africa's share of world population	World population	Eastern Africa Rank within Africa
2018	433,643,132	2.75%	11,606,889	−166,013	18.1	4.80	65	25.4%	115,265,762	5.68 %	7,632,819,325	1
2017	422,036,243	2.78 %	11,398,256	−166,013	18.1	4.80	63	26.1%	110,357,119	5.6%	7,550,262,101	1
2016	410,637,987	2.8 %	11,179,961	−166,013	18.1	4.80	62	25.7%	105,614,849	5.5%	7,466,964,280	1
2015	399,458,026	2.86%	10,494,106	−190,400	18.0	4.89	60	25.3%	101,034,466	5.7%	7,383,008,820	1
2010	346,987,498	2.91%	9,277,558	−145,400	17.4	5.35	52	23.2%	80,635,931	5.3%	6,958,169,159	1
2005	300,599,706	2.86%	7,897,163	−103,100	17.1	5.76	45	21.7%	65,108,773	4.9%	6,542,159,383	1
2000	261,113,890	2.93%	7,018,571	67,600	16.9	6.12	39	20.5%	53,484,357	4.5%	6,145,006,989	1
1995	226,021,034	2.62%	5,475,019	−415,700	16.8	6.41	34	19.4%	43,843,113	4.2%	5,751,474,416	1
1990	198,645,937	3.03%	5,504,571	−32,000	16.6	6.80	30	17.9%	35,563,628	4.1%	5,330,943,460	1
1985	171,123,080	3.01%	4,720,838	−104,200	16.6	7.02	26	16.1%	27,578,835	3.8 %	4,873,781,796	1
1980	147,518,890	3.01%	4,062,924	−94,200	16.8	7.10	22	14.6%	21,468,109	3.6%	4,458,411,534	1
1975	127,204,270	2.89%	3,382,440	−114,500	16.9	7.14	19	12.4%	15,757,081	3.4%	4,079,087,198	1
1970	110,292,070	2.82%	2,859,083	−28,200	17.1	7.11	17	10.5%	11,587,492	3.3%	3,700,577,650	1
1965	95,996,653	2.66%	2,362,837	−28,700	18.2	7.05	14	5.1%	4,851,920	3.2%	3,339,592,688	1
1960	84,182,469	2.45%	1,919,333	−35,000	17.8	7.05	13	7.6%	6,362,149	3.0%	3,033,212,527	2
1955	74,585,806	2.24%	1,565,510	−28,700	18.2	7.05	11	6.5%	4,851,920	2.9%	2,772,242,535	2

Elaboration of data by United Nations, Department of Economic and Social Affairs, Population Division. World Population Prospects: The 2017 Revision. (Medium-fertility variant).

Source: Reproduced from Worldometers (www.Worldometers.info). http://www.worldometers.info/world-population/eastern-africa-population/.

FIGURE 18.3

East Africa Community.

Source: Courtesy researchgate.net.

and Zimbabwe) are within the Eastern African subregion. Because Eastern African countries belong to different economic blocks, they create institutional and policy fragmentation due to the secretariats operating in different political, normative, and geographic contexts. This is elaborated in the cases below.

The East African Community

EAC comprises six countries (Burundi, Kenya, Rwanda, South Sudan, Tanzania and Uganda) (Fig. 18.3) with a common development aspiration. Their development blueprint, *East African Community Vision 2050: regional vision for socio-economic transformation and development*, aims "to dove-tail the tactical regional and individual state initiatives into a larger framework for transformation and development …" (EAC, 2015) (p. 12). The vision has an emphasis on agriculture and rural development through improved practices including mechanization, irrigation, improved seeds, and the use of fertilizers. It recognizes the effects of climate change and environmental degradation on agriculture and food security through extended droughts and flooding in parts of the region, hence the need for water management. A separate pillar addresses natural resource and environment management with its "goal for effective and sustainable use of natural resources with enhanced value addition and management" (EAC, 2015) (p. 57). Water and energy are considered to be part of the pillar on infrastructure, with goals on (1) affordable and efficient energy and (2) management of transboundary water resources.

By framing its vision around specific pillars, the EAC's Vision 2050 policy fails to reflect on the nexus approach to sectoral issues such as food, energy, water, land, or climate change, meaning they will be addressed through different mandates and governance systems both regionally and nationally. The vision acknowledges the important role of smallholder farmers in food production and their occupation of the majority of land used for crop and livestock, but again, this linkage between food and land is not considered in a nexus perspective. Instead, low productivity among small hold farmers is perceived to stem only from the lack of access to markets, credit, and technology. In the vision, RWH as a technology is neither discussed nor mentioned, despite the emphasis on technology transfer as an opportunity for the region. The vision also states that "effective governance at the local, national and regional levels is critical for advancing sustainable development" but does not provide a framework for the implementation of such a structure.

The Intergovernmental Authority on Development in Eastern Africa

IGAD has eight members (Djibouti, Ethiopia, Eritrea, Kenya, Somalia, South Sudan, Sudan, and Uganda) and was established to address issues of development and drought control (https://igad.int/about-us/what-we-do). Since 70% of the IGAD region comprises drylands receiving less than 600 mm in annual rainfall and experiences recurrent drought, the organization developed the IGAD Drought Disaster Resilience and Sustainability Initiative (IDDRSI) aimed at "addressing the effects of drought and related shocks in the IGAD region in a sustainable and holistic manner" (http://resilience.igad.int/index.php/functions-test/programming/rpp/projects/rplrp). Implementation of this program is undertaken independently by each country. This approach allows regional policies to be integrated into national development plans. However, institutional fragmentation at the national level can create artificial barriers to success. For example, in Kenya, directorship of IDDRSI is based at the IGAD Center for Pastoral Areas and Livestock Development, operating under the Division of Agriculture and Environment, but project activities are executed by staff of the Conflict Early Warning and Response Mechanism and the IGAD Climate Prediction and Application Center.

Although IGAD does not use the FEW nexus approach in addressing its mandate, there is opportunity for using a nexus approach in its "Regional Pastoral Livelihoods Resilience Project", which "aims at enhancing the sustainable management and secures access of pastoral and agro-pastoral communities at natural resources (water and pasture) with trans-boundary significance" (http://resilience.igad.int/index.php/functions-test/programming/rpp/projects/rplrp). While the strategy does not refer explicitly to RWH as a technology, it includes plans for developing infrastructure for 458 water points to create improved access to water resources by 50% for pastoral and agropastoral communities in the project area.

The Southern Africa Development Community

SADC has three separate policy frameworks for food, water, and energy: (1) Protocols on Energy (1996) (http://www.sadc.int/files/3913/5292/8363/Protocol_on_Energy1996.pdf), (2) Dar es Salaam Declaration of Agriculture and Food Security in the SADC Region (2004) (http://www.sadc.int/files/6913/5292/8377/Declaration_on_Agriculture__Food_Security_2004.pdf), and (3) Regional Water Policy (2005) (SADC, 2005). The three frameworks provide a basis for a nexus approach, with examples including: (a) Article 5(c) of the agriculture policy which focuses on developing water harvesting technologies; (b) Articles 9(b) (iv, v, vi and vii) of the Regional Water Policy which address RWH and efficient use for food and livestock productions; and (c) Articles 7 and 8 of the Protocol on Energy which address the participation of citizens and communities and the need for environmentally sound energy development and use. These policy frameworks provided a foundation for the eventual explicit mandate for an FEW nexus approach in Article P8.2 of the "Regional Strategic Action Plan on Integrated Water Resources Development and Management Phase IV (2016—20)" (http://www.sadc.int/files/9914/6823/9107/SADC_Water_4th_Regional_Strategic_Action_Plan_English_version.pdf).

This latter strategy focuses on water as a driver in energy production to produce food through irrigation and stock watering. SADC's aim is to use the FEW nexus to confront challenges posed by rapid population growth, increased urbanization, and climate change and variability. As with the above two regional bodies, SADC's strategy on water fails to adequately address technological advances in RWH as an option to address the key challenges that have been identified. SADC's FEW nexus approach nonetheless highlights the need for clear determination and establishment of governance systems at all levels, from the Secretariat to river basins and Member States (http://www.sadc.int/files/9914/6823/9107/SADC_Water_4th_Regional_Strategic_Action_Plan_English_version.pdf). Although the strategy is focused on water resources, they acknowledge other sectors in the implementation processes and the need to coordinate responsibility with other sectoral entities. Critically, they provide for a framework to build consensus on the nexus lead coordinating entity at regional and Member States levels while facilitating the establishment of a SADC Secretariat cross-sector working group on nexus at Member States level. An end goal is transformative governance that will "involve institutional change and joint implementation by the public and private sectors".

A comparison of the three regional development entities in Eastern Africa shows that (1) with exception of SADC, FEW nexus perspectives are not major considerations in development policies and strategies; (2) none of the regional bodies considers technological advances in RWH as a strategy to address food, energy, and water needs or as an adaptation to climate change among small hold farmers; and (3) the EAC and IGAD fail to provide a clear governance structure for implementation of regional policies, while the SADC provides a structure that cascades implementation to the level of Member States.

Conclusion and recommendations

Technological innovations and governance structures critical for building drought and climate change resilience among small hold farmers have not been adequately addressed in existing regulatory frameworks at the national and regional levels in Eastern Africa. In particular, policy frameworks pay inadequate attention to RWH as a technology that can address food security, increase access to water and biomass energy, and contribute to climate adaptation strategies. To further develop such frameworks, an FEW nexus approach should be taken. An urgent need exists for resources to support planning and decision-making processes, particularly in light of burgeoning populations, increasing urbanization, and food—energy—water insecurity in a changing climate. Despite the progressive policy approach by SADC, it has failed to incorporate RWH technologies into its strategies on food, energy, water, or climate change at all levels. The EAC and IGAD particularly need to rethink their existing policies to become more aligned with SADC's policy approach. However, the development of new frameworks will require a collective approach across regional entities to foster coordination at the Member States level in Eastern Africa, irrespective of the development entity.

Policies will need to include appropriate evaluation of trade-offs and synergies as key aspects of planning. However, mobilizing financial, institutional, technical, and intellectual resources will be key to creating the enabling environment that is required to implement new policies with FEW nexus perspectives (Yumkella and Yillia, 2015). Appropriate sustainable governance strategies will require transdisciplinary approaches and meaningful stakeholder engagement where the small-scale farmer is a part of the change. This will entail the development of innovative ways to unite stakeholder efforts and capacities at all levels.

References

AfDB, 2011. Cost of Adaptation to Climate Change in Africa. African Development Bank, p. 41.

Africa Renewable Energy Access Program, 2011. Wood-based Biomass Energy Development for Sub-saharan Africa: Issues and Approaches. https://siteresources.worldbank.org/EXTAFRREGTOPENERGY/Resources/717305-1266613906108/BiomassEnergyPaper_WEB_Zoomed75.pdf.

Barron, J., et al., 2009. Rainwater Harvesting: A Lifeline for Human Well-Being. United Nations Environment Programme/SEI.

Bazzaz, F.A., Sombroek, W.G. (Eds.), 1996. Global climatic change and agricultural production: An assessment of current knowledge and critical gaps, Global climate change and agricultural production. Direct and indirect effects of changing hydrological, pedological and plant physiological processes. FAO, Rome & John Wiley & Sons, Chichester, ISBN 92-5-103987-9.

Cairns, R., Anna, K., 2016. Anatomy of a buzzword: the emergence of 'the water-energy-food nexus' in UK natural resource debates. Environmental Science and Policy **64**, 164−170.

Chaudhury, M., 2017. Strategies for reducing vulnerability and building resilience to environmental and natural disasters in developing countries.

Daron, J.D., December 2014. Regional Climate Messages: EastAfrica. Scientific report from the CARIAA Adaptation at Scale in Semi-Arid Regions (ASSAR) Project.

De Trincheria, J., et al., 2017. Transnational Policy and Technology Transfer Recommendations on the Use of Rainwater for Off-Season Small-Scale Irrigation in Sub-Saharan Africa: Fostering Innovation and Replication of Rainwater Harvesting Irrigation Strategies in Arid and Semi-arid Areas of Ethiopia, Kenya, Mozambique and Zimbabwe. AFRHINET Project. Hamburg University Applied Sciences, Hamburg, Germany.

De Trincheria, J., et al., 2017. Training Materials for Local Communities on Rainwater Harvesting Irrigation Management: Capacity Building on the Use of Rainwater for Off-Season Small-Scale Irrigation in Arid and Semi-arid Areas of Sub-Saharan Africa. AFRHINET Project. Hamburg University of Applied Sciences, Hamburg, Germany.

De Trincheria, J., et al., 2017. Best Practices on the Use of Rainwater for Off-Season Small-Scale Irrigation: Fostering the Replication and Scaling-Up of Rainwater Harvesting Irrigation Management in Arid and Semi-arid Areas of Sub-saharan Africa. AFRHINET Project. Hamburg University of Applied Sciences, Hamburg, Germany.

EAC, 2015. East African Community Vision 2050: Regional Vision for Socio-Economic Transformation and Development. The East African Community, Arusha, Tanzania, p. 142.

Easterling, D.R., Meehl, G.A., Parmesan, C., Changnon, S.A., Thomas, R., Karl, T.R., Mearns, L.O., 2000. Climate Extremes: Observations, Modeling, and Impacts. Science 289, 2068. https://doi.org/10.1126/science.289.5487.2068.

Fabiola, R., Dalila, D.R., 2016. How the nexus of water/food/energy can be seen with the perspective of people well being and the Italian BES framework. Agriculture and Agricultural Science Procedia 8, 732−740.

IPBES, 2018. Summary for policymakers of the regional assessment report on biodiversity and ecosystem services for Africa of the Intergovernmental Science-Policy Platform on Biodiversity and Ecosystem Services. In: Archer, E., Dziba, L.E., Mulongoy, K.J., Maoela, M.A., Walters, M., Biggs, R., Cormier-Salem, M.-C., DeClerck, F., Diaw, M.C., Dunham, A.E., Failler, P., Gordon, C., Harhash, K.A., Kasisi, R., Kizito, F., Nyingi, W.D., Oguge, N., Osman-Elasha, B., Stringer, L.C., Tito de Morais, L., Assogbadjo, A., Egoh, B.N., Halmy, M.W., Heubach, K., Mensah, A., Pereira, L., Sitas, N. (Eds.), IPBES secretariat, 49. Bonn, Germany.

IPCC, 2013. Climate change 2013: the physical science basis. In: Stocker, T.F., et al. (Eds.), Contribution of Working Group I to the Fifth Assessment Report of the Intergovernmental Panel on Climate Change. Cambridge University Press, Cambridge, United Kingdom and New York, NY, USA. https://doi.org/10.1017/CBO9781107415324, 1535 pp.

IWA, 2015. Nexus Trade-Offs and Strategies for Addressing the Water, Agriculture and Energy Security Nexus in Africa. Riddell Associates, London.

Kurian, M., 2017. The water-energy-food nexus: trade-offs, thresholds and transdisciplinary approaches to sustainable development. Environmental Science and Policy 68, 97−106.

Mdee, A., 2017. Disaggregating orders of water scarcity − the politics of nexus in the Wami-Ruvu river basin, Tanzania. Water Alternatives 10 (1), 100−115. Retrieved from: http://www.water-alternatives.org/index.php/alldoc/articles/vol10/v10issue1/344-a10-1-6/file [Web of Science [®]].

Millennium Ecosystem Assessment, 2005. Ecosystems and Human Well-Being: Synthesis. Island Press, Washington, DC.

NASCA, 2015. Climate Change Adaptation and Resilience in Africa: Recommendations to Policymakers. Network of African Science Academies, p. 52.

Othieno, H., Awange, J., 2016. Energy Resources in Africa. https://doi.org/10.1007/978-3-319-25187-5_2.

Oweis, T., Prinz, D., Hachum, A., 2001. Water Harvesting: Indigenous Knowledge for the Future of the Drier Environments. ICARDA, Aleppo, Syria, 40 pages.

Pardoe, J., et al., 2017. Climate change and the water−energy−food nexus: insights from policy and practice in Tanzania. Climate Policy. https://doi.org/10.1080/14693062.2017.1386082.

Parry, M.L., Canziani, O.F., Palutikof, J.P., van der Linden, P.J., Hanson, C.E. (Eds.), 2007. Contribution of Working Group II to the Fourth Assessment Report of the Intergovernmental Panel on Climate Change. Cambridge University Press, Cambridge, United Kingdom and New York, NY, USA.

Pascual, U., et al., 2017. Valuing nature's contributions to people: the IPBES approach. Current Opinion in Environmental Sustainability 26−27, 7−16.

Reddy, P.P., 2016. Micro-catchment rainwater harvesting. In: Sustainable Intensification of Crop Production. Springer, Singapore.

SADC, 2005. Regional Water Policy. Gaborone, Botswana, p. 77.

Schreiner, B., Baleta, H., 2015. Broadening the lens: a regional perspective on water, food and energy integration in SADC. Aquatic Procedia 5, 90−103.

Shepherd, A., Mitchell, T., Lewis, K., Lenhardt, A., Jones, L., Scott, L., Muir-Wood, R., 2013. The geography of poverty, disasters and climate extremes in 2030. ODI, London, UK.

Trincheria, J., et al., 2016. Fostering the Use of Rainwater for Small-Scale Irrigation in Sub-saharan Africa. AFRHINET Project. Hamburg University Applied Sciences, Hamburg, Germany, ISBN 978-3-00-054353-1.

Weitz, N., Strambo, C., Kemp-Benedict, E., Nilsson, M., 2017. Closing the governance gaps in the water-energy-food nexus: insights from integrative governance. Global Environmental Change 45, 165−173.

WWF, 2006. East Africa Climate Impacts. WWF-World Wide Fund For Nature, Gland, Switzerland, 12.

Yumkella, K.K., Yillia, P.T., 2015. Framing the water-energy nexus for the post-2015 development agenda. Aquatic Procedia 5, 8−12.

Drought management in the drylands of Kenya: what have we learned?

Caroline King-Okumu[a,b,d,*], **Victor A. Orindi**[c], **Lordman Lekalkuli**[e]

[a]*Geodata, University of Southampton, Southampton, United Kingdom,*
[b]*The Borders Institute (TBI), Nairobi, Kenya,*
[c]*Ada Consortium and National Drought Management Authority (NDMA), Nairobi, Kenya,*
[d]*Centre for Ecology and Hydrology, United Kingdom,*
[e]*National Drought Management Authority, Isiolo, Kenya*
[*]*Corresponding author*

Introduction

A strong international agenda to reduce the risks of droughts and other disasters (ISDR, 2005) was reflected in Kenya's first Medium Term Plan 2008—12 (GoK, 2008 p. 104) for its Vision 2030. This called for a shift in emphasis from disaster response to preempt and prevent drought emergencies. However, two more Medium Term Plans later, the translation of this aspiration into a reality is still not completed in Kenya and the surrounding Horn of Africa region. In light of this, it is important to take stock of what lessons have been learned about the intended transition to drought preparedness over the decade 2008—18. These lessons should identify how best to end drought emergencies before 2022.

There is widespread agreement that the ability of societies to reduce the impacts of future disasters depends on successfully extracting, analyzing, sharing, and internalizing these lessons (Birkland, 2009; Deverell, 2009; Staupe-Delgado et al., 2018). On the other hand, past efforts from around the world to learn lessons from experiences of managing droughts and other disasters are recognized often to have resulted in misdiagnoses, as well as occasionally also some more profound evidence-based reflections (Glantz, 1977; Egner et al., 2015). It is therefore important to consider carefully which are the lessons that may be of most use, and to learn from them effectively.

The Global Water Partnership (GWP) (2015) has already observed a shift occurring across the Horn of Africa region from a focus on emergency/crisis response toward integrated drought management, including preparedness, drought mitigation, and early warning. This is needed urgently as growing economic water demand and extractions may already exceed predicted increases in precipitation and the costs of hydrological droughts are increasing. This chapter focuses on lessons that can be identified in the Kenyan experience in relation to three key regionally shared challenge areas identified previously by GWP:

1. Limitations in the human resource and institutional capacities
2. Lack of natural resource information and weak early warning systems (EWSs)
3. Limited financial resources for drought risk management (DRM) and resilience building

Drought Challenges. https://doi.org/10.1016/B978-0-12-814820-4.00019-5

In the following section a brief contextual and conceptual background is provided. We then explore what lessons have been learned in relation to the three challenge areas. The discussion provides recommendations for the intended shift for the arid and semi-arid lands (ASALs) of Kenya and the Horn of Africa region out of recurring crisis mode and onto an accelerated path to sustainable development by 2022. It builds on and deepens reflections that have emerged through an internal process which focused on lessons learned through the achievement of National Drought Management Authority's (NDMA) strategic objectives for 2013−17 (NDMA, 2018b). The present review considers a longer timeframe and some more fundamental lessons that are emerging which may be of interest for the wider region as well as in Kenya.

Background: what is drought management and preparedness?

In the 21st century, drought management has been firmly built upon an understanding of drought as a manageable phenomenon. This is based on the observation that the hydrometeorological aspects of drought and flood are familiar and occur on an almost predictable cyclical basis (Glantz and Katz, 1985). At the policy level in Kenya, drought preparedness is most often addressed as part of a globally supported climate change adaptation agenda. This reflects the well-explored relationships between the changing climate and patterns of drought events that have been observed across the region. However, there are also other anthropogenic reasons for shortages of water and the resulting life-threatening drought emergencies. Preparedness[1] is now widely considered to include a broad range of measures, which should be introduced proactively early on while the drought cycle is still under "normal" phase conditions.

There is a strong conviction among the international development community that in order to be effective, a drought management system must concern itself not just with the immediate cause of droughts (the deficit of water) but rather, it should engage with the underlying human and institutional factors that leave people vulnerable to drought effects, and their capacities to adapt. This reflects a longstanding—and still growing—recognition that the principal causes of loss and damages during droughts in the Horn of Africa have often been driven by institutional failures, insecurity, marginalization, and lack of normal State services for some groups of the national populations (Sen, 1981; Waal, 1989).

The regional challenge in the Horn of Africa

The Horn of Africa region includes eight countries: Djibouti, Eritrea, Ethiopia, Kenya, Somalia, South Sudan, Sudan, and Uganda, each with its own unique and different institutional context. However, these countries share a common vulnerability to droughts that occur on a regional level (IFRC, 2011; GWPEA, 2015). The 2008−11 regional drought crisis killed a quarter of a million people in Somalia alone and created 955,000 refugees who sought relief in neighboring countries (GFDRR, 2018). About 80% of livestock in the pastoral areas on the Ethiopia−Kenya−Somalia border were

---

[1]For UN definition see UN, 2016. Report of the open-ended intergovernmental expert working group on indicators and terminology relating to disaster risk reduction in *Seventy-first session Agenda item 19 (c) Sustainable development: disaster risk reduction* 41. United Nations General Assembly.

destroyed and mass migration of pastoralists out of the drought-affected areas caused major international alarm (FAO, 2011; Headey et al., 2012).

Already in 2008, there was a recognition that food aid generally arrived late and was not ever an adequate solution to the problem of the drought emergencies. This had already been observed during the previous 2006 drought. Despite warnings that came as early as July 2005, major donor interventions still did not start until February 2006. Furthermore, there was already a generalized recognition that dependence on disaster relief perpetuated a debilitating situation of increasing destitution (GWPEA, 2016). The shift away from crisis management was sought to build resilience and to reduce reliance on relief aid over the longer term.

The heavy dependence on external aid delivered through the drought management system in Kenya's arid northern areas was established in the 1980s. During this period, relations between the Kenyan government and the populations across Kenya's Northern Frontier District and the wider region were conflict-prone. In 1984, Wagalla which today is part of Wajir County, was a zone of conflict between the Kenyan government and suspected Somali separatists. However, with development partner support, drought coordination structures were established in the late 1990s at both the national and (then) district levels: the Kenya Food Security Meeting (KFSM) and Kenya Food Security Steering Group (KFSSG), and the District (now County) Steering Group (D/CSG). Government ownership of the drought management system in the north of Kenya deepened over time.

For Kenya and its neighbors, social, economic, and political developments can only be achieved in a climate of enhanced security and stability for all. Events in one country affect the others (IBRD, 2014). The critical terms of trade for the pastoral livestock systems, and the mobility strategies that vulnerable communities employ during times of drought both operate on transboundary and regional scales.

At a Regional Summit held in 2011, governments and development partners agreed on a declaration highlighting the importance of developing long-term sustainable solutions to end drought emergencies. Leaders from the region, under the coordination role of the Intergovernmental Authority on Development (IGAD) secretariat, developed the Horn of Africa Regional Disaster Resilience and Sustainability Strategy Framework with the following overarching motto, as stated in the Nairobi Strategy: "While droughts may be an unavoidable natural phenomenon in the Horn of Africa, their impact can be mitigated by human action." It was strongly emphasized that "Droughts need not, and should not, lead to famine and other disasters" ("The Nairobi Strategy: Enhanced partnership to eradicate drought emergencies," 2011, Art. 71). Regional initiatives for lesson learning are essential and have recently begun to seek out and draw together the country experiences (GWPEA, 2016; Duguma et al., 2017).

Institution building in the Kenyan arid lands 2008–18

Following an extended period of negotiation and discussions, in 2010, Kenya adopted a new constitutional settlement. This seeks to overcome the internal divisions which underpin persistent vulnerability and risk within Kenya. The Constitution and subsequent reformed national legislation include recognition of the need to address the particular marginalization of the drought-prone former Northern Frontier populations. The 2010 constitutional reforms included, for example, the devolution of both funds and functions from the national level to the county level governments across Kenya, starting in 2013. It also provided for new mechanisms such as an Equalization Fund, which would have a primary objective to transfer funds solely for provision of basic services to marginalized areas.

The 2010 Constitution identified DRM as a responsibility shared by both national and county governments (IBRD, 2018). A vision strategy dedicated to the acceleration of development in the arid lands (GoK, 2011) focused on the newly introduced devolution process as a key means to overcome drought, flood, and other development challenges faced in Northern Kenya:

> *Devolution will now bring services closer to people. In theory it will allow a greater degree of self-determination, as county governments plan and legislate in response to local needs and concerns. For an area such as Northern Kenya, which requires policy solutions tailored to its unique ecological and social realities, this is a major opportunity. (p. 107)*

However, it was well recognized in the strategy (p. 107) that the ASALs would face challenges in managing devolved governance. These were expected to include the need to facilitate the cross-boundary sharing of resources (including water), to include marginal groups and women (who are often excluded from customary clan-based resource management institutions), and to source local technical expertise and bureaucratic experience.

In 2012, a new policy for the ASALs charted a direction for drought management which embedded it in the development process (GoK, 2012). It focused on two key strategies: first, investing in human development and economic growth so that those living in the ASALs can better withstand shocks; and second, establishing permanent institutional mechanisms—specifically the NDMA and a dedicated drought contingency fund—which enable action much earlier in the drought cycle, carried out in ways which reinforce rather than undermine people's livelihoods. These two strategies are at the core of the Common Programme Framework (CPF) (GoK, 2015a) for Ending Drought Emergencies (EDE), launched in November 2015 (Fig. 19.1).

The CPF is now the principal reference document for coordination. It refers to six pillars (as numbered in Fig. 19.1). The sectors and counties are responsible for implementing the first four EDE pillars, while the NDMA is responsible for pillars five and six (which are aligned with its strategic plan) and for steering and coordinating the EDE initiative as a whole. The creation of the NDMA as a state entity with its own national budgetary allocation channeling and coordinating donor support—rather than as a project—was seen as a departure from previous crisis management approaches in Kenya. This was an important step away from over-reliance on project-based approaches and opened the way for more coherent long-term national programs.

The Government of Kenya Medium Term Plan for EDE for 2013–17 committed the government to end the worst of the suffering caused by drought by 2022, by strengthening the foundations for growth and development, and by strengthening the institutional and financing framework for DRM. This commitment has been further strengthened in the 2018–22 medium term plan (GoK, 2017a).

The implementation of the four foundational pillars of the CPF at the devolved county level is a major coordination task involving local communities, and international partners, as well as the emerging institutions of the new devolved government. Two rounds of 5-year integrated development planning have now been developed at the county level: 2013–17 and 2018–22.[2] These document the status

---

[2]See https://roggkenya.org/story-suggestions/kenya-county-cidp-kenya-five-year-plan-on-track/download-page-cidp-county-integrated-development-plans/.

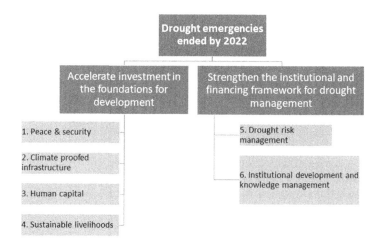

FIGURE 19.1

Ending Drought Emergencies framework (GoK, 2017a, p. 9).

of drought preparedness, levels of access to climate-proofed infrastructure in the ASAL counties, and ongoing approaches to improvement (amongst other development concerns). Establishing a precedent for devolved county governments and communities to take charge of their own access to resources is taking time. In 2016, a legal basis for the county governments to manage their water resources was introduced through a new Water Act (GoK, 2016). The NDMA, through the county drought coordinators is continuing to work closely with the county governments so that they could progressively assume more of this responsibility.

Local level coordination connecting the myriad of local institutions and actors to the devolved county government processes and institutions at other scales is key to drought response and preparedness (see example and discussion in Jarso et al. (2017). The inclusion of local associations for natural resource use and management is particularly key. NDMA's 2018 strategic plan (NDMA, 2018b) observed lessons learned over the period 2013—17 in relation to coordination issues—particularly that there were competing coordination structures at county and national levels as well as constraints on coordination at county and lower levels. The NDMA concluded that the CSG had proved to be the most recognized coordination structure at the county level. However, CSGs are coordinated by the NDMA. Rather than continuing in this way, the CSGs have been working hard to transfer more of the coordination role to the county government officers.

Under the new Water Act (GoK, 2016), for 2018, the devolved institutions in Kenya are expected to drive catchment planning from the bottom-up (for further discussion of the institution-building challenges at the local level, see Tari et al., 2015; Hujale, 2015; King-Okumu, 2015a; Ifejika Speranza et al., 2016). Due to the downstream location of many of the most vulnerable ASAL populations in catchments that are already water-stressed, drought preparedness requires coordination and planning upstream at the catchment level as well as the local level (GoK/WRMA/JICA, 2012; WRMA, 2013b, see discussion in King-Okumu et al., 2017).

Early warning system and information

The weakness of EWSs across the region was highlighted as a key challenge by GWP (2015). Ineffective early warnings and lack of confidence in their predictions may partly explain slow and inadequate government and donor responses. During the 2011 drought, early warnings of poor rainfall were noted as early as May 2010. In February 2011, the Famine Early Warning Systems Network (FEWSNET) issued a further warning that poor rains were forecasted for March to May. However, humanitarian funding did not increase significantly until the UN declared a famine in Somalia in July 2011. By then, thousands had already suffered and significant long-term damage had been done to the national economies, including Kenya's own national economy (PDNA, 2012).

Since 2011, considerable effort and investment has gone into the drought EWS to trigger effective drought management actions in Kenya. Monitoring of drought conditions and factors affecting vulnerability and resilience to drought is central to the drought management approach. The Kenyan monitoring system is community-based, involving face-to-face interviews carried out by NDMA staff on a monthly basis with residents in their homes as well as direct observation of field conditions. This is then complemented by information from other sources, particularly remotely sensed analysis currently provided through a service contract between the Government of Kenya and Boku University with support from the European Union. This includes a colour-coded assessment of vegetation conditions that is generated on a monthly basis (Fig. 19.3). When combined with the other available indicators and local knowledge, the vegetation condition index helps to provide an indication of the comparative severity of stress at different times of the year.

The results of this monitoring contribute to a suite of indicators that determine the drought phase for a particular area in a particular season. The drought phases include the following five: normal, alert, alarm, emergency, and recovery. For all indicators, current conditions are compared with the long-term average for the same time of year, based on compiled records since 2001. This is important because it shows whether the situation in a particular month is better or worse than might normally be expected, taking customary seasonal fluctuations into account. However, it is important to note that the socioeconomic indicators refer only to selected households that are permanently resident in the target areas. It is not possible for the EWS to capture the numbers or conditions of vulnerable households and livestock that migrate into each area from across the region during dry seasons and drought periods. Nor is it possible to identify and register the needs for government services of those nonresident uncounted populations that remain at other times.

Within 5 days of the month's end, the analysis that supports the selection of the drought phase and a related set of recommendations are published in a county early warning bulletin which is discussed with, and validated by, the County Steering Group (CSG).[3] The county bulletins are then consolidated into a national bulletin. Both the county- and national-level bulletins are published on the NDMA website: http://www.ndma.go.ke/index.php/resource-center/early-warning-reports. On occasion, gaps or delays to the system may still occur. For example, the comparison of trends in the types of water sources used by communities (e.g., as in Fig. 19.2) is often not published. However, the progressive reduction in the level of flows through the largest of the ASAL catchments toward Ethiopia and Somalia has

---

[3]The CSGs include heads of departments of both county and national government departments/institutions, NGOs, and other actors active in that county. They are chaired by the County Governor with NDMA providing the secretariat.

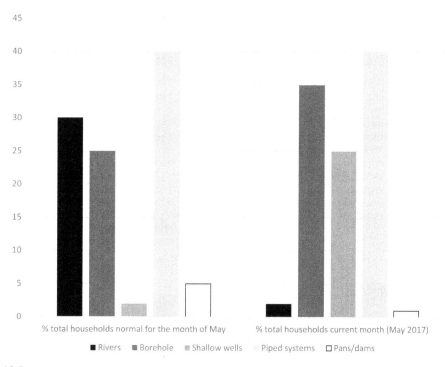

FIGURE 19.2

Household water sources in Isiolo County, May 2017[4] (% of households using them compared to normal).

Based on NDMA, 2017. Early Warning Bulletin, Isiolo, May 14, 2017. National Drought Management Authority, Nairobi.

been widely observed by local community members and scientists (Wiesmann et al., 2000; Aeschbacher et al., 2005; Liniger et al., 2005).

Using the data from the EWS and climate forecasts, the NDMA coordinates biennial multistakeholder food security assessments involving the international donor community through its leadership of the Kenya Food Security Steering Group (KFSSG). The long rains assessment takes place between July and August, while the short rains assessment takes place around February. These assessments are carried out in partnership with the county governments, specifically the CSGs. The purpose is to determine the impact of the short and long rains on food security. The assessments use a global set of standardized tools called the Integrated Food Security Phase Classification (IPC) which allows the situation to be compared over time and ensures that international standards are applied. The IPC uses five phases (Table 19.1).

Over the period 2008−18, considerable investments have been made in enhancing climate information services in Kenya.[5] However, information and predictions of water resource availability is still poor, particularly in areas of the catchments where populations must rely increasingly on

---

[4]May 2017 was preceded by a relatively dry period—see Fig. 19.3. A national drought emergency was declared in February 2017.

[5]See http://www.adaconsortium.org/index.php/our-approach/achievements/95-blogs/150-climate-information-service-for-reducing-kenya-s-vulnerability.

Table 19.1 Integrated Food Security Phase Classification.

Phase 1 None	Phase 2 Stressed	Phase 3 Crisis	Phase 4 Emergency	Phase 5 Catastrophe
Households are able to meet essential food and nonfood needs without engaging in atypical, unsustainable strategies to access food and income, including any reliance on humanitarian assistance.	Households are able to afford minimally adequate food consumption but are unable to afford essential nonfood expenditures without engaging in irreversible coping strategies.	Households are marginally able to meet minimum food needs but only with accelerated depletion of livelihood assets that will lead to food consumption gaps.	Households have large food consumption gaps resulting in very high acute malnutrition and excess mortality or extreme loss of livelihood assets that would lead to food consumption gaps in the short term.	Even with humanitarian assistance, households have an extreme lack of food and other basic needs where starvation, death, and destitution are evident.

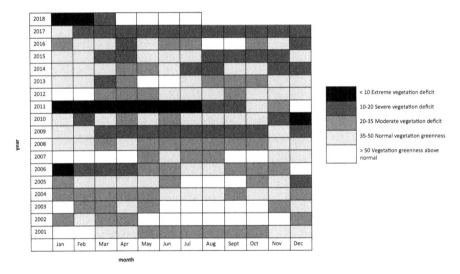

FIGURE 19.3

Three monthly vegetation condition index for Isiolo County 2001–18 (NDMA, 2018a).

groundwater (WRMA, 2013a, 2016a, 2016b). In addition to continuing with the systematic compilation of available climate information in the EWS, the county steering committees and drought coordinators have obtained access to different types of complementary information. In counties taking part in the Hunger Safety Nets Programme (HSNP),[6] more detailed databases on household socioeconomic characteristics have been compiled (Venton, 2018a). SMART surveys provide additional

---

[6]See a recent program evaluation at https://www.opml.co.uk/projects/evaluation-kenya-hunger-safety-net-programme-hsnp.

Table 19.2 Review experience, lessons, and challenges in providing drought and climate information to facilitate concerted action by relevant stakeholders 2013–17 (NDMA, 2018, p. 19–20).

Achievements	Challenges	Lessons
a) Drought early warning system rolled out in 23 counties, and monthly early warning bulletins produced and disseminated b) Climate information consolidated and disseminated c) Biannual food security assessments undertaken to inform drought preparedness and response by all stakeholders d) National Drought Management Authority's capacity on GIS and remote sensing strengthened	a) Livelihood zoning is out of date b) Inadequate funding for field monitors undermines data transmission	1. Appropriate technology mix is key to an objective drought early warning 2. Linkage between early warning and drought response is essential for drought risk management (DRM) 3. Establishment of agreed drought response triggers is essential for early action 4. Sector leadership and ownership is essential for effective DRM coordination

Reproduced from NDMA, 2018. National Drought Management Authority Strategic Plan, 2018–22. National Drought Management Authority, Government of Kenya.

information about health issues. These have reconfirmed that water access and quality challenges influence child nutrition levels—since diarrheal diseases prevent children from gaining the nutritional value that they should obtain from food that they consume.

Participatory resource mapping techniques (GoK, 2015c; Jarso et al., 2017) have enabled communities in some counties to document the status of their water resources and to highlight priorities for improved management. Versions of these techniques have been in use since the 1980s and have demonstrated relevance to enhance the level of information that is currently available in the EWS. However, water resource information has still not yet been incorporated systematically to the drought management systems across the ASAL counties. This is a matter of concern in light of the ongoing progressive depletion of water resource availability, and its effects in increasing the exposure of vulnerable communities to drought risk.

The NDMA Strategic Plan has identified major achievements and lessons learned in the improvement of the EWS over the period 2013—17 (Table 19.2). But it has not picked up on the opportunity to supplement the early warning information with additional information on natural resource conditions, including information on the condition of water resources. Remaining challenges in this area are not mentioned in the strategic plan for 2022. Nor is the opportunity to make greater use of the EWS and other information on natural resource availability to assess value for money and impacts achieved through investments made in drought response, preparedness, and resilience building.

Financing of national drought management in Kenya

Financing drought management is a widely recognized and shared challenge for countries across the Horn of Africa region and the international community (Caravani et al., 2017). A major reason for the lack of preparedness observed during the 2008—11 drought was recognized to have concerned over-reliance on project-based development partner funding. Although developed nations have set a goal of mobilizing US$100 billion per year by 2020 to support climate change mitigation and adaptation activities in developing countries according to the Paris Agreement, this should not encourage continued reliance on relief aid projects alone. Climate finance is rather defined differently from conventional aid/relief funding. Development partner countries have expressed concern that Kenya should aim to provide basic services to populations in the ASALs and make use of more of its own resources to cope better with drought (e.g., McVeigh, 2018). In line with this view, a new phase of the HSNP beginning in March 2019 will be largely funded by the Kenyan government. This contrasts to the previous phase which was primarily funded by the United Kingdom Department for International Development (DfID). The Kenyan Government has also established its own National Drought Emergencies Fund (NDEF) to allow contribution of funds from both the Kenyan Government and from development partners.

Kenya's Nationally Determined Contribution to climate change adaptation and mitigation (GoK, 2015b) estimated that US$40 billion will be required to finance adaptation and mitigation interventions across six sectors until 2030. The national government is expected to contribute KSh 1.9 billion (US$18.5 million) through the NDMA to implement EDE in Kenya between 2014 and 2018 (DI, 2017). Other national funds will come from sectoral budgets with a component of drought risk or response interventions. For example, the National Climate Fund is being operationalized to deal with priority Climate Change interventions as highlighted in the NCCAP II-2018−2022. A National Drought Emergency Fund was allocated around KES 2 billion in 2018−19. Development partners are expected to contribute more than three quarters of the EDE implementation budget during the 4-year period.

According to the World Bank (IBRD, 2018), the Kenyan government already crowds in approximately US$65 million in additional private sector risk capital every year to help manage climatic risk. This includes microlevel agricultural insurance programs. There are also Contingency Emergency Response Components in World Bank projects and contributions to various funds, such as the NDEF, a proposed National Disaster Management Fund, and sovereign-level drought insurance through the African Risk Capacity (ARC) (2014/15 and 2015/16).

It is also important to consider the investments that are routinely made by the populations who live in the ASALs. Although these are rarely assessed comprehensively, they are believed to dwarf external investments (Tari et al., 2015; King-Okumu, 2015b).

Throughout the decade 2008−18, actors at all levels have continued to highlight needs for increased external assistance to respond to the losses and suffering caused by droughts. But at the same time, the channeling of Contingency Funds to affected communities in the drought-affected areas through the new permanent structure encountered practical problems, and because of this, commentators observed that the absorption of external funds was ineffective (Duguma et al., 2017). This is an area where a steep learning curve has been necessary to put in place accounting systems that are required both by development partners and by the Kenyan government and its people.

The 2018 Strategic Plan placed financing issues at the top of its list of challenge areas for lesson learning. However, the main lesson learned during the period 2013−17 that was listed in relation to these consisted in the observation that cash transfers empower beneficiaries to make own decisions on investment options. The review did not explicitly mention any lessons learned concerning the establishment of funds at other levels of government, although it did observe that capacity building had helped communities in prioritization and planning.

The Kenyan government and its partners have worked to establish the Contingency Fund, in line with Article 18 of the NDMA Act, 2016 (Box 19.1). This took time, so in the short term the gap still had to be filled by grant funding from the European Union, through the Kenya Rural Development Programme (KRDP), which supports both preparedness and early response and a range of other

---

Box 19.1 Article 18, (1) and (3) of the NDMA Act, 2016

(1) There is established a fund to be known as the National Drought Emergency Fund which shall be administered by the Secretary with the overall guidance and supervision of the Board.

(3) The objective of the Fund shall be to—
   a) facilitate timely response to drought during its different stages;
   b) provide for a common basket emergency fund in order to minimize the negative effects of drought;
   c) provide funds for capacity and technical expertise development to improve on drought management; and
   d) finance the establishment, management and coordination of projects, activities or programmes to further the foregoing purposes.

---

project-based interventions by NGOs continuing to work with bilateral donors. However, in 2017, the Kenyan Parliament adopted the Kenya National Policy on Climate Finance which aims to enhance national financial systems and institutional capacity to effectively access, disburse, absorb, manage, monitor, and report on climate finance in support of national sustainable development goals. By the end of the decade a National Drought Emergency Fund had at last secured budgetary provisions from July 2018. Moving forward, this could enable faster implementation of the strategic and institutionalized contingency planning process.

Next steps are anticipated to include increasing the size of the Contingency Fund (current annual allocation is KSh 5 billion) and incentivizing county governments to allocate money toward emergency response and relief funds ahead of time (GoK, 2018). The treasury has also been working with the NDMA during 2018 to operationalize the National Drought Emergency Fund and explore sovereign risk transfer instruments as part of its overall Disaster Risk Finance portfolio.

Over the period 2008−18, the Kenyan government has learned how to establish financing systems to channel funds to drought affected areas (Mahadi, 2018; Bonaya and Rugano, 2018). It is fair to say that the government is increasingly putting in place different financing vehicles/structures for sustainable financing of drought and climate-related challenges. However, there are still some ongoing challenges in assessing value for money and impacts achieved through the adaptation funds (Siedenburg, 2016; King-Okumu 2016, 2017; Macgregor, 2018). One of the challenges concerns identification of the benefits obtained from the investments in adaptation. A second important question concerns the identification of counterfactual scenarios where expenditures have been made (i.e., what would have happened *without* the funding). In order to weigh the benefits versus the costs of the adaptation finance and as an alternative to disaster relief, it is necessary to establish confidence in both pieces of information, even if one or other case must inevitably be hypothetical.

For the most part, the available estimates of losses and damages that would have occurred without investments that have been made in adaptation to drought in Kenya are characterized by wide ranges of uncertainty. As a result of this, they are generally treated with a high level of skepticism. Modeling of counterfactual scenarios to show drought impacts on hydrology, agricultural and livestock production, and household incomes with and without interventions (as demonstrated by Gies et al., 2014) could offer a means to obtain the required benefit cost calculations to determine the effectiveness and efficiency of adaptation investments. However, the development of robust hydrological models itself requires similar investments in human capacities, data collection, and processing to those that would be required for water resource planning and management. These are still generally not yet prioritized by development partners or governments, despite substantial funding available for infrastructure and climate information.

Discussion of scope for further learning in Kenya

A decade of drought management experience (2008–18) has clearly not yet been enough to end drought emergencies in the Horn of Africa region. While the region may be more adept at managing droughts than it was in 2010/11, a drought crisis was still declared in 2017 in Kenya and the surrounding region. The most affected areas included most of Somalia, southeastern Ethiopia, northeastern and coastal Kenya, and northern Uganda. The United Nations Food and Agriculture Organization (FAO) estimated that 17 million people were impacted.[7] Yet when the 2018 March–May long rains were high, and the alarm phase in Kenya deescalated (see Fig. 19.3), still no major participatory scientific program to assess and recharge reserves for the next drought was instituted and the national EWS still includes no hydrological monitoring nor any use of available modeling tools to support improved management and sharing of the available water resources. In light of this, further scope for improved learning from Kenya's drought experiences still remains.

Kenya has made significant strides in a positive direction toward improved disaster preparedness through the establishment of the NDEF. However, preparedness requires more than finance: proactive water resource management actions should be taken early on while the drought cycle is still under "normal" phase conditions. There have been some successful localized investments in improving water pans and sand dams (NDMA, 2014).[8] But these local level successes have still not yet been enough to achieve the intended shift for the arid portion of the country or the region out of recurrent crisis mode and onto an accelerated path to sustainable development. Higher rainfall has been predicted for Kenya over coming years (WRMA, 2013b) and high rainfall events are already occurring. However, due to weak catchment management and increasing upstream water extractions, flows through the major catchments are still reducing, water tables are falling, and the water deficit is growing (WRMA, 2013a, 2013b, 2016a, 2016b). As a result, the majority of the population in the ASALs is still unable to access sufficient supplies of clean water during droughts. This is increasing exposure and vulnerability to hydrological drought.

Over the decade 2008–18, Kenya has succeeded as never before to build institutions, transfer funds together with development partners, and track the impacts of the drought on its households, even while frequently hosting in-migrating transhumant people from across the region. As the NDMAs early warning databases have grown, they have begun to enable some interesting insights concerning the effectiveness of the Kenyan drought management actions. For example, a recent study (Venton, 2018a) was able to include an analysis of the long-term trends in the remotely sensed data on vegetation conditions in the EWS. From 2013 onward, reduced correlation between periods of fodder shortage and child malnutrition impacts have been suggested to indicate successful drought management and resilience building interventions (Venton, 2018a).[9] This is a rather striking observation with potentially major significance that could be worthy of further interrogation.

[7]https://news.un.org/en/story/2017/01/550422-warning-dire-food-shortages-horn-africa-un-agriculture-agency-calls-urgent.

[8]See also the County Integrated Development Plans at https://roggkenya.org/story-suggestions/kenya-county-cidp-kenya-five-year-plan-on-track/download-page-cidp-county-integrated-development-plans/.

[9]Interestingly, from 2013 international donor commitments did not increase (though they did spike in 2011 and some delays in disbursement may have occurred). However, since 2013, there has been a philosophical shift in the language of international assistance toward investments in building resilience. But is not yet possible to assess the practical significance and impact of this shift.

Other interesting observations that the EWS datasets partially support concern periods when the VCI analysis has suggested that pastures did not become as badly degraded as they might usually do during dry seasons and droughts. This might be understood to suggest that investments in local rangeland management had been effective (Tari et al., 2015). This observation and those of other rangeland management groups also require further investigation and more evidence-based evaluation (King-Okumu, 2015a). This type of learning about the material impacts of drought preparedness programs on the ground requires the establishment of participatory scientific processes and monitoring systems. Learning systems of this kind should build local and regional capacities for resource management.

At the regional level, it has been argued that 30% of humanitarian spending on drought relief could be saved through earlier and more proactive local scale actions such as fodder production and water harvesting (Venton, 2018b). This would be equivalent to savings of US$1.6 billion when applied to US government spending over the last 15 years in Kenya, Ethiopia, and Somalia alone. Furthermore, it has also been argued that taking the proactive response a step further by protecting people's income and assets could be even more cost-effective in enabling households to manage the effects of shocks. If these benefits could be incorporated, the overall savings from the avoided disaster relief burden increase to US$4.2 billion over the 15-year period. Put another way, according to Venton (2018a, 2018b), "every US$1 invested in building people's resilience in the Horn of Africa could result in up to US$3 in reduced humanitarian aid and avoided losses."[10]

Greater use of the EWS to assess the potential and actual achievements of adaptation interventions could offer a considerable boost to learning, donor confidence, and local confidence in public decision-making. This could include an assessment of the likely return on investment that could be obtained from improved catchment management (see discussions in King-Okumu et al., 2016; Jillo et al., 2016; GoK 2017b). However, such analyses of the compiled datasets are few and cannot be conducted independently. An initiative to make the EWS dataset more accessible to researchers and the public has not yet been achieved.

Summary of lessons and recommendations

Reflecting on the Kenyan experiences in relation to the three key challenges discussed in this chapter, the lessons learned over the past decade and their implications for the next drought in the Horn of Africa can be summarized as follows:

Challenge 1: limitations in the human resource and institutional capacities
Lesson 1
Drought preparedness requires coordination connecting local institutions to government processes and institutions and partners at other scales.

---

[10]This estimate is more modest than one presented previously at the global level by Judith Rodin who suggested a return of 4:1 in her book *The Resilience Dividend*.

Recommendation 1

Awareness raising and capacity building are needed to encourage the devolved local institutions to fulfill their new roles and drive improved catchment planning to prepare proactively before droughts hit (whether they are socioeconomic droughts, hydrological droughts, or just temporary meteorological phenomena).

Challenge 2: lack of natural resource information and weak early warning systems

Lesson 2

Over the period 2008—18, the Kenyan government and development partners have learned how to manage systematic community-based monitoring systems providing monthly updates for early warning.

Recommendation 2

There is an opportunity to use the early warning information together with additional information on natural resource conditions to transition through drought response, preparedness, and resilience building toward more sustainable and drought-resilient development planning.

Challenge 3: limited financial resources for DRM and resilience building

Lesson 3

Over the period 2008—18, the Kenyan government and donors have learned how to establish financing systems to channel funds to drought-affected areas ensuring budgetary allocation and disbursement on a timely basis for early response/action.

Recommendation 3

There is an opportunity to make further use of participatory resource accounting and hydroeconomic decision support systems to assess the material effects of drought preparedness and management improvements, including value for money. Other nonmonetary impacts also require further consideration and assessment.

Conclusions

The ability of societies to reduce the impacts of future disasters by effectively preparing for them depends on successfully extracting, analyzing, sharing, and internalizing lessons from experience. At a time when the future of donor support for drought relief is uncertain, it is all the more important to effectively identify and learn lessons from significant investments made to enhance drought preparedness in Kenya. In this chapter, we have identified three lessons and three recommendations that should be of interest not only in Kenya but also for the surrounding Horn of Africa region to transition out of recurring crisis mode and onto an accelerated path to sustainable development by 2030.

The GWP (2015) has observed a shift occurring across the region from a focus on emergency/crisis response toward integrated drought management, including preparedness, drought mitigation, and early warning. This chapter identifies scope for Kenya to enhance institutions and capacities for coordination at the local level, make use of the richness of its early warning databases to boost evaluation of investments in resilience building, and connect investments in drought preparedness planning to sustainable development planning more broadly. In light of the long periods of underdevelopment/marginalization that occurred in the arid lands, good strides have been made toward ending drought emergencies in Kenya. These need to continue and be enhanced for meaningful impact.

The Kenyan approach to drought preparedness has been based on the observation that drought management must concern itself not *just* with the immediate cause of droughts (the deficit of water) but with the underlying human and institutional factors that leave people vulnerable to drought effects. Ultimately, however, *if the system does not make a credible and informed effort to rebalance the deepening water deficits in the ASAL catchments, it will not end the drought emergencies.* This requires concerted action during periods of predicted high rainfall—not just following the announcement of drought emergencies. This will require continued learning; increased attention to evaluation; and the use of participatory scientific methods and capacities to monitor, model, and manage the hydrological systems in the ASALs in the precrisis drought phases.

As water extractions for economic uses increase, the risks of hydrological and socioeconomic droughts are also increasing. In other regions and parts of Kenya, hydrometeorological information is more routinely available to support the private sector in their projections of the anticipated returns on investments in infrastructure or other projects. But this is not yet possible in the Kenyan ASALs and across the Horn of Africa. It still remains to be seen whether the private or public sector will insist that more robust hydrometeorological projections should be used to ascertain and mitigate the extent of drought risks in the Horn of Africa.

Looking ahead for 2018—30, the devolved local institutions that have been put in place in the Kenyan ASALs and across the IGAD region over the period 2008—18 could achieve the end of drought emergencies. Needs highlighted for investment to build local and national capacities to improve monitoring and modeling systems by combining participatory resource accounting with hydrometeorological tools are of relevance not only in Kenya but across the region as a whole. Potential savings on drought relief that could then be verifiable should be of interest to national and local governments across the region as well as to international donors.

References

Aeschbacher, J., Liniger, H., Weingartner, R., 2005. River water shortage in a highland-lowland system: a case study of the impacts of water abstraction in the Mount Kenya region. Mountain Research and Development 25, 155—162.

Birkland, T.A., 2009. Disasters, lessons learned, and fantasy documents. Journal of Contingencies and Crisis Management 17, 146—156.

Bonaya, M., Rugano, P., 2018. Gender Inclusion and the CCCF Mechanism: Increasing the Voice and the Benefits for Women. Ada Consortium, Nairobi.

Caravani, A., Greene, S., Trujillo, N.C., Amsalu, A., 2017. Decentralising Climate Finance: Insights From Kenya and Ethiopia. In Working Paper. ODI, London, 115 pp.

Deverell, E., 2009. Crises as learning triggers: exploring a conceptual framework of crisis-induced learning. Journal of Contingencies and Crisis Management 17, 179—188.

DI, 2017. March 2017 Analysis of Kenya's Budget 2017/18 What's in It for the Poorest People? Report. Development Initiatives, 28 pp.

Duguma, M.K., Brüntrup, M., Tsegai, D., 2017. Policy options for improving drought resilience and its implication for food security the cases of Ethiopia and Kenya. In: Studies Deutsches Institut für Entwicklungspolitik 98. Deutsches Institut für Entwicklungspolitik gGmbH, Bonn, Germany, 103 pp.

Egner, H., Schorch, M., Voss, M., 2015. Learning and Calamities — Practices, Interpretations, Patterns. Routledge, Abingdon, UK.

FAO, 2011. Drought-Related Food Insecurity: a Focus on the Horn of Africa Drought Emergency. Emergency Ministerial Level Meeting. FAO, Rome, 7 pp.

GFDRR, 2018. Somalia Drought Impact & Needs Assessment - Volume 1 - Synthesis Report. GFDRR, UNDP, EU, World Bank, 160 pp.

Gies, L., Agusdinata, D.B., Merwade, V., 2014. Drought adaptation policy development and assessment in East Africa using hydrologic and system dynamics modeling. Natural Hazards 74, 789–813.

Glantz, M.H., 1977. Nine fallacies of natural disaster: the case of the Sahel. Climatic Change 1, 69–84.

Glantz, M.H., Katz, R.W., 1985. Drought as a constraint to development in sub-Saharan Africa. Ambio 14, 334–339.

GoK, 2008. Sector Plan for Drought Risk Management and Ending Drought Emergencies First Medium Term Plan 2008-12. Republic of Kenya, 218 pp.

GoK, 2011. Vision 2030 Development Strategy for Northern Kenya and Other Arid Lands. Republic of Kenya, Nairobi, 121 pp.

GoK, 2012. Sessional Paper No. 8 of 2012 on the National Policy for the Sustainable Development of Northern Kenya and Other Arid Lands. Office of the Prime Minister, Republic of Kenya, Nairobi, Kenya, 42 pp.

GoK, 2015a. Ending Drought Emergencies Common Program Framework, April 2015. National Drought Management Authority (NDMA), Republic of Kenya, Nairobi, Kenya, 189 pp.

GoK, 2015b. Kenya's Intended Nationally Determined Contribution (INDC) 23 July 2015. Ministry Of Environment And Natural Resources, Nairobi, Kenya, 7 pp.

GoK, 2015c. Resource Atlas of Isiolo County, Kenya: Community-Based Mapping of Pastoralist Resources and Their Attributes. IIED/ADA, Isiolo County, Kenya.

GoK, 2016. The Water Act, 2016. In: Kenya, G.O. (Ed.), Kenta Gazette Supplement No. 164 (Acts No. 43). National Council for Law Reporting with the Authority of the Attorney-General, Nairobi, Kenya, 108 pp. www.kenyalaw.org.

GoK, 2017a. Sector Plan For Drought Risk Management And Ending Drought Emergencies Third Medium Term Plan 2018–2022. Nairobi, Republic of Kenya, 56 pp.

GoK, 2017b. In: NEMA (Ed.), Submission in the Area of Ecosystems, Interrelated Areas Such as Water Resources and Adaptation under the Nairobi Work Programme. NEMA/NDMA/Ada Consortium, p. 6.

GoK, 2018. Disaster Risk Financing Strategy 2018–2022, Republic of Kenya. The National Treasury and Planning, Government of Kenya, Nairobi, 4 pp.

GoK/WRMA/JICA, 2012. The Project on the Development of the National Water Master Plan 2013 Final Report. Water Resources Management Authority, Japan International Cooperation Agency, 281 pp.

GWPEA, 2015. Assessment of drought resilience frameworks in the Horn of Africa. In: Integrated Drought Management Program in the Horn of Africa (IDMP HOA). Global Water Partnership Eastern Africa (GWPEA), Entebbe, Uganda, 32 pp.

GWPEA, 2016. Building Resilience to Drought: Learning from Experience in the Horn of Africa. Integrated Drought Management Programme in the Horn of Africa. Global Water Partnership Eastern Africa (GWPEA), Entebbe, Uganda.

Headey, D., Taffesse, A.S., You, L., 2012. Enhancing resilience in the Horn of Africa an exploration into alternative investment options. In: IFPRI Discussion Paper 01176. Development Strategy and Governance Division, IFPRI, Washington D.C., 48 pp.

Hujale, D., 2015. The Role of Somali Local Institutions in Building Resilience in the Arid and Semi-arid Lands (ASALs). Oxford Brookes, Oxford, 58 pp.

IBRD, 2014. World Bank Regional Integration Department Africa Region, October 23. World Bank, Washington, DC, 101 pp.

IBRD, 2018. Disaster Risk Management Development Policy Credit with a Catastrophe Deferred Drawdown Option (Cat DDO) (P161562). The World Bank, Nairobi, Kenya, 56 pp.

Ifejika Speranza, C., Kiteme, B., Wiesmann, U., Jörin, J., 2016. Community-based water development projects, their effectiveness, and options for improvement: lessons from Laikipia, Kenya. African Geographical Review 1—21.

IFRC, 2011. Drought in the Horn of Africa: Preventing the Next Disaster. International Federation of Red Cross and Red Crescent Societies, Geneva, 24 pp.

ISDR, 2005. Summary of the Hyogo Framework for Action 2005—2015: Building the Resilience of Nations and Communities to Disasters. International Strategy for Disaster Reduction, Hyogo, Japan, 2 pp.

Jarso, I., Tari, D., King-Okumu, C., 2017. Recommendations to the County Government of Isiolo for preparation of a strategic plan on water, energy and climate change. In: IIED Report, vol. 62. IIED, London.

Jillo, B., Adaka, V., Jarso, I., Shandey, A., Kinyanjui, J., Lekalkuli, L., Tari, D., Okumu, C.K.-, 2016. Cracking the Climate-Water-Energy Challenge in the Drylands of Kenya. IIED, London, 4 pp.

King-Okumu, C., 2015a. Inclusive Green Growth in Kenya: Opportunities in the Dryland Water and Rangeland Sectors. International Institute for Environment and Development, London, 52 pp.

King-Okumu, C., 2015b. Rapid assessment of investments in natural resource stewardship in comparison to the value of returns. In: IIED Working Paper. International Institute for Environment and Development (IIED), London.

King-Okumu, C., 2016. Distilling the Value of Water Investments. IIED, London, UK, 2 pp.

King-Okumu, C., 2017. Rethinking Cost/benefit Assessments of Decentralised Investments in Resilience Building. BRACED, London, UK, 2 pp.

King-Okumu, C., Wasonga, O.V., Jarso, I., Salah, Y.M.S., 2016. Direct Use Values of Climate-dependent Eco-system Services in Isiolo County, Kenya. IIED.

King-Okumu, C., Jillo, B., Kinyanjui, J., Jarso, I., 2017. Devolving water governance in the Kenyan arid lands: from top-down drought and flood emergency response to locally driven water resource development planning. International Journal of Water Resources Development.

Liniger, H., Gikonyo, J., Kiteme, B., Wiesmann, U., 2005. Assessing and managing scarce tropical mountain water resources: the case of Mount Kenya and the semiarid Upper Ewaso Ng'iro Basin. Mountain Research and Development 25, 163—173.

Macgregor, J., 2018. Assessing the "Business Case" for Investing in the CCCF Mechanism. Ada Consortium, Nairobi, 4 pp.

Mahadi, Y., 2018. Strengthening the CIDP Process and Contribution of CCCF Mechanism. Ada Consortium, Nairobi.

McVeigh, K., 2018. Cuts to UK Aid: Package to Drought-Hit Kenya to End in 2024 the Guardian Newspaper.

NDMA, 2014. Isiolo County Adaptation Fund: Activities, Costs, Impacts after the 1st Investment Round. Kenya National Drought Management Authority (NDMA), Nairobi/London, 32 pp.

NDMA, 2017. Early Warning Bulletin, Isiolo, May 2017. National Drought Management Authority, Nairobi, 14 pp.

NDMA, 2018a. Early Warning Bulletin, Isiolo, July, 2018. National Drought Management Authority, Nairobi, 13 pp.

NDMA, 2018b. National Drought Management Authority Strategic Plan, 2018-22. National Drought Management Authority, Government of Kenya, 72 pp.

PDNA, 2012. Kenya Post-Disaster Needs Assessment (PDNA) for the 2008-2011 Drought. Republic of Kenya with Technical Support from the European Union, United Nations, and World Bank and Financial Support from the European Union and the Grand Duchy of Luxembourg, 188 pp.

Sen, A., 1981. Poverty and Famines an Essay on Entitlement and Deprivation. Clarendon Press, Oxford.

Siedenburg, J., 2016. Community-based cost benefit analysis (CBCBA). Findings from DFID Kenya's arid lands support programme. In: Evidence on Demand, vol. 62. Landell Mills, London, UK.

Staupe-Delgado, R., Kruke, B.I., Ross, R.J., Glantz, M.H., 2018. Preparedness for slow-onset environmental disasters: drawing lessons from three decades of El Niño impacts. Sustainable Development 1−11.

Tari, D., King-Okumu, C., Jarso, I., 2015. Strengthening Local Customary Institutions: A Case Study in Isiolo County, Northern Kenya. Ada Consortiumj, Nairobi, 52 pp.

UN, 2016. Report of the open-ended intergovernmental expert working group on indicators and terminology relating to disaster risk reduction. In: Seventy-first Session Agenda Item 19 (C) Sustainable Development: Disaster Risk Reduction, vol. 41. United Nations General Assembly.

Venton, C., 2018a. Economics of Resilience to Drought - Kenya Analysis. USAID, Washington DC, 43 pp.

Venton, C.C., 2018b. Economics of Resilience to Drought. USAID, 43 pp.

Waal, A. d., 1989. Famine that Kills: Darfur, Sudan, 1984−85. Oxford University Press, Oxford, UK.

Wiesmann, U., Gichuki, F.N., Kiteme, B.P., Liniger, H., 2000. Mitigating conflicts over scarce water resources in the highland-lowland system of Mount Kenya. Mountain Research and Development 20, 10−15.

WRMA, 2013a. Final Report - Surface and Groundwater Assessment and Planning in Respect to the Isiolo County Mid Term ASAL Program Study Volume 1 Main Report. In Report No. 47/2013. Earth Water Ltd, 462 pp.

WRMA, 2013b. The Project on the Development of the National Water Master Plan - Final Report: Volume I: Executive Summary. 219 + Figures. Water Resources Management Authority, Nairobi, Kenya.

WRMA, 2016a. Surface Water and Groundwater Resources Assessment in Wajir County for Decision Making: Final Report. Geekan Kenya Ltd for the Kenyan Water Resource Management Authority (WRMA), Nairobi, 209 pp.

WRMA, 2016b. Water Resources Assessment for Decision Making in Garissa County Final Report June 2016. In Contract No. WRMA/GOK/MTAP2/3/1/2015-2016 - LOT 1. MTAP, Nairobi, 203 pp.

Achieving policy coherence for drought-resilient food security in SSA—lessons from Ethiopia and Kenya

20

Michael Brüntrup

Program B, Transformation of Economic and Social Systems, Deutsches Institut für Entwicklungspolitik (German Development Institute) DIE, Bonn, Germany

Introduction

Drought is the most far-reaching of all natural disasters (WMO and GWP, 2014). Its impacts extend from effects on water availability for people, animals, domesticated and wild plants, and industry to fires, transport, electricity, tourism, infrastructure, supply chains, and social relations (Cai et al., 2017; Van Loon et al., 2016). Competition over dwindling resources can exacerbate or create local and regional conflicts but can also create subsequent pushes for democratic developments (Maystadt and Ecker, 2014; Aidt and Leon, 2016). Migration may also increase, too, which in turn can exacerbate local conflict outside the affected region (Raleigh et al., 2008; Obokata et al., 2014).

As the impacts of drought are manifold, so are the policy measures required to assist people, communities, and nature to become more resilient, i.e., to "anticipate, absorb, or buffer losses, and to recover" (UN, 2015). The policies and policy measures are not only various, they may also partially or temporarily conflict with each other, in different phases of an individual drought and in successive droughts, so timeliness, coordination, and cooperation between policies and their actors are required—"drought policy coherence" (Nilsson et al., 2012). This bundle of (coherent) policies and measures, their institutional setup and funding, should be the objective of a national drought strategy and policy. General guidelines on how to establish such drought policies have been elaborated on various occasions (e.g., WMO and GWP, 2014; Tadesse, 2016).

While larger drought events result in stress for all kinds of economies and societies, they have a particularly severe effect on food security in poor countries. Whatever definition of food security is used—and there have been many over the last decades (Jarosz, 2011)—in poor countries, a severe drought certainly threatens it. The difference between rich and poor countries is critical. Although economically more advanced countries may be affected by drought in various ways, their food security is (at least at present) barely affected as there are safeguards on two sides. On the demand side, the overall economic losses (at national and local level) are relatively low due to the limited importance of rainfall-dependent agriculture, the existence of unemployment insurance and social safety nets, and the relatively low share food makes up in overall household expenditures even of the poorer people

(10%–30%), which means food price fluctuations can typically be absorbed. On the supply side, local food markets are well integrated into regional and international markets, there are large supply buffers of commodities normally used for animal feed and energy which can be used for food and fodder during a drought (even if of lower quality), and food market transaction costs are low, thereby limiting the translation of drought-depressed production into high price increases for food.

Such conditions tend not to exist in poor, developing countries, nor are they likely to emerge in the short and medium term: their absence is a defining feature of underdevelopment. Thus, droughts strongly impact on food security in rural areas of poor countries. Crop production becomes insufficient for subsistence, let alone market sales, while livestock lose weight and die, or have to be sold, at low prices due to oversupply (e.g., McPeak and Barrett, 2001; Devereux, 2007). There are significant reductions in income for most households, not only due to direct production and income effects but also because agriculture and livestock are crucial to other income-generating activities in rural areas (Christiaensen et al., 2011). Food prices increase sharply due to weak market integration, and people have limited food or cash reserves (Devereux, 2007). It is unrealistic to expect people and governments to fundamentally change the dependencies and weaknesses of these markets and income patterns in the short and medium term. A more realistic approach would be to tackle the challenges posed by droughts in ways compatible with local capacities and capabilities, while in the long term striving to achieve more structural resilience in food systems, as described for developed countries.

Historically, drought-induced famines and food security crises were portrayed as nature-driven failures of subsistence agricultural systems, but political ecologists have argued convincingly in recent decades that under functioning governments, drought alone does not provoke famine—it is rather the catalyst; it is governments who—willingly, willfully, or ineptly—exacerbate or even provoke droughts to become disasters (Sen, 1982; Devereux, 1993; de Waal, 2002; Tschirley and Jayne, 2010). If this is indeed so, what are the implications for building drought resilience through governance? Is resilience maximized in circumstances where a household is able to cope with drought on its own? Should governments intervene only when private means do not suffice? Is resilience achieved if household members simply survive, or should the measure of resilience include the protection of productive assets, so that the longer-term survival, recovery, or even development is not handicapped?

The answers to these questions are anything but trivial and can have far-reaching consequences for drought policies and their coherence. For instance, if (bottom-up) household self-reliance is not deemed crucial (costly and long-term), external efforts to support bottom-up resilience can be avoided in favor of a well-functioning, top-down food-aid system (or cash handouts plus efforts to improve food markets). If short-term survival is the measure of resilience, protection of productive assets will not be accounted for in assessing the success of drought-resilience strategies. Another principle to be discussed in relation to any region is whether drought is a rare event or is a frequently recurring one. If droughts are rare, the mixture of measures to grow and to protect should be different from cases where droughts strike frequently.

General approaches to dealing with disaster risks, such as the Hyogo framework (UNISDR, 2006) and its successor, the Sendai framework (UNISDR, 2015), are based on the concepts of risk management (Wilhite et al., 2000). The emphasis lies on resilience building and preparedness; early warning systems (EWSs); vulnerability assessments; multisectoral approaches; intertwining of normal sector emergency policies; personal, institutional, and financial capacity building; and, infrastructural "rebuilding better," i.e., rebuilding with a better protection against risks. Particularly in the case of recurrent droughts, where the intervals between droughts are so close that preparing for, going through and rebuilding after drought potentially overlap, a drought cycle management approach is the right way to further conceptualize drought risk-management and drought policy (see Fig. 20.1).

Such a coherent drought policy is anything but easy to establish, since it has to stretch over several policy domains, must embrace short-term and long-term solutions and perspectives, bring together very different stakeholders, and work in a very timely manner. In this context, lessons from the recent past of Ethiopia and Kenya, both situated in the Horn of Africa, one of the most drought-stricken regions in the world (Mengisteab, 2013), will be compared in the remainder of this chapter.

Drought policies in Ethiopia and Kenya

Ethiopia and Kenya provide useful examples for policy learning since they have similar ecological settings but different economic and political situations. Both have some high-potential humid and semi-humid areas as well as some arid and semi-arid lands (ASAL). Both area types are affected by recurrent droughts, but the ASALs are more vulnerable. The topography diversity of the agroecological zones ensure that even a very large drought does not usually affect the whole region, and that food reserves can usually be found within each country. The two countries' ASALs share a long common border, ecological zones, and (pastoralist and agropastoralist) production (and livelihood) systems. Sometimes the same ethnic groups live on both sides of the border, and the traditional knowledge systems can therefore be assumed to be comparable.

However, economically and politically the countries are quite different. Kenya is one of the more prosperous countries in this part of the world, with strong economic growth since 2005, a long

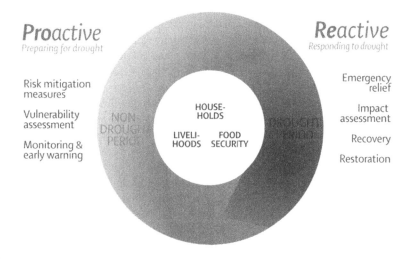

FIGURE 20.1

Drought cycle management.

Source: Adapted from Brüntrup, M., Tsegai, D., 2017. Drought Adaptation and Resilience in Developing Countries, Briefing Paper 23. German Development Institute/Deutsches Institut für Entwicklungspolitik (DIE), Bonn. https://www.die-gdi.de/briefing-paper/article/drought-adaptation-and-resilience-in-developing-countries/.

post—colonial (deficient) democratic history, and, since the constitutional reform of 2010, one of the most rapid and ambitious devolution processes worldwide. Ethiopia is one of the poorest countries in the world, although its economy, too, has grown rapidly over the past 15 years. It is characterized by weak democratic tradition and strongly centralized policy making. Both countries have been members of the Intergovernmental Authority on Development (IGAD) since its inception in 1996 and, as such, share the common IGAD Drought Disaster Resilience and Sustainability Initiative (IDDRISI, IGAD, 2013), which provides a modern framework for learning national action and regional cooperation on drought policy (Duguma et al., 2017).

The evolution of drought policies in Ethiopia and Kenya

Ethiopia has been almost synonymous with a high vulnerability to droughts, with historic and catastrophic droughts having been registered for many centuries, and with national and international policy failures having contributed to widespread starvation during the 19th century (Pankhurst 1961, 1968; de Waal, 1991; Pflanz, 2011; Venton et al., 2012; de Perez et al., 2019). A formal Relief and Rehabilitation Commission was first established after the 1973/74 famine (which was one of the immediate reasons for the overthrow of Emperor Haile Selassie) with a focus on organizing relief operations more effectively. This institution was later abused by the succeeding DERG regime and could not prevent, or might even have caused, some of approximately 1 million deaths during the 1983—86 famine. From 1993 onwards, a National Policy on Disaster Prevention and Management (NPDPM) and then a National Disaster Prevention and Preparedness Commission (NDPPC) were established. The focus gradually shifted, as the names imply, from relief to risk management. Key sectoral offices, such as the Ministries of Agriculture and Rural Development, Health and Water Resources, became more actively involved in disaster management through the establishment of emergency task forces in 2003. In 2004, the tasks of emergency response and of chronic food insecurity were split up, with the former being transferred to the Ministry of Agriculture and Rural Development in 2007.

In 2013, a national Disaster Risk Management Policy and Strategic Framework were established, and the system made accountable to the prime minister's office (MOA, 2013). The framework recognizes multiple hazards and associated disasters and is the shared responsibility of multiple ministries and local administrations. It centralizes information flows and coordinates national management for these hazards, but at the same time emphasizes the need for bottom-up approaches to risk management and hazard responses, and therefore calls for participation and decentralization in the planning, execution, monitoring, and evaluation of disaster risk-mitigation measures. Accordingly, a multisector and multiagency national platform entitled the DRM Technical Working Group is supported by a series of task forces from sectors including agriculture; water, sanitation, and hygiene (WASH); health; nutrition; and education; there is also a working group on gender. Similar groups are also established at subnational level.

Kenya also has a long history of drought (Mbithi and Wisner, 1973). While in the 1970s the number of people affected was reported in the tens of thousands, in the 1980s it was in the hundreds of thousands, in the 1990s up to 1.4 million, between 2.3 million and 4 million three times in the early 2000s, and in early 2010 up to 10 million people were at the risk of hunger after harvests failed due to drought (Kandji, 2006; Mateche, 2011). However, in modern times, drought has rarely led to famine and widespread death, but rather to emergency declarations followed by the distribution of food aid, although the livelihoods of people have been heavily affected. In general, drought is concentrated in

ASALs and has become a bone of contention between these regions and the central government. The strong decentralization in Kenya since 2010 is partly to be explained by the lack of strategic response by the central government to recurrent drought stresses (as felt by the ASAL regions), which exacerbated the feeling of unjust repartition of government funding and economic wealth between ASALs and the center (i.e., the capital and central highlands). The dispute finally led to a specific policy for the ASALs, with many drought-related elements, as a special pathway to the overall Kenya Vision 2030.

Accordingly, the new National Drought Management Authority established through the Ending Drought Emergency (EDE) strategy of 2012 is located under the Ministry for Devolution and Planning (Kenyan Ministry of Devolution and Planning, 2015). Interestingly, besides more common elements it has peace and security among its six pillars, acknowledging that in the region the exacerbated competition for resources (particularly during droughts) can lead to additional violence and, vice-versa, violence is preventing effective implementation of drought-related (and other) measures. Measures are classified as being implemented by central government, by decentralized county governments, or as activities by the Drought Authority. Several ASAL multistakeholder forums have been established for implementation at national, regional, and local levels and in different stakeholder combinations. The highest forums are chaired by the president (Intergovernmental Forum) and the secretary of cabinet (Intergovernmental Committee).

Lessons learned

Lessons are classified according to the main pillars of a drought cycle management: early warning and drought (disaster) response during drought; post drought vulnerability assessment; and risk mitigation in preparation for the next drought.

Early warning and drought (disaster) response

It is commonly accepted that the later the response to a drought, the more costly the measures and the greater the damage (Hillier and Dempsey, 2012). This suggests that early warning is a key to resilience. However, unnecessary costs may occur if drought warnings prove unfounded. Unfortunately, it is one of the particularities of droughts (compared to other natural disasters) that their characteristics in terms of duration, depth, geographical coverage slowly evolve over time in ways that are hard to predict. To make things more complicated, the effects are modified by sectoral and local vulnerabilities that need to be taken into account when advancing from meteorological to agricultural/biological drought to socioeconomic drought (WMO and GWP, 2014). An adaptive response is the way forward, starting with low-cost and no- or low-regret measures (e.g., early vaccination, fodder accumulation) and moving to increasingly more specific measures as drought warnings get more precise. For this to happen, good EWSs are a precondition, but have to be continually revised and need to be continually integrated into ongoing and new measures.

EWSs have been set up in the East African region over several decades. In the two countries used as examples here, national meteorological agencies produce short- and medium-term outlooks of up to a whole season. For converting meteorological information into more specific drought warnings, water and soil, vegetation as well as socioeconomic information (see also vulnerability assessments) are also taken into consideration. In Kenya, the Meteorological Department also produces drought bulletins; in Ethiopia this is done by the National Disaster Risk Management Commission. Bottom-up reporting systems are supposed to provide local observations. In Kenya, it is the National Drought Management Authority and the President who declare drought, but it is the counties who finally have to take up this information and convert them into local warnings.

The national systems are supported by the regional IGAD Climate Prediction and Application Centre. Several international agencies, such as the World Meteorological Organization (WMO), Food and Agricultural Organization (FAO), World Food Programme (WFP), and USAID also provide information, directly or indirectly, for example, through the Famine Early Warning Systems (FEWS) Network, to the national responsible organizations. Several international nongovernmental organizations (NGOs), such as Oxfam and Action Aid, also have their own reporting systems or use a locally adapted blending of national and local forecasts, such as Participatory Scenario Planning in the Pastoral Resilience Improvement through Market Expansion (PRIME) project (Singh et al., 2016).

Systems in the past were often accused of being sluggish in declaring drought, which then led to delayed, more costly, deficient, or even harmful responses (Hillier and Dempsey, 2012). Even with improvements in the evolving drought strategies described above, this problem has not stopped completely. The severe drought in the whole of the Horn of Africa in 2010/11, which cost about 200,000 lives (mostly in Somalia) was forewarned in May 2010, when the first signs of an El Nino were noted; in February 2011, the FEWS Network issued a warning, but the UN did not declare a famine until July 2011 (Venton et al., 2012). In Ethiopia, the first signs of the 2015 drought, which affected 10 million people, were visible when spring rains failed and an El Nino—induced reduction in summer rains became likely (Singh et al., 2016). However, the government hesitated to declare a drought until August 2015, when it released USD 700 million from its budget for drought relief measures. In Kenya, too, a common complaint is that all too often information about droughts comes late, is not taken up, or is not converted adequately into action (Kandji, 2006; Venton et al., 2012).

The reasons for weak performance of EWSs are several:

a) There are indications that the quality of the forecasts is not yet very good (although it is better in the case of El Nino droughts than in other situations (de Perez et al., 2019) and sometimes too vague or not site-specific enough (Singh et al., 2016).

b) Local people are unaware, do not understand, or mistrust formal warning systems and warnings (Singh et al., 2016).

c) Governments are reluctant to declare drought in order to maintain an international image (of advancement), to save money in the hope of an early recovery, or simply due to lack of concern (with pastoralists or with particular regions for reasons of political difference) (Cabot-Venton et al., 2012; Cohen et al., 2016).

It is the combination of these, but particularly the indifference of some governments and the frequent delays in declaring drought, combined with (suspicions of) political economy considerations (see above), which limits the usefulness of EWSs —and not only in these two countries (Drèze and Sen, 1991; Batterbury and Warren, 2001; Tschirley and Jayne, 2010; Ncube and Shikwambana, 2016).

Vulnerability assessment

Drought does not strike all people living in a given area in a uniform way. It depends on the type of drought and to an even greater extent on the nature conditions and the predisposition of the people—i.e., their vulnerability (Carrao et al., 2016). For instance, while some crops are very sensitive even to short dry spells of a few days, other varieties of the same crop or other species are less sensitive; the same goes for farming systems, farming households, and animals. Natural vegetation may be more or less adapted to droughts, and thus more or less able to produce fodder and biological products under drought conditions. Groundwater availability and irrigation options can modify that ability substantially, while the availability of groundwater in the medium and longer term also usually depends heavily on rainfall in the same region or in catchment areas. People who rely more on drought-dependent natural resources for subsistence production and consumption, for income generation, and for food markets are more vulnerable to drought than people with other sources of food and income. Women are affected differently than men (Getu and Mulinge, 2013). All these elements should be taken into account when assessing drought vulnerability and when assessing the likely impacts of an emerging drought risk.

Note that vulnerability is not a static concept and will change, even for a given population, depending on the characteristics of the drought and the often variable status of the various elements. For instance, groundwater or household savings, the level of grain and fodder stocks could be in a depleted state as a result of a previous calamity. Thus, one-off and time-bound assessment methods have to be combined. Note also that nature, people, and socioeconomic environments are not only vulnerable to drought but also to other risks (compare the wider risk management frameworks mentioned above).

In Ethiopia, there is a systematic national disaster risk profiling program at the community (Woreda) level. It collects statistical data on a wide range of issues (water, electricity, health services, etc.), survey data, key information, and focus group discussion information. There is a special section on women. The collected data are summarized and the profiles are made available on the Internet (MOARD, 2010). However, the program's financing is only assured by short-term donor funding, profiling activities have been behind schedule, and the more advanced parts of the profiles—the contingency and climate change adaptation plans—have only existed for a minority of communities. In addition, communities often only have the use of outdated IT hardware, services are understaffed, the staff are poorly trained, and turnover is high so that knowledge and training effects are lost. The constant updating of the vulnerability profiles is far from assured, and the merging of bottom-up with top-down information does not work well except where organized by the many NGOs in the region (see, for instance, Singh et al., 2016).

In Kenya, following strong decentralization, the counties are nowadays responsible for disaster risk vulnerability profiling, supported by the drought management authority and line ministries. Significant differences in execution exist between counties. Issues of funding, awareness, weak intergovernmental cooperation (between national and county governments), and capacity gaps (both human and technological) at county level are slowing down the processes. However, it should be taken into consideration that the counties have been given, in a short period of time, a huge number of additional tasks, for which it is impossible to provide trained staff in the short and medium term. Again, NGOs and projects step in a variety of ways, including the distribution and exchange and information.

For both countries, international organizations—for instance, the joint assessments of FAO and WFP, and other large international and national NGOs—have their own systems of vulnerability assessment. Often these assessments are executed only in the wake of a drought. While they are sometimes embedded into national systems, this is not necessarily the case. The reasons for that separation include past bad experiences with government systems (which may have reasons to over- or underreport assistance needs, see above).

(Predrought) risk mitigation

Often, the assessment of vulnerabilities is also used to collect (ideas on) options for reducing them in the future, i.e., resilience-building measures. These can be at the level of individuals, but more often of farms, households, or communities. Wider government-based mechanisms may be added but are designed at higher levels and belong to the wider concept of non-self-reliant resilience. The options for drought mitigation for food security are many and go far beyond water-related issues. Brüntrup and Tsegai (2017) provide a roster for typical sectors and the instruments involved (Table 20.1).

It is important to note that many instruments, particularly agroecological ones, must be carefully fine-tuned to the local context: improved varieties and trees have to fit soil conditions, climate profile, and farm—household system; local infrastructure (ponds, dams, weirs) has to fit local geographical patterns as well as institutional and organizational logics and responsibilities not only for construction but also for maintenance in order to be managed sustainable. This is why it is so important to integrate

Table 20.1 Role of key policy domains/sectors for building up food security, enhancing drought resilience during drought and nondrought times.

Policy domain	Nondrought period	Drought period
Water/landscape	• Landscape/watershed management, water harvesting, and conservation on- and off-farm • Water storage • (Water saving) irrigation • Water contingency planning	• Contingency execution (drinking and livestock first)
Agriculture	• Drought resilience breeding • Cropping system adjustment (new crops) • Fostering livestock markets • Seed (emergency) stocks • Managing pastoralism and crop/live-stock integration	• Irrigation or stop according to drought severity and outlook • Livestock vaccination (as early as possible) and reduction • Protecting key animals, recovery • Seed distribution (recovery)
Finance	• Crop and livestock (weather) insurance • Savings • Cash transfer facilities	• Ease disbursements • Use for emergency cash transfers (private and public)
Social protection	• Establishing social security systems	• Scaling up to drought-affected populations, cash or in kind
Food markets	• Fostering food crop markets (integration, storage, commercial linkages,...) • Establishing food price monitoring systems	• Facilitating commercial food inflows • Situation-sensitive regional food aid
General economic development	• Income diversification • Migration as income diversification measure • Infrastructure (transport, storage, tele-communication, etc.) • Contingency planning	• Infrastructure-building as part of emergency aid and reconstruction (cash/food for work)

Source: Adapted from Brüntrup, M., Tsegai, D., 2017. Drought Adaptation and Resilience in Developing Countries, Briefing Paper 23. German Development Institute/Deutsches Institut für Entwicklungspolitik (DIE), Bonn. https://www.die-gdi.de/briefing-paper/article/drought-adaptation-and-resilience-in-developing-countries/.

drought risk mitigation measures into the existing sectoral and administrative logic and processes, and to have bottom-up participation. On the one hand, sector politics have additional (sometimes overwhelming) nondrought resilience logics which usually prevail over drought logics (see Fig. 20.1 on drought cycle management). On the other hand, many measures only make sense for drought mitigation in a package with other measures (e.g., drought-resilient seeds may require certain inputs, distribution and reserve systems, dams and their water use need regulation for various stages of drought, and opening of borders for food importation at the right time may have to be combined with restoring protection at a later time to help farmers recover).

It goes beyond the scope of this chapter to review in detail recent advancements in each of the sectors in both countries. Duguma et al. (2017) find that great advancements have been in implementing drought mitigation measures, but that a lot is yet to be done, too, both in terms of individual measures and in coordination across the drought cycle (see next chapter, this volume). In Ethiopia, community disaster risk-profiling programs not only describe the local vulnerabilities but also try to analyze the underlying causes of vulnerability and collect propositions for resilience building. However, completion rates have been low, and experience has found that local knowledge has its limits in terms of imagining new, innovative measures. Even more significantly, most proposed measures cost money and need other capacities, both of which are in short supply. In Kenya, the decentralization process and the merging of drought resilience with development planning is considered to be the right approach for the drought-stricken ASAL counties. Again, planning and implementation are far from complete.

Some sectoral highlights of drought mitigation are as follows:

The **Productive Safety Net Programme (PSNP)** in Ethiopia was set up in 2005 as a response to the 2003 drought experience. It provides cash for work for millions of poor people at certain times of the year, while constructing communal assets that build both wealth and resilience to drought, such as new ponds, cisterns, and water harvesting systems to supply drinking water. PSNP is based on systematic poverty assessments and allows (and has been effectively used) for the scaling up of operations in the case of disasters, particularly of droughts. However, the scientific literature shows positive but only limited impact on resilience to severe droughts, and the program is seen as still being unable to make a fundamental positive impact on long-term household-level drought resilience, although it is able to save lives and provide short-term support against asset depletion (Andersson et al., 2011; Béné et al., 2012., Adem et al., 2016). In Kenya, similar schemes on a lower level are operational and are intended to be scaled up.

Natural resource management has been very actively pursued in the Ethiopian Highlands for decades: constructing microbasins and ridges, often combined with tree planting to stabilize ridges, protect soils, and surface water regulation. Many regions in the central and western highlands have profited from large-scale government-donor initiatives, such as the Sustainable Land Management Programme and similar projects; a recent assessment estimates the coverage at 41% of cropland with slopes greater than 8% (Hurni et al., 2015). Farmers face constraints with regard to capacities, incentives, and enabling conditions (Adimassu et al., 2016). The impacts seem to be significantly positive and many, but not very large when measured at household level, are differentiated according to local and site conditions (e.g., Kassie et al., 2010; Nyasimi et al., 2014). The impacts of these often community-based programs must also be analyzed on a larger scale, i.e., watershed level such as in Siraw et al. (2018), where impacts were again positive but small. The ASAL regions had been neglected before, also due to political and security issues, but more recently the programs are extended and modified to some extent to cover them. For instance, the Strengthening Drought Resilience Programme in the Afar region (GIZ, 2013) has a focus on capturing the floodwaters flowing from the highlands into the desert and thereby replenishing groundwater resources. However, this is still in its pilot phase, with technical (weir stability) and institutional (maintenance) challenges still to be addressed in the long run. More generally, it seems that many NRM programs do not take the particularities of ASALs fully into account.

In Kenya, the neglect of the ASAL regions in governmental planning has already been mentioned, although many, particularly NGO, donor-funded projects work on rural development, pastoralism, and drought resilience. Many NGOs prefer to work in isolation from government structures, and this may be one of the reasons why some successful approaches by NGOs have not yet been scaled up in government programs. Other reasons could be possible, such as the difficulty of sectorally organized administrations adopting holistic, cross-sector livelihood support strategies; it may also be as simple as the lack of a systematic assessment of drought resilience projects and programs across the Horn of Africa.

A special mention must be made of the **large dam construction projects**, some of which are often presented as irrigation (and thus drought control) instruments. In Ethiopia, construction costs of the Great Renaissance Dam alone had been projected in 2006 to cost USD 4.8 billion (Getahun, 2014), and there are many more in the making. However, many of these dams are primarily hydroenergy rather than irrigation projects, and it is yet to be seen which utilization will dominate. In any case, there are serious concerns about their effects on the food security of pastoralist and agropastoralists, since irrigation areas in the valley bottoms are exactly those flood recession agriculture and grazing lands that constitute reserves during dry seasons and droughts. In addition, the impact of some dams with international importance such as the Great Renaissance Dam on Sudan and Egypt or the Gibe dam cascade on Kenya is the subject of fierce debate and international dispute (Verhoeven, 2013). Kenya itself has also resumed dam-building projects, many minor ones for storing drinking water, but also the High Grand Falls Dam along the river Tana, which is a multipurpose dam intended also to provide irrigation and seasonal flow control (Gachanja, 2018).

There is a **diversity of household coping mechanisms** that should not be underestimated and could be supported by policies when searching for drought management solutions (compare, e.g., Adams et al., 1998). Personal savings and access to credit to bridge temporary shortages (of fodder, food, cash flow, or income) are more flexible ways to respond to the many needs of the poor, including drought resilience (Rutherford, 2003). Social networks, such as families, friends, and neighbors, provide further personal mechanisms for coping with droughts. Drought insurance and wider crop and livestock insurance are potentially valuable instruments that have received considerable attention but are not yet widely used; their implementation struggles with low willingness to pay and adoption rates by the potential farmer and pastoral clients (Tadesse et al., 2015).

Coordination

Integrated drought strategies require a lot of coordination, harmonization, and negotiation of trade-offs. It is not only about the interaction of sector policies and stakeholders, it is also about temporal alignment and the handing over from one priority area to another, including disaster management, often under time pressure and stress. As has been clearly seen in the two example countries, droughts are highly political, and their consequences can extend well beyond the immediate effects on households. Severe, widespread droughts can undermine politically regimes that cannot manage them well, and even local droughts can undo national integration. A template for the "right" drought coordination is not possible, since it has to work under the conditions of national governance pertaining at the time, be it centralized or decentralized, autocratic or democratic. However, some general lessons emerge from the two study countries.

While both countries have established sophisticated drought coordination mechanisms within the frameworks of wider disaster risk management systems and with the highest level political steering mechanisms, there are still important caveats in implementation. Sectoral ministries often do not fulfill their duty to integrate drought resilience into their long-term activity planning, or are reluctant to coordinate. Multilevel coordination among line ministries is hampered by weak capacities of material, personnel, and other resources. Necessary top-down initiatives for systematic drought cycle management show weaknesses and reluctance to facilitate adoption of bottom-up local opportunities and ideas; people, particularly in structurally weak, historically neglected ASALs, are often unable and unwilling to listen to and communicate with official channels. Relations between central governments and ASAL regions are often complicated by ethnic divides. Development partners, both governmental and nongovernmental, are sometimes blamed by national administrations for not aligning with government priorities and plans. On the other hand, governments are not innocent when it comes to corruption, human rights infringements, and politically motivated mismanagement, including of drought systems, which cause development partners to maintain their own information systems, priority setting and use of approaches, instruments and implementation systems. There is considerable mistrust among stakeholders, which does not ease coordination and harmonization.

Conclusions and policy lessons

This chapter has described how general disaster risk management ideas for droughts and drought cycle management have been adopted in both Ethiopia and Kenya, in line with international frameworks and a wider regional strategy. This is the result of a long history of drought disasters in the region which not only caused a heavy humanitarian toll but has also threatened the political order and long-term development and economic achievements at the national level since the mid-2000s. The countries are different in terms of their level of development, their political systems, and the intensity and pattern of decentralization, all of which are important to drought management. It can be said that Kenya is more advanced in terms of institutional sophistication and inclusion of bottom-up priorities, solutions, and approaches in its drought (and other) policies, while Ethiopia is more advanced in social protection and implementation of natural resource management programs in disadvantaged rural areas (though not in its lowland ASALs). Yet, still a lot is to be done, and the following 10 recommendations would, if implemented, improve policy coherence for drought resilience:

1. **The understanding of what constitutes drought resilience** and priorities between long-term and short-term measures is still not uniform across all stakeholders, with some important negative repercussions. A guiding principle is needed where resilience is initiated at the lowest

possible level (the household) and is progressively opened to resource mobilization at higher-level structures (community, district, region, nation) when increasingly severe drought surpasses individual households' and communities' capacities to withstand drought impacts. This approach respects human dignity, acknowledges individual options and potentials, better protects longer-term development capabilities, and provides more independence from ruthless strategic regime opinions (and their changes). It avoids an overreliance on social protection, which often leads to a reduction in households' own capacities. Wealth accumulation, economic diversification, savings, and credit access are universal resilience builders. On the other hand, too strong an expectation of what can be achieved by self-reliance of poor households and local communities against heavy droughts should not lead to the neglect of top-down emergency readiness. Common understanding can be achieved by learning from the further development of the international frameworks, exchange of the experiences of more advanced economies, frank learning from one's own and from regional experiences, and a clearer exchange between public institutions, the private sector, civil society, and academia. Public awareness campaigns through mass media, schools, informal networks, sectoral education programmes, etc. are needed, too.

2. Based on the need to search for more individual adapted solutions, it follows that **vulnerability assessments should be refined** to include the possibility of adapting continuously to local conditions, specific groups, and changing conditions. They should be stipulated top-down, but must be responsive and adaptive to bottom-up solutions. This can be done through differentiating between subgroups of vulnerable groups (gender, landless youth, people with disabilities) to ensure that interventions benefit the needy, and by using participatory approaches which differentiate between local subgroups such as women, young people, small farmers, and large (agro)pastoralists. It must be further recognized those in ASALs (pastoralists, in particular) face specific threats and options. Cash support allows more flexibility than food aid, which must be restricted to really limited supply situations. Knowledge is a very flexible resource for poor households.

3. It is vital to further promote the **integration of drought risk management approaches into long-term development measures**, especially "no-regret" solutions that prevent and mitigate the impacts of drought, prepare for crises, and respond to them. Many natural resource and livestock management practices, social protection schemes, and measures to increase income and diversification comply with such a double purpose. But trade-offs should also be clearly addressed, for instance, in terms of scarce capacities of administrations, costs of stabilization and risk mitigation versus average income maximization and wealth accumulation, etc. Public and project decisions on the best (mix of) options should be made transparent and revisable. In addition, humanitarian and drought risk management interventions (development measures) should be linked in a way that mutually reinforces the efficiency and effectiveness of each, again while also clearly identifying and negotiating trade-offs. This means, in particular, that humanitarian measures should be planned early on and integrated into development plans, for instance, location, long-term institutional responsibility, and maintenance of local infrastructure created through food or cash for emergency work programs.

4. For **efficient and properly functioning drought EWSs**, effective communication among all relevant stakeholders is critical for vulnerability assessment, preparedness planning, better targeting, and proactive action for emerging droughts. This will require the establishment of a

credible, independent, regional/national platform that consolidates the early warning information from multiple sources. This can be in the form of a consortium of various governments, NGOs, and research institutions with high-profile expertise and reputations. Improved transparency and the provisioning of access to data for all relevant stakeholders would facilitate the process.

5. A **strong and comprehensive connecting institution** is indispensable to allow for multisectoral and multistakeholder communication and coordination, for creating mutual accountability and facilitate interinstitutional and multiactor learning. For this, a coordination unit with a solid authority, clear accountability, and sufficient capacities to carry out its responsibilities should be created, at a very high level within the government hierarchy.

6. Drought knows no geographical or sectoral boundaries, particularly in developing countries with old transboundary linkages and more or less open, uncontrolled, and uncontrollable borders. Drought episodes thus call for strengthened **collaboration between African countries, regional and subregional institutes**, and international organizations in the implementation of drought risk management and implementation plans. In the case of the two case study countries, IGAD is the right level of regional cooperation, but also other neighbors and other African regional organizations should be involved where it makes sense (e.g., Egypt and the Nile Basin Initiative in the case studies). IGAD and other African regional organizations should prioritize and help mobilize resources for cross-border initiatives that enhance cooperation.

7. **Monitoring and evaluation and knowledge management** is vital for effective follow-up, reporting, and documentation of drought resilience efforts and achievements. Therefore, we recommend establishing an independent, strong monitoring and evaluation system under the above proposed coordination unit, which would be responsible for monitoring and evaluation, identifying strengths and weaknesses, and ensuring scale-up of good practices. In addition, mutual accountability among government, nongovernment stakeholders, and development partners should be strengthened through reporting, possibly under common standards. Again, regional organizations could play an important role here.

8. **Emergency funding** is short-term and costly, and becomes more so the later it is initiated. Therefore, development partners and governments should increase funding for anticipatory drought resilience building as opposed to emergency funding. The use of contingency funding should be enhanced to link relief and development and provide easy and quick funding for early action.

9. **Expertise is a critical resource in building drought resilience** within individuals, institutions, and organizations. In poor countries in particular, it is essential to exploit readily available internal expertise and enhance efforts to reduce labor turnover at national level, with a special focus on the subnational level.

10. **Large dams** are an important tool for combating droughts, but they are also very tricky, with potentially very negative consequences (unlike other measures discussed in this chapter) and are thus very controversial. They need particular attention in drought strategies, including careful environmental and social impact assessments, participatory processes, and often crossborder river basin cooperation.

It should be reiterated that, in the long run, income diversification and food market integration are important means of achieving resilience against drought and other natural disasters, and that other sectors affected by droughts (such as energy, health, bioreserves) have their additional complex requirements, lessons, and recommendations. Given its "spider web" character, touching various sectors, stakeholders, and levels of government from local to national and even beyond, drought can be leveraged as a "connector" and can therefore serve as an opportunity for governments to enhance policy coherence, not only in relation to droughts but also other areas.

References

Adams, A.M., Cekan, J., Sauerborn, R., 1998. Towards a conceptual framework of household coping: reflections from rural West Africa. Africa 68 (2), 263−283.

Adem, B.I., Lim, S., Bosi, L., Venton, C.C., 2016. Reality of Resilience: Perspectives of the 2015−16 Drought in Ethiopia. https://reliefweb.int/sites/reliefweb.int/files/resources/51332_resilienceintelethiopiapaperweb.pdf.

Adimassu, Z., Langan, S., Johnston, R., 2016. Understanding determinants of farmers' investments in sustainable land management practices in Ethiopia: review and synthesis. Environment, Development and Sustainability 18 (4), 1005−1023.

Aidt, T.S., Leon, G., 2016. The democratic window of opportunity: evidence from riots in Sub-Saharan Africa. Journal of Conflict Resolution 60 (4), 694−717.

Andersson, C., Mekonnen, A., Stage, J., 2011. Impacts of the Productive Safety Net Program in Ethiopia on livestock and tree holdings of rural households. Journal of Development Economics 94 (1), 119−126.

Batterbury, S., Warren, A., 2001. The African Sahel 25 years after the great drought: assessing progress and moving towards new agendas and approaches. Global Environmental Change 11 (1), 1−8.

Béné, C., Devereux, S., Sabates-Wheeler, R., 2012. Shocks and Social Protection in the Horn of Africa: Analysis from the Productive Safety Net Programme in Ethiopia. IDS Working Paper 395. Institute of Development Studies, Brighton.

Brüntrup, M., Tsegai, D., 2017. Drought Adaptation and Resilience in Developing Countries. Briefing Paper 23. German Development Institute/Deutsches Institut für Entwicklungspolitik (DIE), Bonn. https://www.die-gdi.de/briefing-paper/article/drought-adaptation-and-resilience-in-developing-countries/.

Cabot-Venton, C., Fitzgibbon, C., Shitarek, T., Coulter, L., Dooley, O., 2012. The Economics of Early Response and Resilience: Lessons from Kenya and Ethiopia (Economics of Resilience Final Report). https://www.gov.uk/government/uploads/system/uploads/attachment_data/file/67330/Econ-Ear-Rec-Res-Full-Report_20.pdf.

Cai, X., Shafiee-Jood, M., Apurv, T., Ge, Y., Kokoszka, S., 2017. Key issues in drought preparedness: reflections on experiences and strategies in the United States and selected countries. Water Security 2, 32−42.

Carrao, H., Naumann, G., Barbosa, P., 2016. Mapping global patterns of drought risk: an empirical framework based on sub-national estimates of hazard, exposure and vulnerability. Global Environmental Change 39, 108−124.

Christiaensen, L., Demery, L., Kuhl, J., 2011. The (evolving) role of agriculture in poverty reduction - an empirical perspective. Journal of Development Economics 96 (2), 239−254.

Cohen, M.J., Ferguson, K., Gingerich, T.R., Scribner, S., 2016. Righting the Wrong: Strengthening Local Humanitarian Leadership to Save Lives and Strengthen Communities. https://www.oxfamamerica.org/explore/research-publications/turning-the-humanitarian-system-on-its-head-saving-lives-and-livelihoods-by-strengthening-local-capacity-and-shifting-leadership-to-local-actors/.

de Perez, E.C., van Aalst, M., Choularton, R., van den Hurk, B., Mason, S., Nissan, H., Schwager, S., 2019. From rain to famine: assessing the utility of rainfall observations and seasonal forecasts to anticipate food insecurity in East Africa. Food Security 11 (1), 57−68.

de Waal, A., 2002. Famine Crimes: Politics & the Disaster Relief Industry in Africa. James Currey, Oxford.

de Waal, A., 1991. Evil Days: Thirty Years of War and Famine in Ethiopia. Human Rights Watch, New York & London.

Devereux, S., 1993. Theories of Famine. Harvester Wheatsheaf, New York.

Devereux, S., 2007. The impact of droughts and floods on food security and policy options to alleviate negative effects. Agricultural Economics 37, 47−58.

Drèze, J., Sen, A. (Eds.), 1991. The Political Economy of Hunger: Volume 1: Entitlement and Well-Being. Clarendon Press.

Duguma, M.K., Brüntrup, M., Tsegai, D., 2017. Policy Options for Improving Drought Resilience and its Implication for Food Security. Studies 98. Deutsches Institut für Entwicklungspolitik. https://www.die-gdi.de/uploads/media/Study_98.pdf.

Gachanja, P., 2018. Kenyan Gov't to Construct 57 Dams Across the Country. https://citizentv.co.ke/news/kenyan-govt-to-construct-57-dams-across-the-country-205357/.

Getahun, T., 2014. Renaissance Dam: Is it Worth Building? https://addisfortune.net/columns/renaissance-dam-is-it-worth-building/.

Getu, M., Mulinge, M.M. (Eds.), 2013. Impacts of Climate Change and Variability on Pastoralist Women in Subsaharan Africa. African Books Collective.

GIZ, 2013. Measures to Fight Drought in the Lowlands of Ethiopia. https://www.giz.de/en/worldwide/23119.html.

Hillier, D., Dempsey, B., 2012. A dangerous delay: the cost of late response to early warnings in the 2011 drought in the Horn of Africa. Oxfam and Save the Children, Oxford, UK.

Hurni, K., Zeleke, G., Kassie, M., Tegegne, B., Kassawmar, T., et al., 2015. Economics of Land Degradation (ELD) Ethiopia Case Study: Soil Degradation and Sustainable Land Management in the Rainfed Agricultural Areas of Ethiopia: An Assessment of the Economic Implications. http://www.eld-initiative.org/fileadmin/pdf/ELD-ethiopia_i_06_72dpi-D.pdf.

IGAD, 2013. IGAD Strategy for the Drought Disaster Resilience and Sustainability Initiative - IDDRSI Strategy. IGAD, Djibouti.

Jarosz, L., 2011. Defining world hunger: scale and neoliberal ideology in international food security policy discourse. Food, Culture & Society 14 (1), 117−139.

Kandji, S.T., 2006. Drought in Kenya: Climatic, Economic and Socio-Political Factors. http://www.worldagroforestry.org/downloads/Publications/PDFS/NL06291.pdf.

Kassie, M., Zikhali, P., Pender, J., Köhlin, G., 2010. The economics of sustainable land management practices in the Ethiopian highlands. Journal of Agricultural Economics 61 (3), 605−627.

Kenyan Ministry of Devolution and Planning, 2015. Ending Drought Emergencies Common Programme Framework. Government of Kenya, Nairobi.

Mateche, I., 2011. The Cycle of Drought in Kenya a Looming Humanitarian Crisis. https://issafrica.org/iss-today/the-cycle-of-drought-in-kenya-a-looming-humanitarian-crisis.

Maystadt, J.F., Ecker, O., 2014. Extreme weather and civil war: does drought fuel conflict in Somalia through livestock price shocks? American Journal of Agricultural Economics 96 (4), 1157−1182.

Mbithi, P.M., Wisner, B., 1973. Drought and famine in Kenya: magnitude and attempted solutions. Journal of Eastern African Research and Development 3 (2), 113−143.

McPeak, J.G., Barrett, C.B., 2001. Differential risk exposure and stochastic poverty traps among East African pastoralists. American Journal of Agricultural Economics 83 (3), 674−679.

Mengisteab, K., 2013. The Horn of Africa. John Wiley & Sons.

MOA (Ministry of Agriculture), 2013. Disaster Risk Management Strategic Programme and Investment Framework. Disaster Risk Management and Food Security Sector. FDRE, Addis Ababa, Ethiopia.

MOARD (Ethopian Ministry of Agriculture and Rural Development), 2010. Woreda Disaster Risk Profiling: Methodological Note. Government of Ethiopia, Disaster Risk Management and Food Security Sector (DRMFSS), Addis Ababa. http://profile.dppc.gov.et/Default.aspx.

Ncube, B., Shikwambana, S., 2016. Coping and Adaptation Strategies for Agricultural Water Use during Drought Periods. Cape Peninsula University of Technology. http://digitalknowledge.cput.ac.za/handle/11189/6350.

Nilsson, M., Zamparutti, T., Petersen, J.E., Nykvist, B., Rudberg, P., McGuinn, J., 2012. Understanding policy coherence: analytical framework and examples of sector −environment policy interactions in the EU. Environmental Policy of the Government 22, 395−423.

Nyasimi, M., Amwata, D., Hove, L., Kinyangi, J., Wamukoya, G., 2014. Evidence of Impact: Climate-Smart Agriculture in Africa. https://publications.cta.int/media/publications/downloads/1815_PDF_7ey3GWo.pdf.

Obokata, R., Veronis, L., McLeman, R., 2014. Empirical research on international environmental migration: a systematic review. Population and Environment 36 (1), 111−135.

Pankhurst, R.R.K., 1961. An Introduction to the Economic History of Ethiopia. Lalibela House, London.

Pankhurst, R.R.K., 1968. Economic History of Ethiopia. Haile Selassie I University Press, Addis Ababa.

Pflanz, M., 2011. UN declares first famine in Africa for three decades as US withholds aid. Daily Telegraph, 20 July. London.

Raleigh, C., Jordan, L., Salehyan, I., 2008. Assessing the impact of climate change on migration and conflict. In: In Paper Commissioned by the World Bank Group for the Social Dimensions of Climate Change Workshop, Washington, DC, pp. 5−6.

Rutherford, S., 2003. Money talks: conversations with poor households in Bangladesh about managing money. Journal of Microfinance/ESR Review 5 (2), 43−75.

Sen, A., 1982. Poverty and Famines: An Essay on Entitlement and Deprivation. Oxford University Press.

Singh, R., Worku, M., Bogale, S., Cullis, A., et al., 2016. Reality of Resilience: Perspectives of the 2015−16 Drought in Ethiopia. BRACED Issue No. 6. http://www.braced.org/contentAsset/raw-data/18256c98-2a10-4586-9317-17a68b45c1a7/attachmentFile.

Siraw, Z., Bewket, W., Adnew Degefu, M., 2018. Assessment of livelihood benefits of community-based watershed development in northwestern highlands of Ethiopia. International Journal of River Basin Management. https://doi.org/10.1080/15715124.2018.1505733.

Tadesse, M.A., Shiferaw, B.A., Erenstein, O., 2015. Weather index insurance for managing drought risk in smallholder agriculture: lessons and policy implications for sub-Saharan Africa. Agricultural and Food Economics 3 (1), 26.

Tadesse, T., 2016. Strategic Framework for Drought Management and Enhancing Resilience in Africa White Paper for African Drought Conference (Draft), Windhoek Country Club and Resort, Windhoek, Namibia, 15−19th August 2016.

Tschirley, D.L., Jayne, T.S., 2010. Exploring the logic behind southern Africa's food crises. World Development 38 (1), 76−87.

UN, 2015. Global Assessment Report on Disaster Risk Reduction. United Nations. http://www.preventionweb.net/english/hyogo/gar/2015/en/gar-pdf/GAR2015_EN.pdf.

UNISDR (United Nations Office for Disaster Risk Reduction), 2006. Hyogo Framework for Action 2005-2015: Building the Resilience of Nations and Communities to Disasters. Geneva.

UNISDR, 2015. Sendai Framework for Disaster Risk Reduction 2015-2030. Geneva.

Van Loon, A.F., Gleeson, T., Clark, J., Van Dijk, A.I., et al., 2016. Drought in the anthropocene. Nature Geoscience 9 (2), 89.

Venton, C.C., Fitzgibbon, C., Shitarek, T., Coulter, L., Dooley, O., 2012. The economics of early response and disaster resilience: lessons from Kenya and Ethiopia, *Final report*. http://prime-ethiopia.org/wp-content/uploads/2015/03/Econ-Ear-Rec-Res-Full-Report_20.pdf.

Verhoeven, H., 2013. The politics of African energy development: Ethiopia's hydro-agricultural state-building strategy and clashing paradigms of water security. Philosophical Transactions of the Royal Society A: Mathematical, Physical & Engineering Sciences 371 (2002). https://doi.org/10.1098/rsta.2012.0411.

Wilhite, D.A., Hayes, M.J., Knutson, C., Smith, K.H., 2000. Planning for drought: moving from crisis to risk management 1. JAWRA Journal of the American Water Resources Association 36 (4), 697−710.

WMO (World Meteorological Organization) and GWP (Global Water Partnership), 2014. National Drought Management Policy Guidelines: A Template for Action (D.A. Wilhite). Integrated Drought Management Programme (IDMP) Tools and Guidelines Series 1. WMO, Geneva, Switzerland and GWP, Stockholm, Sweden.

Drought adaptation when irrigation is not an option: the case of Lincoln Co., Colorado, USA

21

Robert McLeman

Geography & Environmental Studies, Wilfrid Laurier University, Waterloo, Ontario, Canada

Introduction

Although the present volume focuses on drought and water scarcity issues in developing regions, drought presents very real challenges to developed nations as well, and the adaptation lessons learned in one may be relevant for the other. As developing countries become more developed, the challenges presented by drought do not disappear, but the potential impacts and range of adaptation responses may evolve. One of the more common drought adaptation trajectories that accompany development is greater utilization of irrigation in agriculture, be it from damming of stream and rivers or drilling wells to tap groundwater. But not all dryland agricultural areas have large reserves of surface or groundwater; what happens then? To explore this question, we include in this book a study of drought impacts and adaptation options in a dryland agricultural region in the United States where irrigation is not an option. Its circumstances are analogous to those that many of the countries and regions already canvassed in preceding chapters may one day find themselves in.

This chapter concerns Lincoln County, Colorado, an agricultural county on the central Great Plains, roughly 100 km southeast of the Denver metropolitan area (Fig. 21.1). The Great Plains are infamously known for the Dust Bowl of the 1930s, a decade when hundreds of thousands of farm families fled extreme drought conditions during the worst years of the Great Depression (Gregory, 1989). This semi-arid region is characterized by low levels of annual precipitation and hot windy summers that are punctuated periodically by severe droughts. Yet despite their dryness, the Plains are very productive in terms of livestock (particularly beef cattle) and agricultural crops (principally wheat, hay, corn, and, in the southern areas, cotton). This productivity is made possible by several key adaptations that were implemented in the decades that followed the Dust Bowl. These include the widespread adoption of irrigation, especially in areas underlain by the massive Ogallala Aquifer (Fig. 21.2); mechanization of production; development of higher yielding varieties of crops; widespread adoption of herbicides and pesticides; and changes in cultivation practices, such as planting wheat in the fall and leaving the land fallow in midsummer (Parton et al., 2007).

A key distinction in agricultural productivity across the Great Plains region is between areas that have sufficient water resources for irrigation, and areas that do not, where agriculture is entirely dependent on local precipitation. Lincoln County is one of many that fit this latter description. There are no large underground water deposits of glacially derived water on the scale of the Ogallala Aquifer,

Drought Challenges. https://doi.org/10.1016/B978-0-12-814820-4.00021-3

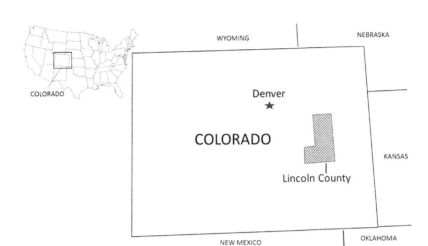

FIGURE 21.1

Map showing location of Lincoln County, Colorado (by author).

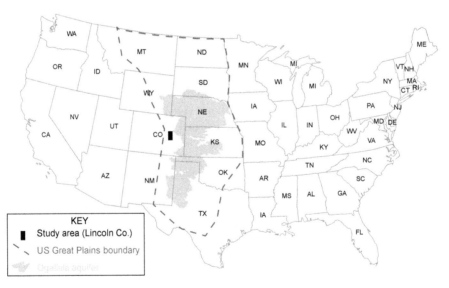

FIGURE 21.2

Location of Ogallala Aquifer (base layer from US Geological Survey).

nor are there perennial rivers or streams of any significance in Lincoln County. Most of the freshwater available to residents is found in shallow alluvial aquifers that are recharged from precipitation that infiltrates from the surface. It is a truly dry area where residents must practice wise and frugal use of their water resources, even as they try to adjust and adapt to larger social and economic pressures that affect lives and livelihoods. Many of the drought adaptation strategies discussed in previous chapters—environmental monitoring, land use planning, insurance, government-led agricultural adjustments, economic diversification, and so forth—have been implemented in Lincoln County with varying degrees of success. Even so, the balance between water and well-being remains a tenuous one, and the county's current experience and future pathways may be similar to those of many peri-urban dryland areas in less developed regions in the not-so-distant future.

The data reported in this chapter were gathered as part of a larger project funded by the Social Science and Humanities Research Council of Canada to model and document the influence of climate and its interaction with other factors on population patterns on the North American Great Plains over the last 50 years. The methods include geospatial regression and "hotspot" modeling of environmental, economic, and social processes in the region using publicly available data from censuses and government agencies, supported by targeted field visits to selected locations to collect secondary data and conduct interviews with local experts and stakeholders. Lincoln County was selected for field visits because it has exhibited population patterns that suggest droughts and extreme heat may have had an influence on population change since the 1970s, primarily by constraining population growth relative to other counties in the central Great Plains. In this chapter, data derived from secondary sources are cited and referenced; data obtained through interviews with local residents are indicated via footnotes. Details now follow.

Lincoln County: location and physical description

Lincoln County, Colorado, is approximately 2500 square miles (6500 square kilometers) in size and is located halfway between the foothills of the Rocky Mountains (known locally as the "Front Range") and Colorado's eastern border with the state of Kansas. Situated in the rain shadow of the mountains, Lincoln County receives an average of only 16 in. (41 cm) of precipitation per year, most of which falls in the summer months (National Weather Service, 2019). Hot temperatures and frequently windy conditions create high rates of evapotranspiration during the summer growing season. Violent thunderstorms that bring hail and/or tornados are common from late spring through early autumn. From late November through to early March, average daily temperatures are at or below freezing for extended periods.

Native vegetation consists primarily of shortgrass prairie—a mix of grasses interspersed with drought-tolerant shrubs, yuccas, and occasional cacti. There are no perennial streams or rivers of any note, although small pockets of wetlands are scattered across the county. Groundwater is found in fragmented pockets in sand and gravel deposits, and only a handful of farms in the very southeasternmost part of the county are able to irrigate. The predominant soils are a mix of loams (silty, sandy, and clay) that are generally quite fertile, with soil moisture being the key limit on their productivity. As elsewhere in the Great Plains, these soils are highly susceptible to wind erosion when vegetative cover is disturbed, an important land management consideration for farmers. Roughly two-thirds of land in Lincoln County is used for grazing, primarily cattle, and one quarter is used for growing nonirrigated field crops, primarily grains (USDA, 2003). Such land use patterns reflect the general scarcity of water in the county.

Brief history of population and land use in the study region

At the time of European contact, the Indigenous people of eastern Colorado—Arapaho, Cheyenne Comanche, and Kiowa—pursued mobile livelihoods centered on the harvesting of massive herds of wild bison that were eradicated by non-Indigenous settlers in the 19th century (Abbott et al., 2013). Eastern Colorado held little attraction for the first Europeans, being primarily an area traveled through by wagonloads of people destined for gold mining sites in the western mountains or for settlements on the Pacific coast. In the 1860s, the US government initiated a concerted program to remove Indigenous people from Colorado, beginning with an infamous 1864 massacre by the US Army of Cheyenne women and children at Sand Creek, Colorado, just south of present-day Lincoln County. A decade-long series of violent conflicts ensued between the army and Indigenous peoples of eastern Colorado until the latter were defeated, with survivors being forcibly relocated to reservation lands in the states of Wyoming and Oklahoma. In the remaining decades of the 19th century, land in Lincoln County was used primarily as rangeland for grazing cattle, a use that declined with the expansion of railways in the Western United States (which connected previously inaccessible grazing areas to markets) and a serious economic recession in 1893 that caused the price of livestock to tumble (Mehls, 1984).

In the early 1900s, large areas of Lincoln County were settled by "homesteaders"—families who came from eastern states to establish farms on 160-acre allotments of former grazing land, encouraged by government policies and railroad companies (Durrell, 1974). Lincoln County's population grew rapidly into the 1920s (Fig. 21.3). During the Great Depression of the 1930s, wide swathes of the Great Plains, including eastern Colorado, were hit by extended periods of severe drought that led to wide-scale outmigration and abandonment of farms, especially in dryer, marginally productive areas (Gregory, 1989). Between 1930 and 1940, Lincoln County's population fell by one quarter. As agricultural production became increasingly mechanized and returned to a greater emphasis on livestock production after World War II, the county entered a multidecadal period of slow population decline. The number of farms fell in the 1950s and 1960s, stabilizing at an average of between 450 and 500

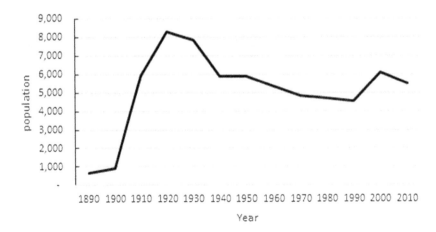

FIGURE 21.3

Population of Lincoln County since 1890 (US Census counts).

farms in total, while the size of remaining farms grew, stabilizing at an average size of between 2500 and 3000 acres (USDA, 2012). The county's population decline was reversed in the 1990s when a penitentiary was constructed near the town of Limon, which now holds nearly 1000 prisoners[1] and created employment opportunities that attracted new residents.

Current demographic patterns

Lincoln County has a current population of just over 5500 residents (all statistics this section from US Census Bureau (2019)). The county's largest population center is the town of Limon, home to just under 2000 people. The county has an uneven ratio of men to women (58:42), reflecting the male prisoner population at the Limon penitentiary and, to a lesser extent, the types of nonfarm employment opportunities available in the county (see below). There are few visible minorities; over 90% of the population is white. Roughly 18% of the population is over the age of 65 years and 20% is under the age of 18 years. There are approximately 2500 housing units in the county, of which 70% are occupied by the owner. The average value of owner-occupied housing units is US$132,000 (for comparison, the average value of housing in Denver is $323,000). There are only 1500 households in the county, implying a high vacancy rate, which was observed firsthand during site visits. Vacancy rates are most visible in smaller villages and hamlets, from which outmigration, especially of young people, has been greatest. Many of these smaller settlements have few, if any businesses or services anymore, and there is concern among remaining residents that abandoned homes may attract transient people from Denver, along with their attendant social problems.

Population patterns in Lincoln County are, for the moment, only lightly influenced by the urban sprawl of Denver and its rapidly growing population of 700,000 people. Denver International Airport, on the easternmost margins of the city, is almost exactly 100k from the westernmost boundary of Lincoln County. A major highway—the four-lane interstate I-70—runs directly from Denver to the Kansas border and beyond, cutting across northern Lincoln County. As Denver continues to sprawl eastwards, it is likely that commuters may look to Lincoln County as a potential option for residence and take advantage of I-70. However, at the moment, the available housing stock in Lincoln County is older and of generally poorer quality in comparison with that which is available closer to Denver. The current average travel time to work of residents of Lincoln County is just over 17 min, implying that most residents work within the county. By comparison, residents of Elbert County, the next county to the west, have an average travel time to work of 40 min, reflecting a large number of residents who commute to Denver.

Local economy and employment in Lincoln County

Agriculture is the predominant land use in Lincoln County. Roughly 1.475 million acres (roughly 600,000 ha) of county land is in agriculture, with 60% of it in grazing and the remaining 40% in crops, mostly unirrigated (all statistics in this section is from USDA (2012) unless otherwise indicated). The

---

[1]For purposes of the US Census, a prisoner's place of residence is counted as being the county that hosts the prison in which he or she is incarcerated.

total value of the county's agricultural production in 2012 (year of the most recent agricultural census) was approximately US$75 million, with $38.5 million represented by grain crops (mostly wheat) and $35 million from beef cattle sales. A common cropping practice of county farmers is to plant wheat in the fall, harvest early the following summer, and then leave the field fallow until fall or the following year. Other crops such as sorghum, commercial birdseed, and sunflower are sometimes mixed into the rotation. Many farmers who grow crops also maintain herds of beef cattle. Although cannabis is a recently legalized and lucrative crop in Colorado, Lincoln County prohibits its production and sale.

Despite having a large spatial area (the 10th largest of 63 counties in Colorado), Lincoln County ranks 22nd among Colorado counties in terms of value of agricultural production. By way of comparison, smaller Cheyenne County to the east, which has one-third less farmland than does Lincoln County, ranks 20th, thanks to greater crop yields made possible by the availability of water for irrigation. Most farms in Lincoln County are family owned and operated, with the majority being more than 1000 acres in size (Fig. 21.4). Local farmers describe continuous pressure to increase the size of their land due to rising production costs and falling profit margins; any farm smaller than 5000 acres in size is seen as having poor long-term prospects.

Of the 464 farms surveyed in Lincoln County in the 2012 agricultural census, 266 operators reported their principal occupation as being farming; the remainder reported a different occupation. This is reflective of the distribution shown in Fig. 21.4 and of reports from local interviewees in Lincoln County, who advised that a farm must be thousands of acres in size to be the principal source of income to the family operating it. Although agriculture is a significant contributor to the overall dollar value of the economy of Lincoln County, it is not significant as a source of employment. In 2015, only 71 people were employed in wage labor jobs in agriculture in Lincoln County (US Census, 2018), representing less than 4% of total employment in the county (Fig. 21.5). The largest sources of nonfarm employment are public sector jobs in administration, health care and education (cumulatively representing 50% of all jobs in the county), and low-paying jobs in retail service, accommodation, and

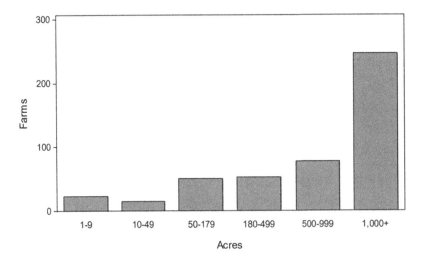

FIGURE 21.4

Number of farms, by size, in Lincoln County (USDA, 2012).

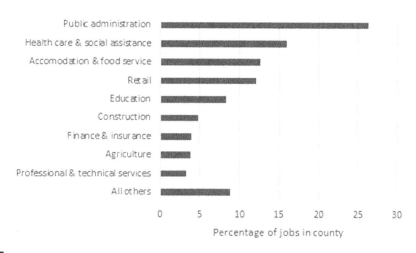

FIGURE 21.5

Main categories of employment in Lincoln County (US Census, 2018).

food services (25% of all jobs). More than half of all jobs are held by women (54.5%). Approximately 60% of jobs in the county pay less than US$40,000 annually, well below the average statewide annual wage of US$54,000 (US Census, 2018; BLS, 2017).

Most private sector jobs and nonfarm economic activity in Lincoln County are centered on the town of Limon, which sits at the junction of an east—west running Interstate I-70, a major north—south highway that runs from Texas to the Canadian border, and three regionally important secondary highways. Its location has attracted a large number of hotels, fast-food restaurants, gas stations, and similar services that cater to travelers and to long-haul truck drivers. The prison at Limon is also a significant source of local employment. The county seat of Hugo, which has a population of about 750, is home to municipal offices and a hospital, which are the key employers in that town. It is also the site of the County Fairgrounds, where a culturally and socially important agricultural fair is held every August. There are secondary schools in Limon and Hugo, but there are no college or university campuses in the county. Most young people who wish to pursue post-secondary education leave the county, many of whom do not return given the lack of employment that requires advanced education.

Beginning in 2012, wind turbines have been erected across extensive areas in northern Lincoln County, providing royalty payments to landowners and additional revenues for the county government. There are patches of oil and gas development across the county as well, again providing royalty payments to landowners and governments. The scale of oil and gas development is, however, much smaller than in other Great Plains states such as Oklahoma and Texas. Both wind and oil development have created construction and technical jobs, although in many cases these are for fixed time periods and are filled by people from outside the county.

How water scarcity limits future development

Water scarcity creates ongoing restrictions and barriers to agricultural productivity and to nonfarming economic development in Lincoln County. A key limit on agricultural productivity is that farmers are limited in terms of the choice of crops they can grow. Crops ideally suited for dryland areas, such as sorghum, millet, and commercial birdseed, have relatively low market values. Wheat, which has a higher market value, is the predominant crop choice, and new varieties are increasingly more drought tolerant and less moisture dependent than in the past. Nonetheless, wheat yields and dollar/acre productivity are simply much lower on nonirrigated land than on irrigated land. Farmers farther east in Colorado and in neighboring western Kansas are able to irrigate extensively, and diversify into higher dollar-value crops like corn (maize) and potatoes. Lincoln County farmers have experimented with sunflower crops, but report finding that these degrade the soil quite rapidly. The USDA (2017) reports that irrigated cropland in Colorado is on average nearly four times more valuable than nonirrigated land (US$4600/acre vs. $1250/acre in 2016). Meanwhile, the costs of production—equipment, fossil fuels, crop inputs, and so forth—are nearly the same on irrigated and nonirrigated farms, meaning that the only way a dryland farmer can improve his or her net income is to expand the size of the farm.

For cattle operations, availability of water and pasture is an ongoing operational challenge. Cattle must have water on a daily basis or they weaken. When grazing is poor due to dry conditions, supplemental hay must be purchased, reducing profit margins. Further, the economic value of the water rights is going up, with ranchers increasingly competing with other users for water rights (see Box 21.1). Although large numbers of cattle are grazed in Lincoln County, there are no large feedlots for fattening cattle prior to slaughter (known as concentrated animal feeding operations (CAFOs)), and there are no meat slaughtering facilities in the county. Such operations are common in other parts of the Great Plains, but limited water supplies likely preclude their development in Lincoln County. The last such operation, a small CAFO south of Limon, closed in the 1990s due to the costs of bringing in corn from elsewhere.

Water scarcity also limits urban development in Lincoln County.[2] With relatively cheap land available for development, proximity to Denver, favorable tax rates, and excellent highway access, it might be expected there would be more warehousing, light manufacturing, and similar businesses setting up in Lincoln County. However, limited water supplies and concerns about contaminating the shallow alluvial groundwater deposits preclude most such development. As it is, the prison and the large concentration of hotels cause Limon to have much greater per capita water demands than comparable towns of its size. Local economic development agencies have been focusing on attracting smaller scale business, such as modest-sized warehouses and services for the trucking industry. The resulting job growth is, however, incremental and concentrated at the Limon highway junction. In Lincoln County's smaller communities, independently owned retail and service businesses have been steadily closing, the result of a declining agricultural workforce and an aging population that has modest spending power. This creates a feedback loop where young people leave the smaller communities in search of better opportunities elsewhere, further shrinking demand for small town goods and services. Anecdotal reports were received of some rural household wells in the county having run dry in recent drought years, but the author was not able to confirm this directly.

---

[2]Data in this paragraph were obtained from interviews with town managers and economic development staff.

Box 21.1 Water rights in Colorado

In Colorado, 85% of water use is for agricultural purposes, with the remainder used by towns, cities, and commercial enterprises (State of Colorado, 2012; Colorado Division of Water Resources, 2019). Water rights in Colorado are complicated, subject to a considerable array of laws and regulations. A common principle in many other jurisdictions, known as "riparian law," is that a landowner has the right to make use of water that is on, under, or adjacent to his or her land. This does not apply in Colorado. Instead, the state developed what is widely known as "the Colorado doctrine"; namely, that all water, whether on the surface or underground, is a publicly owned resource. Private individuals and companies can hold a transferable right to use some portion of this public resource, but the state of Colorado ultimately owns and controls all water, regardless of where it is found. Water rights are therefore separate from landowners' rights. A water rights holder is legally able to build or do work on someone else's land if the purpose is to extract or transport water. A water rights holder may also use existing streams and rivers to transport or store water to which he or she has the right of use. If a stream flows across private land, the landowner may not draw water from it unless he or she also owns the water right. Landowners are also not allowed to drill wells to access groundwater without first obtaining a permit, the only exceptions being wells for household use and shallow wells to water grazing animals. Water rights must be used for a beneficial purpose; a rights holder cannot hoard water. If water is not used for a period of 10 years, the rights are deemed to have been abandoned, and new users may step in to acquire those rights.

The state of Colorado is organized into water management divisions, each of which has a "bank" where water rights holders who are unable to make beneficial use of their complete allotment may deposit their rights without losing them (they may then be temporarily allocated to other users). In times of water shortage, the principle of "prior appropriation" applies. That is, when two or more rights holders are in competition for water and there is not enough to go around, the holder of the water right who was issued first takes priority (this is also known as "first in time, first in right"). New users can obtain a new water right only if they can demonstrate there is unallocated water in the area in which they intend to operate; if not, they must purchase or lease an existing water right from another holder. So, for example, someone wishing to develop a new tract of housing in an area previously used for farming must acquire sufficient water rights to transfer to the purchasers of the homes, before they can be occupied. This has created a phenomenon known locally as "buy and dry," whereby municipalities and developers will purchase a farm or ranch simply to obtain its water rights. The land may be resold but is no longer useful for farming unless the new owner can acquire water rights by another means. The town of Limon has taken this approach, having acquired sufficient water rights from surrounding ranches to support a doubling or tripling of current water use. Limon is currently unable to use all these water rights, and the excess is leased out to ranchers.

Impacts of droughts and extreme events

In addition to the general conditions of dryness, Lincoln County also often experiences wind and dust storms, violent thunderstorms that produce hail and tornados, and periodic droughts. Although the previously described 1930s "Dust Bowl" droughts live large in the popular imagination, older farmers recall extremely severe droughts hitting the county in the 1950s and 1970s as well. More recently, 2010 and 2012 were drought years, and 2014, 2016, and 2018 were relatively dry years, with the intervening years being wetter than average (National Weather Service, 2019). 2018 was a particularly hit-or-miss year; severe summer thunderstorms brought sufficient rain to farms in some parts of the county, while others experienced hail that damaged crops and buildings, and still others were missed entirely by the storms and were left dryer than usual.

The most common impact of recent droughts on farmers has been crop yields below what is hoped for and well below farmers with irrigated land.[3] This in turn leaves crop farmers with little financial liquidity or working capital, often forcing them to take out new loans or draw more heavily on existing lines of credit. Although many county farmers participate in various crop insurance schemes, the payouts from crop insurance decline when yields are depressed over the course of multiple years, for future payout amounts are determined by recent past performance. Smaller operations and people who are relatively new to farming (and therefore have greater debts, higher net costs, and lower profit margins) struggle most during droughts. The options available to drought-stricken farmers experiencing financial trouble are to:

- postpone maintenance on buildings and equipment;
- ask lenders for new loans or to restructure existing debts (i.e., defer repayment and extend the length of the loan);
- ask someone (typically a relative or friend) to cosign for loans, and therefore take on part of the associated default risk;
- sell off some land;
- seek employment off the farm;
- exit farming altogether.

There is also a range of federal government support programs for crop farmers, including ones that incentivize dryland farmers to switch from growing grain crops to hay and fodder. All of the above adaptation options, most of which contribute to the trend toward there being fewer, larger farms, are currently being implemented by county farmers.

Droughts also result in there being less grass available for cattle grazing and, in some cases, ponds for watering animals may run dry. The principal short-term adaptation is to buy feed and/or truck in water, both of which increase significantly the cost of production and rapidly make a herd unprofitable. During the 2012 drought, many county livestock operators sold off large numbers of cattle. In the dry year of 2018, there was a significant reduction in the number of yearling cattle (calves between 1 and 2 years old) being kept on county farms, as operators became more strategic about herd size. Cattle operations have historically tended to be harder hit than crop farmers by droughts because there were fewer government programs to assist them. The last decade has seen an expansion of government programs to assist cattle operators. In 2014, the USDA's Livestock Forage Program provided drought-stricken livestock operators in the county with $25 per month per animal for 3 months to purchase supplemental feed. The federal government has also increased the amount of public pasture land available for grazing in eastern Colorado. Cattle operators for whom these adaptations are inadequate have few options but to exit the business entirely.

The impacts of droughts on towns and villages vary from one place to another. As noted earlier, Limon has more water rights than currently needed, and water use restrictions have therefore not been needed. By contrast, the county seat of Hugo is in a separate irrigation district from Limon, one that is linked with counties to the south that are within the headwaters of the Arkansas River system. The Arkansas, which flows into Kansas, has struggled with low flows in recent years, with Kansas

---

[3]Data in this section were collected from interviews with local farmers, local government officials, and loans officers at county financial institutions specializing in lending to farmers and ranchers.

farmers blaming irrigation withdrawals by farmers in southeastern Colorado. As a result, this irrigation district now has greater restrictions on water use, especially during times of drought.

Local sales tax revenues are an important source of revenue for towns and villages in Lincoln County, and these generally decline during droughts, as residents tend to spend less. An exception is Limon, where most businesses are not directly linked to agriculture, and a large portion of retail sales are made to travelers from outside the county. By contrast, smaller towns where businesses cater to local residents have seen revenues drop by up to one quarter in dry years. In response, affected towns defer major capital expenditures and reduce the amount of road work and municipal maintenance. Local charitable and nonprofit organizations also experience declining contributions during droughts and must scale back their activities accordingly. Drought impacts on county government revenues in recent years have been offset to a certain extent by the coinciding expansion in wind energy, oil, and natural gas development.

Extreme weather events present another set of climate-related risks to residents. Lincoln County is prone to violent thunderstorms from spring through early autumn, often accompanied by hail and tornados. In 1990, much of the central business district of Limon was destroyed by a direct hit from a tornado, which injured 14 people. The town received significant financial assistance from the state government and from private donors. When businesses were rebuilt, they tended to relocate closer to the interstate highway access points, leaving much of the historic town center empty and vacated. Hailstorms can cause considerable damage to crops, buildings, and vehicles; rarely a summer goes by without at least one such event in Lincoln County. Being highly localized in nature, it is often the case that one farm or neighborhood in a town experiences severe hail damage while the next is unscathed.

Future challenges

Lincoln County will experience a range of environmental, social, and economic challenges in coming decades that will test the adaptive capacity of residents and local governments. The US National Climate Assessment (Melillo et al., 2014) projects that, in the absence of significant global greenhouse gas emission reductions, temperatures will rise across the central Great Plains, leading to increased rates of evapotranspiration, necessitating greater rates of irrigation (of which Lincoln County has little) and reduction of livestock production on dry and marginal lands (of which Lincoln County has lots). Increased heat and higher maximum temperatures would also lead to more intense summer thunderstorms, exacerbating wind, hail, and tornado risks. In such a scenario, crop yields in Lincoln County will decline significantly and inevitably result in greater conversion of current cropland to grazing, fewer farms, abandonment of dryer pastures, fewer livestock operations, and more people exiting agriculture altogether. This would in turn lead to greater depopulation of rural areas and smaller towns and villages. Greater demands would be placed on crop insurance programs, and without greater financial support from higher levels of government, Lincoln County's agricultural economy will dwindle.

Climate-related depopulation could be offset by influxes of new residents who commute to Denver and Colorado Springs for work, especially if land values in the "Front Range" continue to grow as they have in recent years (Markus, 2018). An influx of younger families is needed to provide long-term stability for the county's population; however, few commuters would settle in more rural areas, and most growth would be expected to take place in and around Limon. Any such influx would require development of new housing stock, which would in turn generate greater competition with farmers for water rights. Meanwhile, rural populations in the county will continue to age, requiring more seniors' accommodation than currently exists, and for which there is an insufficient tax base to develop more. Although there is little documentation, there is anecdotal evidence of growing opioid drug use in the county by existing residents and by recent arrivals from Denver. The overall long-term socioeconomic picture is one of greater socioeconomic inequity between haves and have-nots within communities and greater economic separation between Limon and the rest of the county.

Current adaptive capacity at household, community, and local government levels is adequate (barely) to meet existing climatic and socioeconomic challenges in the Limon area, and to a lesser extent in rural parts of the county. However, even marginal increases in the intensity of climatic risks or small economic downturns in the regional or national economy could make for considerable trouble for residents of Lincoln County, for there are few (if any) opportunities to increase economic prosperity and adaptive capacity that are not already being pursued.

Conclusion: relevance for developing regions

The climatic, demographic, and socioeconomic trajectories of Lincoln County are similar to those emerging in many less developed countries. Around the world, dryland areas are seeing growing outflows of young people, consolidation of farming and grazing lands into larger holdings owned by fewer individuals, and greater competition for scarce water resources. Water management regulations and procedures are central to socioeconomic development in Colorado. The Colorado Doctrine (that water is a shared resource owned by the people and administered by the state for the common good) is a good starting point for successful water management in a hotter, dryer future and for dryland regions is much preferable to riparian law. The fact that Colorado has a highly regulated, comprehensive, and strongly enforced system for administering water use rights is essential for a dryland area and is something to be emulated in countries where surface and groundwater resources are loosely defined and administered. However, a key drawback is that most water is already allocated, and water rights created in the past may persist in perpetuity. This creates considerable inequity in water access, amplifying the dollar-value costs of water and potentially impeding more compelling uses that emerge over time. To get a water right is becoming increasingly difficult for individuals, farmers, and small businesses, and pits rural communities against urban centers.

A first step in building drought adaptation in less developed countries is therefore to develop comprehensive water management systems that treat water as a public good and ensure that the allocation of water is fair, equitable, and managed with a view to maximize the benefits of the resource for all, and not just a wealthy or fortunate few. This is particularly important for areas that are consistently water scarce. A second step is to develop and implement crop insurance systems along with programs that provide short-term assistance to farmers and livestock operators during times of drought. These have so far been successful in helping many farmers in Lincoln County get through recent droughts. However, there are limits to their utility in keeping farmers and ranchers financially solvent, and these should be seen as components of a wider drought adaptation strategy and not a single solution. A third step is increased flexibility in agricultural practices. Farmers and ranchers must continually adjust crop choices and animal stocking rates as soil moisture and grazing conditions vary from one year to the next and, as the climate changes, will in areas that lack irrigation water need to evolve from food crops to fodder, from crop farming to livestock rearing, and reducing stocking rates on marginal lands.

However, such changes will cause significant demographic disruption in rural areas, promoting higher rates of depopulation that will necessitate strategies to assist rural out-migrants destined to new, primarily urban, locations in finding new labor market opportunities, and to address growing socioeconomic inequities within the remaining rural population. There are no easy answers to be learned to these latter questions from the experience of Lincoln County, for residents are still in the process of discovering the best adaptation pathways.

References

Abbott, C., Leonard, S.J., Noel, T.J., 2013. Colorado: A History of the Centennial State, fifth ed. University of Colorado Press, Boulder.

Bureau of Labor Statistics, 2017. Occupational Employment Statistics: May 2017 State Occupational Employment and Wage Estimates: Colorado. US Department of Labor. https://www.bls.gov/oes/2017/may/oes_co.htm.

Colorado Division of Water Resources, 2019. Water Rights Website. Colorado Department of Natural Resources. http://water.state.co.us/SurfaceWater/SWRights/Pages/default.aspx.

Durrell, G.W., 1974. Homesteading in Colorado. The Colorado Magazine 51 (2), 93−114.

Gregory, J.N., 1989. American Exodus: The Dust Bowl Migration and Okie Culture in California. Oxford University Press, New York.

Markus, B., 2018. Denver's housing crunch has many starting to think outside the metro. Colorado Public Radio News, 23 April. https://www.cpr.org/news/story/denver-s-housing-crunch-has-many-starting-to-think-outside-the-metro.

Mehls, S.F., 1984. The Rancher's Frontier. In: BLM Cultural Resources Series (Colorado: No.16) "The New Empire of the Rockies: A History of Northeast Colorado". US National Parks Service.

Melillo, J.M., Terese, T.C., Richmond, Yohe, G.W. (Eds.), 2014. Climate Change Impacts in the United States: The Third National Climate Assessment. U.S. Global Change Research Program. https://doi.org/10.7930/J0Z31WJ2, 841 pp.

National Weather Service, 2019. Denver-Boulder Forecast Office NOAA Online Weather Data. https://w2.weather.gov/climate/xmacis.php?wfo=bou.

Parton, W.J., Gutmann, M.P., Ojima, D., 2007. Long-term trends in population, farm income, and crop production in the Great Plains. BioScience 57 (9), 737−747.

State of Colorado, 2012. Guide to Colorado Well Permits, Water Rights and Water Administration. Department of Natural Resources. https://www.colorado.gov/pacific/sites/default/files/wellpermitguide_1.pdf.

US Census, 2018. Longitudinal Employer-Household Dynamics Data Visualization Site. https://lehd.ces.census.gov/.

US Census Bureau, 2019. QuickFacts: Lincoln County, Colorado. https://www.census.gov/quickfacts/lincolncountycolorado.

USDA (US Department of Agriculture), 2012. Census of Agriculture: County Profile: Lincoln County, Colorado. https://www.nass.usda.gov/Publications/AgCensus/2012/Online_Resources/County_Profiles/Colorado/cp08073.pdf.

USDA, 2003. Soil Survey for Lincoln County. https://www.nrcs.usda.gov/Internet/FSE_MANUSCRIPTS/colorado/CO073/0/Lincoln-Web.pdf.

USDA, 2017. Colorado Agricultural Statistics. National Agricultural Statistics Service Mountain Region, Colorado Field Office. https://www.nass.usda.gov/Statistics_by_State/Colorado/Publications/Annual_Statistical_Bulletin/Bulletin2017.pdf.

Drought challenges and policy options:
lessons drawn, and the way forward

22

Daniel Tsegai[a],*, **Michael Brüntrup**[b]

[a]*United Nations Convention to Combat Desertification (UNCCD), Bonn, Germany,*
[b]*Program B, Transformation of Economic and Social Systems, Deutsches Institut für Entwicklungspolitik (German Development Institute) DIE, Bonn, Germany*
*Corresponding author

The preceding chapters of this volume have detailed the challenges and various policy options to mitigate the impacts of droughts that are becoming more frequent, severe and long lasting as a result of climatic and ecological system changes. Not only does climate change increase droughts, the vulnerability of the landscapes and of human land-based activities is exacerbated through degradation of vegetation and soils through inter alia lower water retention capacities, more vulnerable plants and microclimate changes. The impacts include threats to water supplies needed for domestic and livestock consumption, reduced agricultural production, energy, transportation, sanitation, human health, tourism, and other sectors. Drought-induced water scarcity adversely affects food security in developing countries through direct impacts on food availability and reducing agricultural incomes, and indirectly by exacerbating food price levels and volatility. This can in turn raise migration pressures and trigger resource conflicts and civil tensions in high-risk areas.

Despite these well-documented impacts, progress on drought preparedness has been slow. With the impacts of climate change and land degradation being increasingly felt, a paradigm shift from "reactive" and "crisis-based" approach toward a more "proactive" and "risk-based" drought management approach is indispensable. The critical components of a *proactive* approach to drought management combine monitoring and early warning systems with impact and vulnerability assessments and risk mitigation measures. These enhance not only the capacity to forecast and monitor the onset, location, and intensity of droughts but also improve communications and coordinate responsive actions between governments, organisations, and vulnerable communities. Although good progress has been made in drought monitoring and management, there remain many challenges. This book has identified and described many of the more pressing of these challenges and readers will hopefully have benefitted from the empirical insights from examples from across the globe and the many policy options that have been suggested, which may be taken up and adapted for more effective management of drought in the future, and contribute to sustainable development and enhanced well-being of vulnerable populations. In the remainder of this chapter, we summarize the key messages in preceding chapters and offer some concluding remarks for moving forward.

Drought Challenges. https://doi.org/10.1016/B978-0-12-814820-4.00022-5

Lessons drawn

After an introductory section, the second chapter reviews existing scholarship on the links between drought, migration, and conflict in the context of sub-Saharan African countries. It was a conscious decision by the editors to have the first substantive chapter tackle one of the more contentious issues that needed to be addressed in this book. When sub-Saharan African droughts feature in Western media, which does not happen often, it is almost invariably in the context of famines, violence, conflicts, and humanitarian emergencies. And while droughts have sometimes been a contributor to political instability, violence, and conflict, the progression from drought to conflict or from drought to displacement is almost never automatic. Rather, such processes are invariably influenced by a multiplicity of environmental, socioeconomic, and political factors. The multistaged, progressive nature of the causal chain from drought to violence and displacement means there are many opportunities for intervention to break the links in the chain, and for proactive measures to prevent the progression from occurring in the first place. The authors highlight integrated early warning systems, weather-based crop insurance schemes, drought mitigation tools, and disaster risk reduction measures—things we need to be doing in drought-prone countries anyhow, as part of wider development initiatives—as being important ways to reduce conflict risks. The authors also note that migration is often an adaptation strategy in dryland and drought-prone regions, and that drought migrants are rarely the source of violence; they are more often the victims of it.

El Niño—induced droughts are a recurrent climate hazard in Southern Africa, and in Chapter 3, the authors describe their impacts: reduced agriculture and livestock productivity; food security and health problems; electrical shortages due to lack of water for hydroelectric power plants; economic slowdowns; depreciation of currencies; greater reliance on food imports, thereby raising food prices; increased migration; and higher rates of school dropouts and child labor. The authors provide a detailed and revealing account of past responses by various countries and the international community. They note that, in the case of the 2015 El Niño drought, the time lag between the early warnings and subsequent action ranged from 3 months in Lesotho to 12 months in Madagascar, reflecting the political dimensions of making an official drought declaration. As the authors note, "In many cases, due to broader consequences related to declaration of drought, such as loan write offs to farmers, associated insurance implications, costs to central Governments and impacts to the broader economy appear to govern how and when governments declare disasters. The bottom line is that countries normally do not want to rush into declaring disasters, even when all other non-government sectors see it as apparent." The Regional Interagency Standing Committee (RIASCO) that was eventually established, developed a drought response plan that hinged on three pillars. The first is a humanitarian response that focuses on issues of food security, agriculture, livelihoods, health, and nutrition. The second pillar focuses on building resilience, while the third pillar emphasizes the potential economic impacts and solutions to the impacts of droughts. The latter pillar situates drought disaster response and mitigation within the context of wider economic development in the region; a logical approach given the regional climate and the economic importance of the agricultural sector. The Chapter ends with a list of lessons learned that will occupy the Southern Africa region for some time:

- improving government leadership
- timely decision-making
- inclusion of resilience building into emergency measures
- improving coordination at all levels of governance
- emphasizing regional level, multisectoral approaches in emergency and longer-term planning
- improving early warning systems and information dissemination

- strengthening institutional and local adaptive capacities
- more active inclusion of development partners in drought planning
- use of standardized indicators
- devising measures that protect productive livelihood assets from being lost or sold at discounted rates during droughts
- better use of social safety nets and cash transfer systems

For all of the preceding goals to be achieved, better data are needed.

Chapter 4 describes the impacts of droughts in dryland Latin America, focusing on the northeast of Brazil. This area has been plagued by recurrent, severe droughts over many centuries, and there are indications that drought frequency has been increasing over time. At the same time, population growth, land and biosphere degradation have exacerbated drought hazard and helped contribute to a decline in cattle stock and loss of economic assets in the region. To some extent, the vulnerability of residents to droughts may have declined: dependency on agriculture has shrunk to 7% of GDP, incomes from other sources have increased accordingly, more people are living in urban areas, and they are better connected through improved infrastructure and transport. Several drought-related strategies have been implemented in the region, including construction of small, medium, and large dams to provide anywhere from 1 to 5 years of water storage; groundwater pumping; desalination plants; irrigation and water distribution systems by trucks and pipelines; and social protection programs that have evolved from simply providing occasional work to more permanent measures. These are all a product of growing incomes and greater government tax revenues that enable Brazil to better fund its own drought resilience programs. Yet, there remains a gap in terms of international, bilateral, and multilateral cooperation. The High-Level Meeting on National Drought Policies has identified a need to better integrate the three pillars of drought preparedness, namely, (1) monitoring, prediction, and early warning; (2) impact and vulnerability studies; and (3) drought mitigation and response. A national drought policy being designed in line with these pillars is expected to contribute to further reducing vulnerability, while improving the quality of assistance to drought-stricken populations.

Chapter 5 looks at indigenous communities in the Yucatan area of Mexico, focusing on the incorporation of local knowledge as a critical component to drought adaptation as part of wider climate change adaptation. Indigenous groups are easily and often overlooked by governments and multilateral agencies when drought management plans are made. This chapter provides an example of why this needs to change. The gradual disappearance of milpa-based practices in Yucatan threatens the cultural survival of the Maya. Milperos of Xuilub, the case study community, are confronted with serious challenges to their agricultural livelihoods, and as such must find ways of adapting to drought and other climate change—related impacts. It is thus envisaged that an inclusion of local knowledge would not only increase the likelihood of indigenous cultural survival but would also contribute to governmental recognition of the territorial sovereignty of its indigenous groups.

Staying with Latin America, the adaptive capacities of rural households to El Niño—triggered droughts in Colombia are examined in Chapter 6. This ethnographic case study seeks to better understand household level decision-making processes in the face of protracted droughts in a village off the Caribbean Coast. It finds that vulnerability and household choices are influenced by the social, cultural, and financial capital endowments of households; the local natural and socioeconomic environments; land ownership; local conflicts over land; and general civil conflict and armed fighting between governments and rebels, which was particularly intense in this region. The chapter provides particularly interesting insights into migration decision-making during times of drought. In areas with reliable water infrastructure, no one migrated during drought. In the rest of the study area, resource-endowed families resorted to temporary migration, while the most vulnerable families adapted in situ, sending individual household members out to other rural areas or urban centers to earn money and remit. However, in situ adaptation to drought often tends to deplete the capital endowments of vulnerable families, particularly if the drought extends multiple years, further exacerbating the prevailing inequalities. It is at this stage that the poorest migrate permanently out of the study area, under destitute conditions.

Chapter 7 highlights an often-forgotten fact: drought impacts have a strong gender differentiation. Men and women respond differently to the impacts of drought on their access to assets critical to their resilience and coping capacities. In explaining the need for a gendered understanding of drought impacts, the authors show that the disadvantaged position of women in some societies heightens their vulnerability and requires a gender-specific innovation and contextualized response to droughts. The recommendation is thus, on one hand, to adopt a more holistic approach to addressing gender inequality by challenging gender stereotypes from an early age, through education improvements that go far beyond specific drought policies. On the other hand, a call is made for livelihoods and gender to be better integrated into vulnerability assessments and resilience building strategies. The authors point out the lack of gender-disaggregated data that are critical to informing coping mechanisms for women, men, youth, and other gendered categories, which hinder progress toward developing best fit solutions to minimizing drought risks and impacts in developing countries.

Chapter 8 discusses the recurring droughts, notably irregular monsoon rains, as a major cause of crop losses, water scarcity, and the mass migration of people and livestock in India, particularly in arid and semiarid regions. Regional disparities with regards to drought are high on the subcontinent. Despite the recurrent adverse impacts of droughts in many parts of the country, the government's approach to droughts has been mostly reactive. The Government of India Manual for Drought Management outlines a series of indicators from multiple sources in declaring the occurrence of drought and seeks to develop relief and crisis management measures such as financial assistance, debt relief, and temporary labor opportunities for affected people. However, the authors point out the lack of proactive measures to reduce vulnerabilities and build long-term resilience, and recommend the elaboration of a more comprehensive drought monitoring and management strategy to facilitate the regular and timely dissemination of relevant data in enhancing drought preparedness. They identify a pressing need to systematize pilot studies for fine-tuned, geographically differentiated vulnerability indexes, and for planning of mitigation measures adjusted to the integrated assessment of vulnerability to drought, in relation to physical, climatic, and social factors.

Chapter 9 reviews the potential of satellite remote sensing for drought monitoring and early warning. While direct remote sensing of precipitation is difficult and not very exact, the measurement of related impacts on the ground, particularly on vegetation, is making substantial progress. It can take local conditions such as soil water content, irrigation, cropping practices, and in certain cases, even crop-specific reactions into account. Combinations of different sensor and archived satellite data, often in combination and adjusted with local data, or refined in crop growth models, can provide timely information on the progress of droughts from plot to large scale, and allow for comparison to past events. Together with information on population and economic activities, such a system can be used to identify affected people, livelihood losses, and economic and social impacts. It is readily available even in inaccessible regions or when governments are unable or unwilling to collect data on the ground. The authors provide a broad overview of the various instruments, methods, resolution, availability, and time span of many indicators and products developed to assess drought-related (and other) effects. This also points to the current weaknesses and gaps, inter alia, the need for validation to distinguish between effects of drought and other influences such as cropping practices, data gaps, and the political economy of classification and declaration of drought.

Chapter 10 presents the Integrated Drought Severity Index (IDSI), an innovative variation on traditional methods for the monitoring and assessment of drought, which has been tailored for use in South Asia. While the chapter may seem more technical than what might ordinarily appear in a volume of this type, its value is to demonstrate how standard drought indices, many of which have been developed for use in other climatic zones, benefit from being adjusted to reflect regional conditions, to take advantage of emerging data sources, and the type of work that goes into doing so. The IDSI aggregates meteorological data, vegetation conditions from satellite imagery, and targeted collection of ground-truthed moisture and crop-yield data, and provides system users with (1) a daily update of meteorological drought indices, using satellite-based rainfall estimates with quality checks and bias correction; (2) an estimation of 10-day soil moisture using ESA's ASCAT data to derive Soil Water Anomaly Drought Index; (3) an integrated drought severity index that reflects vegetation conditions, precipitation, temperatures, and soil moisture; and (4) a warning classification ranging from "watch" to "extreme." An important value of this approach is that, of the eight South Asian countries, only India and Pakistan possess drought monitoring systems along with the necessary experience and technical capability for constant updating and maintenance; and even in these two countries there exists a vast difference in capabilities across regions and governments. The system described in Chapter 10 addresses many knowledge and methodological gaps, while providing the much-needed evidence to manage climate-related risks for smallholder farmers to effectively adopt climate-smart agricultural production. The developed knowledge and capacities described in this chapter could hold lessons for other drought-vulnerable regions.

In Chapter 11, we return to a more detailed focus on links between drought and conflict, with the authors looking at the strengths and weaknesses of early warning systems for drought and conflict. Although both types of systems exist, they are often siloed or isolated from one another, and often fail to incorporate important socioeconomic factors that underlay conflict in drought-affected regions. The authors provide a compelling case for integrating conflict event datasets and conflict early warning systems with drought early warning and response systems as part of a more holistic approach to risk management, reduction, and mitigation.

Crop insurance is a relatively new instrument to help farmers in developing countries manage drought, but it has been in India for some time now, as is described in Chapter 12. India's Weather-based Crop Insurance scheme (WBCI) provides farmers insurance against a number of climate hazards for a range of common field crops, but its effectiveness is constrained by limited access and quality of weather data in the Reference Unit Areas (RAU) used in the scheme. The authors provide a detailed explanation of the procedural steps for an experimental WBCI that trigger payouts to farmers who experience losses of cotton, rice, and/or chili crops due to excessive rainfall and/or high temperatures. The authors outline the very fine-tuned knowledge that needs to be incorporated into insurance contracts to make them effective, describing steps on how to improve these knowledge needs. The authors also describe how region-specific training programs help farmers in understanding the crop and weather index insurance products, and how timely payouts of claims facilitate farmer adaptation and adoption of weather-based and agricultural insurance products.

Chapter 13 also addresses the concept of drought insurance, focusing on how to design innovative insurance products that help farmers cope with drought and which simultaneously tackle land degradation. The Sustainable Land Management (SLM) practices described here are envisaged as a condition/

contractual obligation to provide farmers access to a premium-free insurance scheme. The authors describe options for funding various incentives that ease the financial burden on governments and on local communities to cover premiums. For a drought insurance scheme to be financially sustainable in the long run, it is critical to develop strategies that place the least strain on public resources. The lessons from Chapters 12 and 13Chapter 12Chapter 13 will be of use to planners and decision-makers creating or updating crop insurance programs in other developing regions.

A comparative assessment of drought monitoring and early warning systems in Tanzania, Kenya, and Mali is provided in Chapter 14. Of these, the authors find that Kenya has the most advanced system. It also has more advanced responsive strategies, such as crop insurance based on an area yield approach; livestock insurance schemes based on remote sensing of vegetation cover; and an enhanced livestock reduction program that encourages reduction of herd sizes before animal quality suffers and prices become deflated. More importantly, the authors identify key challenges and opportunities for making monitoring systems across all three countries more precise and ultimately more effective, ones that are likely equally germane to systems in other countries in sub-Saharan Africa. Some of these challenges include proper timing when making financial investments in drought resilience; strategies for allocating funds; and decision-making processes in drought early warning. The authors note that indigenous knowledge is vital to effective drought monitoring and early warning systems (which harkens back to the lessons from Mexico described in Chapter 5) and emphasize the need for national and local stakeholders to work together. Extension agents need to be equipped with the knowledge and resources and ideally would work alongside interdisciplinary experts on a regular basis, to help with the integration of local knowledge and dissemination of scientific forecasts.

In Chapter 15, another important aspect of long-term drought resilience in agriculture, as well as improving drought resilience of crops and cropping systems are tackled. The authors describe the contributions of drought-tolerant maize and maize-legume intercropping to the climate resilience of rural households in northern Uganda. These two technologies are examples of climate-smart agriculture, an approach that seeks to increase agricultural productivity sustainably and enhance resilience of households while reducing emissions of greenhouse gases. So far, most studies have concentrated on the productivity aspect, which can improve resilience through higher incomes and income diversification. This chapter goes beyond this, using a theory-based resilience measurement approach to estimate the probability of a household remaining above a low-income threshold given various household characteristics and exposure to external shocks such as droughts. The authors describe an increase in drought resilience among farmers who adopted drought-tolerant maize and for those who combined this with maize-legume intercropping. Resilience dropped, however, among farmers who adopted intercropping technology with normal maize varieties, showing that when adjusting agricultural practices, careful analysis and empirical evidence is essential, and sweeping recommendations as to what constitutes "best practice" should be made cautiously. The study also reveals the many factors affecting the adoption of new agricultural technologies and practices, such as kinship networks, previous experience of drought, the prevailing trends in food security, and knowledge bases. The authors conclude that climate-smart agriculture should be promoted in vulnerable and marginal areas to enhance resilience to drought impacts, ideally in combination with social safety nets.

Chapter 16 looks at how social protection schemes can potentially enhance drought resilience in both emerging and advanced economies. Usually, these are not established specifically for drought challenges, but they can be used to rapidly scale up support to drought affected people, and to build long-term resilience. In poor countries, such programs are still often an exception, or experimental in nature. As part of its long-term rural development program, Ethiopia is one of the few poor countries to have introduced a country-wide social protection scheme, known as the Productive Safety Net Program (PSNP). The PSNP has made strides in filling food gaps, reducing livelihood asset depletion, and creating assets both at household and community levels. This has provided long-term improvements through the creation of community assets and (to a lesser degree) household assets. Nonetheless, a number of challenges persist, including weak coordination and harmonization of activities among government institutions, development partners, and NGOs; and limited decentralization and capacity gaps (institutional, organizational, technological, and financial) at multiple levels. These weaknesses undermine the potential of PSNP in enhancing drought resilience in the country. The authors describe potential ways for the PSNP to improve its internal implementation capacity while enhancing its linkage with other drought resilience programs, to bring a stronger and meaningful impact on drought resilience at national, subnational, and community level. As with many other chapters, though the details are particular to Ethiopia, the broader lessons may be relevant to other developing countries.

Chapter 17 assesses the drought preparedness and responses of pastoralists in dryland Northern Kenya, a hotspot of drought vulnerability in Africa. The authors describe a number of lessons for better pastoralist rangeland management, early warning systems, modern livestock management strategies, and better water management. In this case, distance to markets is a critical determinant of both exposure to climate hazards and in the adoption of livestock management strategies. With official early warning systems not yet widely implemented in the area, the authors point to community leaders as an effective means of disseminating information to rural households. To overcome the threat of drought-induced food insecurity, it is also proposed that permanent water points be developed to enable pastoral communities to adequately water their stock, and to better manage common natural pasture areas and trekking routes, thereby contributing to conflict reduction.

Climate change scenarios suggest a future increase in precipitation levels in many parts of Africa. Despite this, water scarcity may worsen in East Africa due to increased effects of heat and changes in the timing of precipitation. Chapter 18 emphasizes the need to enhance resilience to climate change in East African drylands by addressing shortfalls in adaptive capacity and base-level development. Innovations in technology and governance that are vital to building drought and climate change resilience for smallholders are missing from existing regulatory frameworks at national and regional levels across the region. The chapter focuses on the food—energy—water (FEW) nexus and the growing demand for these resources due to climate change impacts and population growth. The bemoans the inadequate attention to rainwater harvesting (RWH) as a technology that can address food security, increase access to water and biomass energy, and contribute to climate adaptation strategies. The author describes a microcatchment approach that involves the collection of runoff from roofs and ground surface, storage methods, and productive uses. There is a whole lot of advantages of RWH going far beyond drought resilience, such as biodiversity and other ecosystem service improvements, health benefits, and others. Limitations in its uptake are linked to information transfer gaps, challenges of technology, right fit to local situations, financial requirements, and maintenance, among others. Government policies and investments are needed to overcome these constraints, with the author encouraging a FEW approach to developing the necessary comprehensive and intersectoral governance. He then screens the policy frameworks of the regional economic communities in the Eastern African region and finds that, except for the Southern Africa Development Community (SADC), FEW nexus perspectives are absent from development policies and strategies.

Chapter 19 takes a deeper dive into drought risk management in Kenya, providing lessons for the wider region of the Horn of Africa. The authors describe how recurrent droughts and new international trends in disaster management have inspired a proactive approach to the management of drought hazards in Kenya's arid and semiarid lands. Weaknesses in early warning systems and information, inadequate financing for national drought management, and inadequate human resource and institutional capacities have historically undermined regional efforts at drought management. For the case of Kenya, an insightful synopsis of the last decade of implementing drought risk management in conjunction with decentralization is provided. Major achievements are national disaster and drought strategies embedded into long-term development plans; institutionalized coordination structures at national, county, and local levels; new disaster and contingency financing structures; and national funding becoming more independent of donors. Important recommendations concern the limitation of human resources and institutional capacities at subnational levels, the lack of natural resource information, weak early warning systems, lack of integration, and the lack of systematic learning on effectiveness and efficiency of implemented measures. On this last issue, the author cites Venton (2018a), who states: "every US$1 invested in building people's resilience in the Horn of Africa could result in up to US$3 in reduced humanitarian aid and avoided losses." This reminds us of the old maxim that "an ounce of prevention is worth a pound of cure," a maxim that might equally be made from evidence in most chapters of this book.

The integration of various drought measures into comprehensive policies and strategies under different macroeconomic and political systems is discussed in Chapter 20. Here, the discussion focuses on how general disaster risk-management ideas for drought management have been adopted in Ethiopia and Kenya, in line with international frameworks and a wider regional strategy. It should not be a surprise to the reader of this volume that multiple chapters analyze different aspects of drought management in East Africa, for this is a region that has continuously suffered from drought impacts over the years. Different levels of development by countries within the region, variations in political systems, and the intensity and pattern of government decentralization have had varying implications for drought management in both countries. By considering and comparing the relative progress and challenges at effective drought management, several recommendations are proposed to improve policy coherence and to enhance drought resilience. The main challenges are to (1) develop a common, national understanding of resilience that considers short-term and long-term interests; (2) weigh the respective merits of bottom-up versus top-down approaches and try to support the former; (3) integrate multiple challenges when planning and seeking synergies (e.g., insurance and social protection as well as natural resource management and livelihood diversification); (4) build up confidence in early warning systems and integrate these with longer-term development programming; (5) actively support and profit from regional integration in national drought strategies; and (6) create early response mechanisms by inter alia incorporating contingency funding in development programs triggered through drought declarations. All of these recommendations dovetail closely with recommendations made in preceding chapters.

In Chapter 21, preceding chapters that focus on drought impacts and adaptation in developing countries are contrasted with a developed-country example from dryland agricultural Lincoln County in Colorado, USA. The chapter emphasizes how economic development alone does not insulate a rural region or its population from drought hazards. Water scarcity in the county has consistently posed a serious challenge to crop irrigation, ranchers, and small urban centers in Lincoln County. Alongside other environmental change impacts and prevailing dry conditions, droughts have had numerous impacts on agricultural productivity and economic activity, and limit future development in the county, despite being only a 100 km from the prosperous booming city of Denver. Many farmers are involved in crop insurance schemes, but the payouts decline over the course of multiple dry years, forcing a transition to other adaptive measures. Smaller farm operations and people who are relatively new to farming (and therefore have higher debts, higher net costs, and lower profit margins) often struggle the most during droughts—characteristics similar to those of vulnerable people in dryland regions the world over. The "Colorado Doctrine" of water management in dry areas—namely, that all water, whether on the surface or underground, is a publicly owned resource and not to the (adjacent) land-owner and can therefore be managed according to common interests and reallocated more easily—is analyzed as an approach that may be worth replicating in water-stressed countries in the Global South, given its tight regulation and strongly enforced system for administering water use rights.

Concluding remarks

From the foregoing empirical findings and discussions in the various chapters of the book, it is evident that drought remains an enduring threat to the well-being of people, in a gendered manner, across the globe. Developing countries are particularly vulnerable, but even in emerging and developed countries, the threat of drought is serious and growing, with climate change, deforestation, land degradation, and excessive water extraction exacerbating the matter.

In countries with higher development status, diversified incomes, integrated food markets, and well-functioning government social services, droughts no longer constitute an immediate risk to food security to the same degree as in many poor developing countries, where droughts are still an important cause of food insecurity, malnutrition, and loss of livelihoods. Populations in arid and semiarid rural areas, notably smallholder farmers and pastoralists, are particularly vulnerable. Drought-induced water scarcity accompanied by rapid population growth has contributed to increasing competition and conflicts in many water-stressed areas. Both people and nature suffer during droughts, when vulnerable people frequently draw down natural resources stocks as they try to mitigate the immediate impacts.

With climatic changes and anthropogenic environmental degradation likely to further alter ecological and hydrological systems, droughts will increase in frequency, severity, and duration in many parts of the world. At the same time, rural populations will remain strongly dependent on rainfall-driven production for a long time to come. International food markets, even if more resilient on a global scale compared with isolated local markets, are not immune from the impacts of mega droughts in important agricultural production areas. This will eventually threaten food security in urban areas through price hikes and subsequent political unrest that exacerbate human insecurity. Conflict and involuntary migration are worst case outcomes that unfortunately still occur.

There is thus an urgent need for strategic interventions to promote efficient drought management systems to minimize vulnerability while enhancing resilience and human security. This is especially true for dryland areas, but also in more humid regions. Drought resilience needs to be built at multiple levels, beginning with households, as well as at community, local, national, and regional levels. This volume's look at the three main pillars of drought management—monitoring and early warning, vulnerability and risk assessment, and risk mitigation measures—identifies a number of broad advances that need to be pursued.

Drought monitoring provides the necessary basis and information for early warnings, for ongoing vulnerability assessments, and for design of appropriate responses. Greater use of remote sensing tools is one obvious option. Many drought-vulnerable regions in developing countries are in politically sensitive areas that are prone to conflict. This often makes drought interventions more difficult or costly and, in some instances, the intervention itself may have the potential to exacerbate conflict. In such cases, the policy emphasis on drought monitoring should be geared at strategic integration of both drought and conflict early warning. This should also involve the integration and interchange with local and indigenous knowledge systems, local leaders, and extension agents. This can reduce potential conflict, improve acceptance and uptake of information and alerts, improve understanding of the system as a whole, and generate locally appropriate responses. By being transparent and independent from political influences, early warning systems and the corresponding recommendations gain credibility over time.

A close link between early warning systems and drought risk mitigation can be drawn through drought insurance schemes. Comprehensive and enhanced weather-based index and agricultural insurance schemes provide drought relief to crop and livestock farmers, but these schemes often require farmers to pay a substantial premium that increases with the more drought-prone an area and activity is. Whether farmers and governments are ready to pay, and how this may influence their risk and attitudes activities depends on location, specific activities, details of the schemes, drought history, and past experiences. These schemes could be expanded to include a wider range of crops and climate risk parameters, but then contracts and premiums would also change. In facilitating a greater acceptance and adoption of such programs, smooth and timely payouts and greater farmer education of how schemes work are needed. Products must be tailored to needs and capacities of farmers, and more trial schemes are necessary.

Part II of this book provides many examples for building drought resilience at the micro and meso level: better drought-resilient crops and cropping systems, soil and catchment area improvements to harvest and store water locally, improved livestock management practices (including vaccination, pasture management, fodder storage, destocking strategies with all the associated activities), and forestry and agro-forestry measures, among others. Irrigation and general water management are obviously and particularly important components of nature-based resilience strategies. Some of these measures can be taken individually, while others only make real sense if applied at the larger community, local, or catchment level. An important element of resilience is income diversification, which includes local development of non−rainfall-dependent activities, and in some locations may entail partial migration. Other factors in resilience include savings and credit schemes that enhance economic buffers and flexibility, along with social protection schemes that provide support when individual and household level resilience is exhausted. Finally, in the case of severe droughts, the international community must still step in when poor people and countries lack the capacities to cope. Both proactive and reactive measures must be part of holistic, well-integrated strategies that avoid costly overlaps and misallocations of resources. We must also be alert to the possibility of counterproductive outcomes; for instance, insurance and social protection can reduce proactive engagement in other forms of resilience; temporary water and fodder transportation schemes may reduce willingness to destock and exacerbate vegetation damages. On the other hand, permanent waterholes for famers may attract herds, which could degrade surrounding soil and vegetation with the potential to stimulate conflict. A particularly delicate issue is large-scale dam construction and management, only marginally touched in this volume, due to the many potential negative social and environmental side-effects and risks. After a long period of neglect, large dams are in construction and/or planned in many parts of Africa now, and they merit special attention.

Effective governance regimes for droughts and drought-cycle management are key. Due to the heterogeneous nature of drought responsibilities across ministries, agencies, and sectors, there is a strong need for political commitment at the highest levels to push lower levels to cooperate and harmonize activities and insist upon accountability at all levels. The strongly integrated nature of drought strategies and the need for multistakeholder interaction make the whole system prone to free-riding and neglect or constant shuffling of responsibilities. Thus, roles and responsibilities need to be clearly spelled out in any drought policy, and interactions have to be enforced and monitored. Most importantly, drought-risk management should be incorporated into long-term development plans and governmental disaster mitigation frameworks to generate synergies. The exact form of actions and interactions will depend on the institutional landscape of a given country. Many advocate for "bottom-up" resilience building, but this can only be achieved through locally adapted forms of economic, social, and political activities geared toward local ecological settings. National strategies must therefore be shaped to support these bottom-up solutions as far as possible, while providing top-down support and allowing for decision-making to be devolved to the appropriate level (an awkward and uncomfortable way of doing things for many governments, to be sure).

The way forward

Many of the chapters in this book focus on drought as it relates to poverty and food security in less developed countries, especially the impacts on agricultural and rural populations. However, other domains and sectors are also affected by drought, such as tourism, river transportation, electricity production, fire hazards, forestry, and infrastructure. These issues become more important as economies develop, with increased diversity in the economic sectors, infrastructure, and other investments.

Interactions between drought measures, policy domains, stakeholders, and their responsibilities as described in preceding chapters are rarely well documented and understood, and vary according to local circumstances. The political economy of drought strategies and responses is a key influence on their success or failure. The merits and limits of regional cooperation warrant further investigation and discussion; beyond examples of cooperation in early warning, other issues such as mutual learning, common resource management of water resources, cross-border movements of animal and people induced by droughts, and other regionally important processes are at stake.

There are many ways by which drought policies interact with other policies, and this volume only goes so far in identifying these. This needs to be done widely and often. For example, policies that seek to maximize agricultural productivity—an important policy objective in a rapidly growing, hungry world—can conflict with drought resilience measures. The trade-offs and cost-effectiveness calculations can be complicated and require long-term data. The best mix will also depend on other factors, such as the willingness of people to participate, and the willingness to save resources during good years to put toward adaptation in bad years. Insurance is an important tool, but drought is only one of a whole range of potentially insurable hazards, meaning that more careful analysis and data are needed. In regard to social safety net policy, the need to harmonize recurrent and drought-induced interventions may require changes in the way such programs are conceived and delivered. Large-scale dams for irrigation—a subject not investigated in great detail in this book—are contentious and often also connect with energy and other policy sectors, warranting coordination. Migration because of drought, and as a means of resilience, is a topic of high and growing importance in many developing regions strongly intertwined with other migration-inducing factors and demands greater research and policy consideration.

A whole body of existing literature is devoted to questions of who pays for losses and damages caused by drought and other climate change impacts, as well as costs of prevention and adaptation. For drought policies, this is an important question, since many drought measures require financial and other resources, which could be provided through adaptation funds. But at the same time adaptation funding may alter drought policy formation by forcing proponents to tailor policies to better fit the parameters of external funding mechanisms. How climate change adaptation funds are channeled into national systems may have a significant difference in effectiveness, efficiency, and integrity of larger systems. More profoundly, in this context, questions may be raised as to whether climate change or local and natural resource degradation is more responsible for loss and damage experienced during droughts, which may in turn shape future policy directions.

There is also a greater need to harmonize drought risks with other disaster planning, such as for flooding, geotechnical hazards, and so forth. Each has its own dimensions (e.g., speed and early warning spans, likelihood, persons and assets at risk, concentration of damage, communication and transportation needs, etc.) and yet there are likely commonalities as well. The Sendai Framework for disaster risk reduction demands for a common agenda for preparedness, and at a certain political level may require common frameworks, even when different responses and resilience measures are needed at other levels. Much more research is needed to understand these issues.

In conclusion, the authors and editors hope that this volume provides you, the reader, with not only the information you can use, but also inspiration to further your own contributions to this important field of research and policy, which will continue to grow in importance for the future of many people and livelihoods, and for the peaceful development of the global community.

Author Index

Subject Index

Printed in the United States
By Bookmasters